Synchronization in Digital Communication Systems

Do you need to know how to develop more efficient digital communication systems? Based on the author's experience of over 30 years in industrial design, this practical guide provides detailed coverage of synchronization subsystems and their relationship with other system components. You will gain a comprehensive understanding of the techniques needed for the design, performance analysis, and implementation of synchronization functions for a range of different modern communication technologies. Specific topics covered include frequency-locked loops in wireless receivers, optimal OFDM timing phase determination and implementation, and interpolation filter design and analysis in digital resamplers. Numerous implementation examples help you develop the necessary practical skills, and slides summarizing key concepts accompany the book online. This is an invaluable guide and essential reference for both practicing engineers and graduate students working in digital communications.

Fuyun Ling is an independent consultant, the managing member of Twinclouds Consulting, LLC, and an Adjunct Professor at Tsinghua University, China. He spent 14 years at Qualcomm, where he was a Vice President of Technology. Before joining Qualcomm, he worked at Motorola for more than 13 years and served as an Adjunct Professor at Northeastern University. Dr. Ling is a Life Fellow of the IEEE.

Synchronization in Digital Communication Systems

FUYUN LING

Twinclouds Consulting, LLC, San Diego

CAMBRIDGE
UNIVERSITY PRESS

CAMBRIDGE
UNIVERSITY PRESS

University Printing House, Cambridge CB2 8BS, United Kingdom

One Liberty Plaza, 20th Floor, New York, NY 10006, USA

477 Williamstown Road, Port Melbourne, VIC 3207, Australia

4843/24, 2nd Floor, Ansari Road, Daryaganj, Delhi – 110002, India

79 Anson Road, #06–04/06, Singapore 079906

Cambridge University Press is part of the University of Cambridge.

It furthers the University's mission by disseminating knowledge in the pursuit of education, learning, and research at the highest international levels of excellence.

www.cambridge.org
Information on this title: www.cambridge.org/9781107114739
10.1017/9781316335444

First published 2017

A catalogue record for this publication is available from the British Library

ISBN 978-1-107-11473-9 Hardback

Additional resources for this title are available at www.cambridge.org/ling

Cambridge University Press has no responsibility for the persistence or accuracy of URLs for external or third-party Internet Web sites referred to in this publication and does not guarantee that any content on such Web sites is, or will remain, accurate or appropriate.

To my wife, Xiaoyun Ma, and our daughter, Jing

Contents

Foreword

Synchronization is an integral part of any digital communication system that transmits digital information through a communication channel. It is such an important component in the design and implementation of a communication system that numerous books have been written on this subject since the beginning of the digital information era, which now spans over 60 years. As communication system developments have evolved over these past 60 years, especially in the design and implementation of new modulation/demodulation techniques that achieve ever greater transmission rates in wireless communication channels, new synchronization techniques have also been developed to satisfy the more demanding system requirements.

This book provides a novel treatment of synchronization techniques for the design of modern digital communication systems, especially code division multiple access (CDMA) and orthogonal frequency-division multiplexing (OFDM) wireless systems. Included in the book are thorough descriptions on key topics, namely, the design and implementation of methods for initial acquisition of various transmitted signal types, the design and implementation of digital phase-locked loops (PLLs), and the integration of PLLs in synchronization circuits and algorithms for carrier phase recovery and tracking, as well as in circuits and algorithms for obtaining symbol timing and tracking. Also treated in detail is the topic of resampling/rate conversion methods, which are widely used in performing timing phase adjustments in nearly all modern receivers for digital communication systems.

The book is intended for engineers and related technical professionals who wish to acquire in-depth knowledge and understanding of state-of-the-art techniques in the design and implementation of synchronization for modern digital communication systems.

The author of this book has a wealth of hands-on industrial experience in the design and implementation of wireless digital communication systems. His career in the telecommunications industry has spanned over 30 years. The synchronization techniques that are described in this book embody his contributions and those of a number of his colleagues working in this field.

John Proakis

Preface

Synchronization functions are among the most important and critical components of digital communication systems. During my career of over 30 years working in the technical field of digital communications, my fellow engineers and I often spent more time on designing, debugging, and testing the receiver blocks that are related to synchronization than on any other functional blocks. This is especially true in the later stages of system and modem development, such as in the testing the prototypes in the field of real systems.

The reason for the importance and the difficulty of the implementation and debugging the synchronization functions is that they more directly interact with the real-world channel conditions than other receiver functional blocks. For example, in CDMA and OFDM system design and development projects in which I had participated, the most difficult receiver functions, on which engineers spend most of their time, were always related to synchronization.

The main objective of this book is to provide a general treatment of the key synchronization functions in one single resource that is easy to access. Although synchronization is included in almost every textbook on digital communications, this subject is usually treated at a high level and important design details are left out. This is due to page limitation constraints of these books that focus on the many general topics of digital communication systems. Indeed, synchronization functions are well covered in the literature; however, most are published as journal and conference papers that treat their implementation and analysis for specific communication systems. It is difficult to obtain a unified and comprehensive view on this topic from the papers scattered in many different publications. This book supplements the textbooks and scholarly papers by treating synchronization functions, from their theoretical foundations to their analyses, designs, and implementation.

The fundamental theories pertinent to the basics of synchronization are covered in this book because they are important to engineers and researchers in their practice. Similar to other elements of a digital communication system, synchronization has a well-developed theoretical foundation. Even though today's computers are capable of performing efficient simulation studies of various aspects of synchronization, analytical tools can always provide more insight into issues of optimality and other properties of interest. At the same time, the scope of the book is also limited to stating the results and leaving the lengthy proofs to the cited references for the readers.

Any theory is only as accurate as the model on which it is based. Due to the diverse applications of synchronization in different communications systems and their interactions with various environments, it is not possible to establish a single or even a few such models. As all engineers know, approximation is a fact of life. Therefore, this book tries to strike a balance between theoretical and empirical treatments. In addition to establishing a solid theoretical foundation, there are also many practical aspects that need to be considered and some approximations that need to be made. Such practical considerations and approximations are described in various chapters of this book, and we have made sure that these approximations and considerations are consistent with established theories.

This book is also intended to fill the gaps of other books in the same technical area published before the late 1990s. Due to the widespread applications of digital communications, especially wireless communications, synchronization technologies also experienced rapid development. Besides the basic theories and general descriptions, most of the examples given in this book are in the context of communication systems in deployment today, including CDMA and OFDM wireless communication systems. Even though these examples are mainly related to third- and fourth-generation wireless systems, the principles and implementations described are also applicable to the future generation of the communication systems that employ the same basic technologies.

Another objective of this book is to share my experiences gained during working in the communications industry for over 30 years with young engineers and researchers. The implementation details of various algorithms and functional blocks are provided in the examples given in the chapters. However, because conditions vary and environments change for different applications, what is described is only intended as references for the readers to approach their problems at hand rather than universal solutions.

The primary readership of this book is engineers, researchers, and graduate students working on digital communication–related projects in industry and academia. This book is intended for their self-study or as a reference book for them. It can also be used as supplemental materials in graduate-level digital communication courses and a textbook in short courses on synchronization. The readers should have an understanding of undergraduate-level digital signal processing, digital communications, and linear system theories. Knowledge of probability and detection and estimation theories is helpful but not necessary.

This book is written with physical layer system engineers/researchers in mind. However, due to the many implementation details described in various examples, it may also be of interest to hardware, software. and firmware engineers working on related projects.

This book is organized as follows:

Chapter 1 provides an overview of how digital communications systems work. The main components of a communication link, including the transmitter, the channel and the receiver, and their typical operations, are presented. Because a significant portion of this book is about synchronization in DS-CDMA and OFDM communications systems, brief reviews of these two communication technologies are given. More specifically, the components and operations of these two communication systems are presented.

A few important topics in detection and estimation theory that are closely related to synchronization are introduced in Chapter 2. These topics, including the likelihood function of continuous signals and the Neyman–Pearson lemma, are essential for establishing the optimality of synchronization functions discussed in later chapters.

In Chapter 3, the general procedure of initial acquisition based on hypothesis testing is first described. The performance and other theoretical and practical aspects of the procedure are then discussed. The implementation details of the procedure in four wireless communication systems are presented to conclude this chapter.

Chapter 4 provides an introduction to the phase-locked loop (PLL) due to its importance in synchronization functions. Given that most current implementations of PLLs are in digital or mixed signal forms, most of the chapter focuses on the digital PLL. The analog PLLs and their implementations are presented separately and serve mainly as a reference.

Chapters 5 and 6 cover the two key synchronization functions: carrier and timing synchronization. The optimal maximum-likelihood carrier and timing phase estimations are first established in the respective chapters. Their classical estimation algorithms in single carrier systems are then presented. Due to the popularity of wireless communications and CDMA and OFDM technologies in recent years, significant portions of these two chapters are devoted to the synchronization in these communication systems.

Finally, Chapter 7 is dedicated to digital resampling/rate-conversion technology, which is employed for performing timing phase adjustment in almost all modern digital receivers. The design, implementation, and performance analysis of resampling/rate-conversion algorithms are presented and their applications to practical receivers are discussed.

This book grew out of my work in the digital communication industry and is based on my experience with conducting many projects that involved the design, analysis, and implementation of synchronization functions. I also gained knowledge from numerous discussions with my colleagues and friends throughout the years. I would like to express my appreciation to them even though I am not able to list all of their names. Below, I would like to acknowledge the people who have had the most influence on me in my career and in the process of writing this book.

First, I would like to thank Dr. John Proakis, my PhD thesis adviser, and my colleague and friend for over 35 years. John introduced me to the field of digital communications and is always there to provide me with guidance and help whenever I need it. I especially appreciate his help during the process of writing this book by reading and editing the entire manuscript as well as providing me with many invaluable comments. I am also strongly indebted to the late Dr. Shahid Qureshi, my first supervisor and mentor when I started my industrial career. He assigned me to and guided me in a number of projects through which I gained most of my knowledge and experience on synchronization. I especially feel grateful for his encouragement for me to start working in the area of wireless communications. Special thanks are due to Dr. David Forney for the help and advice he gave me when I was working at Codex Corporation, and for encouraging and supporting me during the process of writing this book.

I would like to express my gratitude to my friends and colleagues Mingxi Fan, Rajiv Vijayan, Jun Ni, Shengshan Cui, Danlu Zhang, Wei Zeng, Haitong Sun, Xiaohui Liu, Shrenik Patravali, Tarun Tandon, Hongkui Yang, Luqun Ni, and Qiuliang Xie for their review and feedback on parts of the manuscript and many helpful discussions during the writing of the book. I would also like to thank Haixia Lan for reading and editing the manuscript to correct my grammar and to improve the presentation of the book.

Finally, I would like to thank my wife, Xiaoyun Ma, for her support throughout my career, in particular, in the past two years during the course of writing this book.

1 An Overview of Digital Communication Systems

1.1 Introduction

Until the late 1980s, analog modems were most widely used for data transmission over telephone wirelines. The techniques developed in digital communications from the 1960s through the mid-1980s were mainly targeted for wireline modems. Since then, various forms of digital transmission technologies have been developed and become popular for communication over digital subscriber loops, Ethernet, and wireless networks. The basic principles of synchronization techniques developed in the analog modem era have evolved. They are still being used directly or as the foundation of synchronization in other types of digital communication systems.

Since the mid-1980s, wireless has taken over wireline as the main form of communication connecting our society and permeating citizens' everyday life. Because of the widespread deployment of cellular communication systems, wireless technologies have been making impressive progress in all disciplines including synchronization.

Another important area of digital communications is satellite communication, which also experienced rapid progress during the same time period. While satellite communication has its unique properties, it also has many commonalities with wireline and wireless communications including those in the area of synchronization. In this book, we will focus mainly on the theories and techniques of synchronization for wireline and cellular-type wireless communication systems. However, what is discussed is also applicable to satellite communications.

To achieve bidirectional communications, two communication channels are needed. The two channels can be physically independent or can share the same physical media. Over wirelines, communications in the two directions can be either symmetric or asymmetric. For wireless, such as mobile communications, they are most likely asymmetric in the two directions. Communications from wireless base stations to mobile devices are usually called *forward link* communications. Communications in the other direction, i.e., from devices to the base stations, are usually called *reverse link* communications. In this book, most techniques discussed and examples given for wireless communications are assumed for the forward link, although what is discussed can be adapted for the reverse link as well.

In this chapter, an overview of the communication system and its main functional blocks is provided in Section 1.2. The details of the three major components of a typical communication system, i.e., the *transmitter*, the *channel*, and the *receiver*, are presented in Section 1.3. Section 1.4 provides a high-level view of the main synchronization functions in communication systems as an introduction to the subjects on which this book focuses. Sections 1.5 and 1.6 describe the basics of code division multiple access (CDMA) and orthogonal frequency-division multiplexing (OFDM) technologies to facilitate the discussion of synchronizations in communication systems that employ these technologies presented in later chapters. Finally, Section 1.7 summarizes what has been discussed and provides a summary of the common parameter notations used in this book to conclude this chapter.

1.2   A High-Level View of Digital Communications Systems

A digital communication system has three major components: a transmitter (Tx),[1] a receiver (Rx), and a communications channel. A high-level block diagram of such a typical digital communication system is shown in Figure 1.1.

The objective of a digital communication system is to send information from one entity, in which the information resides or is generated, through a communication channel to another entity, which uses the information. A high-level description of the process of how a communication system achieves information transfer from the source to the destination is given below. Details of their realization in different types of communications systems are provided later in this chapter.

A Brief Overview of Transmitter Operations

To be processed by the transmitter, the information from the source must be in binary form, i.e., represented by bits in the form of zeros and ones. If the information exists in nonbinary forms, such as in quantized audio or video waveforms, it is first converted to the digital form by source coding. The source-coded information or information that is already in the binary form may be encrypted and/or compressed. Such preprocessed source information in binary form is referred to as *information bits*, which are ready for further processing in the transmitter.

Inside the transmitter, the information bits are first processed in digital form. This step is commonly referred to as *digital baseband processing*. Its output is baseband signal samples, which are converted to analog baseband waveforms by a digital-to-analog converter (DAC).

The generated analog baseband waveforms are modulated onto a carrier frequency f_c to become *passband signals*. After being filtered and amplified, the passband signal at the carrier frequency is transmitted over a communication channel. The communication channel can be a wireless channel through radio wave propagation, or it can be a wireline channel such as a twisted pair of wires or cables.

---

[1] In some references, *Tx* is often used as a shorthand notation of transmitter. It can also be used to describe or specify any quantity that is transmitter related. The same usages can also apply to *Rx*.

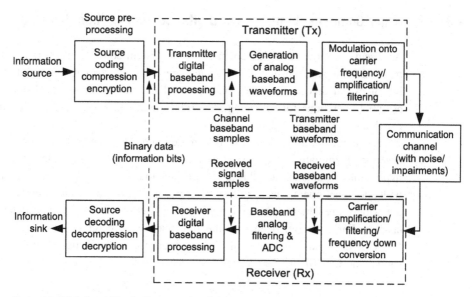

Figure 1.1 High-level block diagram of a digital communication system

A Brief Overview of Receiver Operations

After passing through the communication channel, the transmitted signal reaches the receiver at the other end of the channel. Communication channels always introduce various types of impairments including linear/nonlinear distortions, additive noise, and interference. As a result, the received signal is a distorted version of the transmitted signal. The task of the receiver is to recover the original transmitted information with no or little loss.

The first step of the receiving process is to convert the passband signal at carrier frequency to baseband. After performing passband filtering to reduce as much outband interference and noise as possible, the received signal is *frequency down-converted* or *demodulated* to become a baseband analog waveform.[2] The generated baseband analog signal can be expressed as the convolution of the transmitted baseband waveform and the channel impulse response (CIR) of the equivalent baseband channel.

The baseband analog waveforms of the received signal are analog filtered and are converted to digital samples. Then the baseband digital signal samples are processed by the receiver digital baseband processing block to regenerate the original information bits sent by the transmitter with no or few errors. The original source information is recovered from the regenerated information bits to complete the operations of the communication system.

The details of the operations performed in the blocks as just described are explained in the following.

---

[2] In digital communication terminology, the term *demodulation* commonly refers to the general operation of recovering the original data symbols from their modulated forms. In a loose sense, it may also mean the operations for data symbol recovery including frequency down-conversion, or carrier phase and/or frequency offset correction.

1.3   Major Components of Typical Digital Communication Systems

As discussed above, a digital communication system consists of a source information processing block, a transmitter, a channel, a receiver, and a source-information recovering block. In this section, we consider its three major components, which perform essential operations for achieving successful digital transmission, i.e., the transmitter, the channel, and the receiver.[3]

The functional blocks of the transmitter and the receiver and their operations described below are in the context of single-carrier digital communication.[4] However, in principle, what is discussed is also applicable to other digital communication systems with additional operations as necessary. Some digital communication technologies that are widely used in modern communication systems will be introduced later in this chapter. For example, details of the direct-sequence code division multiple access (DS-CDMA), which is a type of direct-sequence spread spectrum (DSSS) communication, and OFDM will be presented in Sections 1.5 and 1.6, respectively.

1.3.1   Transmitters and Their Operations

In this section, we describe various transmitter functional blocks and their operations for data transmission in single-carrier digital communication systems. The functional blocks considered are channel coding and interleaving, data symbol mapping, data packet forming and control signal insertion, spectrum/pulse shaping, digital-to-analog conversion, and carrier modulation and filtering. The block diagram of such a transmitter is shown in Figure 1.2.

1.3.1.1   Channel Coding and Interleaving

In order to improve the reliability of data transmission over the communication link, the *information bits* to be transmitted are first coded by a forward error correction (FEC) encoder. This step is called *channel coding*. Channel coding introduces redundancy into the input bits. Such redundant information is used by the decoder in the receiver to detect the information bits and to correct the errors introduced when the information was transmitted over the communication channel. Thus, a system with FEC coding can tolerate higher distortion and noise that are introduced during transmission than a system without it. In other words, a coded system can achieve more reliable communication at a lower signal-to-noise ratio (SNR) of the received signal than an uncoded system. The difference between the SNRs required by the two systems to attain the

---

[3] The source information processing and recovery blocks are sometimes viewed as associated with the transmitter and receiver, respectively. However, in order to concentrate on the aspects that are most pertinent to the subject matter of this book, their functions will not be considered here.

[4] Direct-sequence spread spectrum communication is also a type of single-carrier digital communication. To avoid confusion, in this book, the term "single-carrier communication" generally refers to non-DSSS single-carrier communication, unless otherwise specified.

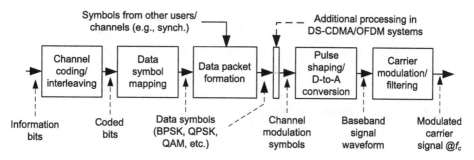

Figure 1.2 Block diagram of a typical transmitter

same error rate is called the *coding gain* with the SNR measured by the ratio of energy per information bit (E_b) divided by noise density (N_0), i.e., E_b/N_0.

Due to the introduced redundancy, the number of coded bits generated by the FEC encoder is always greater than the number of input information bits. Consequently, the ratio of the number of the input information bits divided by the number of the output coded bits, known as the *coding rate R*,

$$R = \frac{\text{number of information bits}}{\text{number of coded bits}} \qquad (1.1)$$

is always less than 1 in order to achieve positive coding gain.

Channel coding is an important component in a digital communication system. Its purpose is to improve the performance of digital communication systems so that these systems can approach the theoretical limit given by the Shannon channel capacity [1]. In earlier years, the most commonly used FEC codes were algebraic block codes, convolutional codes, and trellis codes. Since the mid-1990s, turbo and low-density parity-check (LDPC) codes have become popular in communication system designs because of their ability to approach and/or achieve the Shannon channel capacity.

FEC is one of the most important disciplines of digital communications and has a very well developed theoretical foundation. Due to the scope of this book, it is not possible to cover all of its details. We will only mention some of its aspects related to synchronization when appropriate. Interested readers can find abundant information regarding the coding theory, code design, and code implementation from many references and textbooks including [2] and [3].

In many communication systems, especially those intended for communication over fading channels, the coded bits are interleaved before being further processed. Briefly, an interleaver changes the order of the input bits. As a result, the consecutive coded bits are distributed over a longer time period when transmitted. Interleaving makes the effects of impulsive noise and channel deep fading spread over different parts of the coded bit stream, so that the impact of the noise and fading on the decoding in the receiver is reduced.

Because an interleaver only changes the order of the coded bits, the numbers of the input and output bits are the same. Thus, the interleaver can be viewed as a rate one encoder, which provides no coding gain over static channels.

The most popular interleavers used for digital transmissions are *block interleavers* [4, 5] and *convolutional interleavers* [6, 7]. Similar to error correction coding, interleaving does not directly affect synchronization. We will not elaborate on its functions further. The interested reader can find the relevant information in the available references.

1.3.1.2   Data Symbol Mapping

In order to be modulated and transmitted over a communication channel, the binary coded bits are usually grouped to form *data symbols*. Thus, each data symbol can represent one or more coded bits. Depending on the modulation technique used for transmission, the data symbols may be represented by a symbol constellation, as shown in Figure 1.3. The most popular types of modulation are amplitude modulation (AM), binary phase-shift keying (BPSK), quadrature phase-shift keying (QPSK), multiple phase-shift keying (MPSK), and quadrature amplitude modulation (QAM).

Note that the bits mapped to the constellation points shown in the figure satisfy a special property such that any two adjacent constellation points differ by only one bit. This is called *Gray mapping* or *Gray coding*. This property is important to achieving good system performance. This is because at the receiver a symbol error is most likely to occur when the adjacent constellation points are misinterpreted as the constellation point of the actually transmitted symbol. With Gray mapping, such a symbol error with high probability will result in only a one-bit error, which is easier to be corrected by channel decoding.

Depending on the symbol constellations used, different numbers of coded bits are mapped to one of the constellation points. The simplest modulation constellations are BPSK and QPSK. The BPSK constellation is used if one bit is mapped to a data symbol. A bit 0 is mapped to the real value 1, and a bit 1 is mapped to the real value -1. The QPSK constellation is used if two coded bits are mapped to a data symbol corresponding to a point in a QPSK constellation. For example, 00 are mapped to the complex point $1+j$, 10 to $-1+j$, etc.

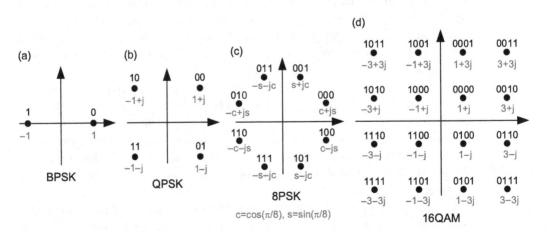

Figure 1.3 BPSK, QPSK, 8PSK, and 16QAM symbol constellations

When each data symbol represents more than two coded bits, MPSK and QAM constellations are often used. An MPSK constellation consists of 2^M points evenly distributed on a circle centered at the origin, and each data symbol corresponds to M coded bits. The 8PSK constellation is shown in Figure 1.3(c). As shown there, the triplet 000 maps to the constellation points of the complex number $\cos(\pi/8) + j\sin(\pi/8)$, 001 maps to $\sin(\pi/8) + j\cos(\pi/8)$, etc.

The most popular QAM constellations have 2^N constellation points on a square lattice grid. Therefore, each constellation point represents a data symbol mapped from an N-tuple of coded bits. The 16QAM constellation with four-bit Gray mapping and the coordinates of its constellation points in the complex plane is shown in Figure 1.3(d).

So far in this section, we have assumed that the output of the channel encoder is coded binary bits, and these bits are mapped to complex-valued data symbols. There are also channel encoders that directly generate complex data symbols from the information bits. One example is trellis coding, which was widely used during the 1980s and early 1990s. However, from the viewpoint of synchronization, only the form of the channel modulation symbols is relevant. Therefore, in the rest of this book, we are interested only in the type of data symbol constellations that is used in the transmission regardless of how the symbols are generated.

In the literature, data symbols as described above are often referred to as data modulation symbols. However, to distinguish the data modulation symbols from the channel modulation symbols to be discussed below, we will refer to them simply as data symbols. For single-carrier communications, data and channel modulation symbols are the same. However, they are different for OFDM and CDMA communication systems.

1.3.1.3 Formation of Data Packets/Streams for Transmission

The data symbol sequence as generated above can be directly used to generate baseband signal waveforms. However, in many digital communication systems, additional information may need to be transmitted together with the basic data. For example, in multiple access systems, there may be data from more than one user that need to be transmitted. Moreover, there are control signals such as signals for synchronization that need to be transmitted as well.

Data symbols carrying signaling/data information are generated in ways similar to those described in Section 1.3.1.2. The data and signaling symbols that are transmitted together may use different symbol constellations. For example, the data symbols may have 16QAM or 64QAM constellation in order to improve the transmission spectrum efficiency. In contrast, the signaling symbols, which are used to facilitate synchronization and for other purposes, such as pilot or reference symbols for performing channel estimation, are often transmitted in BPSK or QPSK. Nonetheless, their symbol rates of transmission are usually the same.

Different types of communication systems may have different data packet/stream transmission formats. For example, in wireline digital transmission, the control signals for training the receiver are usually sent at the beginning of a data communication session. Then regular data are transmitted continuously as a long symbol stream. Because the wireline channel does not change much in a short time duration, no additional training signals are sent during the normal data transmission/receiving session to improve the efficiency of the system.

In contrast, for wireless communication, the channel often experiences fast fading. As a result, the data symbols to be transmitted are usually organized in packets, which are transmitted consecutively to form a long data stream. Each packet has a short time span and is embedded with synchronization signals. In the receiver, these synchronization signals, such as pilot symbols, are used to perform channel estimation for the estimation of data symbols, as well as to achieve carrier and timing synchronization. These signals can also be used to obtain other information from systems that the receiver needs to communicate with.

1.3.1.4 Generation of Baseband Signal Waveform – Spectrum/Pulse Shaping

Transmitter channel modulation symbols, or simply *channel symbols*, are generated from the data symbols described above and converted to analog baseband waveforms. A channel symbol is represented by a complex number belonging to one of the modulation symbol constellations, including BPSK, QPSK, MPSK, QAM, etc. The baseband analog signal is generated by multiplying channel symbols with time-shifted continuous-time pulses that satisfy certain time- and frequency-domain characteristics. The Fourier transform of the most commonly used time pulse, denoted by $g_T(t)$, for generating analog baseband signal $x(t)$, is a square-root raised cosine (SRCOS) function, i.e.,

$$G_T(f) \cong \mathrm{FT}[g_T(t)] = \begin{cases} 1 & |f| \leq \dfrac{(1-\beta)}{2T} \\ \sqrt{0.5 \cos\left[\dfrac{\pi T}{\beta}\left(|f| - \dfrac{1-\beta}{2T}\right)\right] + 0.5} & \dfrac{(1-\beta)}{2T} < |f| \leq \dfrac{(1+\beta)}{2T} \\ 0 & |f| > \dfrac{(1+\beta)}{2T} \end{cases}$$

(1.2)

The time pulse $g_T(t)$ in the transmitter is actually generated by multistage processing, which will be mentioned later. It has a large peak at $t = 0$ and decays to zero when t goes to positive or negative infinity.

In single-carrier communication systems, the data symbols are directly used as the channel modulation symbols to generate the analog baseband signals. Thus, no complex conversion is needed, except maybe a single scaling operation. However, in DS-CDMA and OFDM communication systems, additional processing steps are performed to convert the data symbols to the suitable forms of channel symbols.

Generating baseband signal pulses with the desired spectrum from the channel symbols is often called *spectrum shaping* or *pulse shaping*. This process is the same for all three types of communication systems mentioned above. A block diagram of baseband waveform generation is shown in Figure 1.4(a).

Channel modulation symbols can be viewed as complex-valued random variables. They are converted to the analog form by *digital-to-analog converters*. The output of a DAC may be analog voltage impulses or rectangular pulses. Every T second, two

DAC outputs are generated for every complex channel symbol, one from its real part and the other from its imaginary part. To simplify our discussion, we will assume that the DAC outputs are impulses. Practically, it is more common to use the "sample-and-hold" DACs to generate output in the form of rectangular pulses with a width equal to T. The description given below is also applicable to such DACs with minor modifications.

The power spectrum of a channel symbol sequence has a constant magnitude as it is assumed that the channel symbols are, by design, uncorrelated in most systems. For T-spaced symbols, the spectrum is periodic with a period of $1/T$ [8] as shown in Figure 1.4(b).

It is possible to generate the analog baseband pulses by suitable analog low-pass filtering of the analog impulses generated by DACs directly from the symbol sequence as indicated in Figure 1.4(b). Because the analog filter performs both spectrum/pulse shaping and image rejection, it is quite demanding to implement. In practice, it is more efficient to perform spectrum shaping first by digital low-pass filtering. Once converted to the analog domain, the spectral images can be rejected by using a simple analog filter as described below.

To perform digital spectrum shaping, the channel symbol sequence is first up-sampled by inserting zeros between adjacent symbols. As an example, we consider the simplest case that one zero is inserted between any two adjacent channel symbols. The spectrum of the symbol sequence does not change after zeros are inserted. However, because the sample spacing is now equal to $T/2$, the spectrum of the new sequence has a base period of $2/T$ as shown in Figure 1.4(c). The zero-inserted sequence is filtered by a digital low-pass filter, which has the desired frequency response such as SRCOS previously mentioned. Figure 1.4(d) shows the power spectrum of the sequence at the filter output.

DACs are used to convert the digital sample sequence at the digital low-pass filter's output to $T/2$ spaced analog impulses, which have the power spectrum shape shown in Figure 1.4(d). Given that the spectrum has a period of $2/T$, the image bands are located at $2m/T$ Hz, $m = \pm 1, \pm 2, \ldots$ A simple analog low-pass filter shown in the figure filters the impulse sequence to retain the baseband signal spectrum centered at zero frequency and removes the image spectra. The output of the analog low-pass filter is the desired baseband signal waveform as shown in Figure 1.4(e).

In practical implementations, sample-and-hold circuits are usually incorporated into DACs to generate rectangular pulses. For $T/2$ spaced input digital samples, the widths of the rectangular pulses are equal to $T/2$. Thus, the power spectrum at the DAC output is equal to the spectrum shown in Figure 1.4(d) multiplied by a window of $\sin^2(\pi Tf)/(\pi Tf)^2 = \mathrm{sinc}^2(Tf)$.[5] As a result, the power spectrum of the actual analog waveform is no longer the same as the squared frequency response of the digital spectrum shaping filter.

---

[5] In this book, the sinc function is defined as $\mathrm{sinc}(x) = \sin(\pi x)/\pi x$, commonly used in information theory and digital signal processing, such as in Matlab. It can also be defined as $\mathrm{sinc}(x) = \sin(x)/x$ in some other disciplines in mathematics.

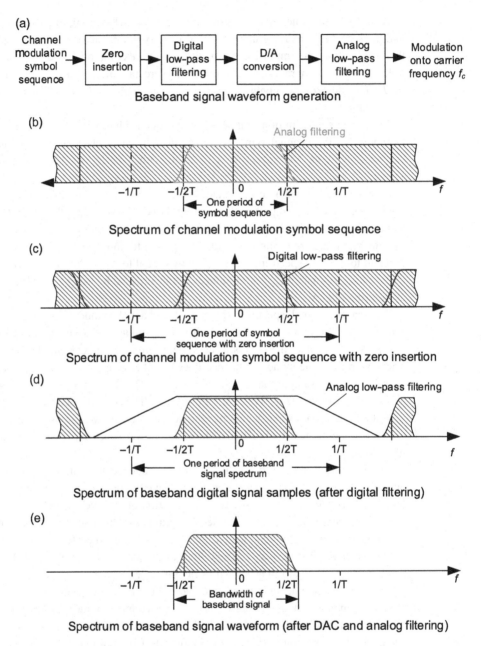

(a)

Channel modulation symbol sequence → Zero insertion → Digital low-pass filtering → D/A conversion → Analog low-pass filtering → Modulation onto carrier frequency f_c

Baseband signal waveform generation

(b)

Analog filtering

−1/T   −1/2T   0   1/2T   1/T f

One period of symbol sequence

Spectrum of channel modulation symbol sequence

(c)

Digital low-pass filtering

−1/T   −1/2T   0   1/2T   1/T f

One period of symbol sequence with zero insertion

Spectrum of channel modulation symbol sequence with zero insertion

(d)

Analog low-pass filtering

−1/T   −1/2T   0   1/2T   1/T f

One period of baseband signal spectrum

Spectrum of baseband digital signal samples (after digital filtering)

(e)

−1/T   −1/2T   0   1/2T   1/T f

Bandwidth of baseband signal

Spectrum of baseband signal waveform (after DAC and analog filtering)

Figure 1.4 Baseband waveform generation

Such distortions in signal spectra can be either precorrected by properly designed spectrum shaping filters or post-corrected by appropriate analog low-pass filters. Practically, the most convenient remedy for the distortion introduced by the rectangular analog pulse is to increase the number of the zeros between the channel symbols, i.e., to increase the upsampling frequency. By such a design, the signal spectrum occupies only a small portion of the main lobe of the sinc filter of the rectangular pulse, where it is relatively flat. Therefore, very little distortion is introduced. Descriptions of the practical baseband pulse generation using DACs with sampling-and-hold circuits can be found in the references (e.g., [9, 10]).

The generated analog baseband signal waveforms can be expressed as

$$x_T(t) = \sum_{k=-\infty}^{\infty} a_k g_T(t - kT) \qquad (1.3)$$

where a_k's are the complex-valued channel modulation symbols at time kT, and $g_T(t)$ is the combined impulse response of the digital low-pass filter, DAC output pulse, and the analog low-pass filter and often has the SRCOS frequency response given by (1.2). In other words, the overall pulse/spectrum shaping filtering is performed by the composite functional block that consists of all of these filtering functions. It is referred to as the *transmitter filter* in the rest of this book.

1.3.1.5 Modulation of Baseband Waveform to Carrier Frequency

In most cases, the baseband signal $x_T(t)$ given by (1.3) cannot be directly transmitted over communication channels. The analog baseband signal is first modulated onto a carrier frequency f_c that is suitable for the application. The modulated carrier signal is generated as

$$x_c(t) = x_{T,r}(t) \cos 2\pi f_c t - x_{T,i}(t) \sin 2\pi f_c t = \text{Re}\left[x_T(t) e^{j2\pi f_c t}\right] \qquad (1.4)$$

where $x_{T,r}(t)$ and $x_{T,i}(t)$ are the real and imaginary parts of $x_T(t)$. In the literature, they are often called *in-phase* and *quadrature* components of $x_T(t)$, denoted by I(t) and Q(t), respectively. The modulated signal can also be viewed as the real part of the product of $x_T(t)$ and the complex sinusoid $e^{j2\pi f_c t}$. The modulator implementation and the spectrum of the modulated carrier signal $x_c(t)$ at the carrier frequency f_c are shown in Figure 1.5.

The modulated carrier signal is then amplified and filtered to meet the regulatory requirements and is ready to be transmitted over the channel.

1.3.2 Channel

In the field of digital communications, a *channel* refers to the physical medium that connects a transmitter and a receiver. The signal sent by the transmitter propagates through the channel before it reaches the receiver. During the transmission process, impairments introduced by the channel distort the signal. Thus, the signal that reaches the receiver will not be the same as its original form when it was generated at the transmitter. The receiver regenerates the original information bits with no or few errors from the impaired received signal.

The impairments introduced by the channel include linear and/or nonlinear distortions, as well as additive noise and interference. Depending on the type of the channel, these impairments have different forms. In most cases, the interference is from many different and independent sources. Thus, it can be modeled as Gaussian noise according to the central limiting theorem.

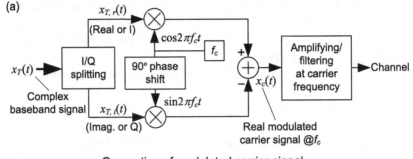

Generation of modulated carrier signal

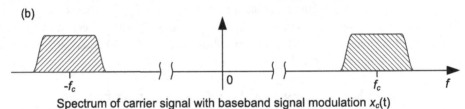

Spectrum of carrier signal with baseband signal modulation $x_c(t)$

Figure 1.5 Modulation of baseband signal to carrier

Based on the nature of the propagation media, communication channels can be broadly classified as wireline and wireless channels. Within each class, they can be further divided into different types as shown below.

1.3.2.1 Wireline Channel

The typical wireline channels include twisted-pair transmission lines and coax cables. These channels are used for transmissions over telephone lines, wired local loops, Ethernet cables, and others. Parts of such transmission channels may also include optical fibers.

Such transmission lines and cables can be viewed as two-port linear networks, which have continuous-time CIRs. The output of the channel is the convolution of the transmitted signal with the CIR. Hence, they may introduce linear distortions into the received signal in the form of intersymbol interference (ISI). In addition, thermal noise and interference from nearby electromagnetic signal sources will also corrupt the signal passing through the wireline channel.

Except for certain types of interference, such as impulse interference, the impairments introduced by the wireline channels have the common characteristic of being slowly time varying. This is because changes in the characteristics of the wireline impairments are most likely caused by changes in the surrounding environment, e.g., temperature variations. The time constant of such changes is most likely in the order of minutes, hours, or even longer. In other words, the tracking requirements for the impairments of wirelines are usually not demanding.

1.3.2.2 Wireless Channel

A number of different types of communication systems use radio waves as the communication media for the transmission of information. As a result, there are many types of wireless channels, such as satellite channels, mobile wireless channels, microwave communication links, Wi-Fi/WLAN channels, etc. Various wireless channels have different characteristics. For example, signals transmitted over mobile wireless channels encounter fast fading at high vehicle speeds, whereas signals transmitted over other types of wireless channels may encounter other type of impairments, such as high nonlinearity in satellite communications. In this book, we will mainly consider wireless channels that are pertinent to synchronizations in mobile and Wi-Fi wireless communication systems.

Wireless communication channels also introduce linear distortions, which create ISI in the signal at its output. In wireless systems, in which the transmitted signals are carried by radio waves from one point to another, the linear distortion introduced by the channel is caused by reflections of radio waves. As a result, unlike wireline channels, which have time-continuous CIRs, the CIRs of wireless channels consist of multiple time-discrete paths. Each of the paths is caused by the reflection from an object hit by a ray of the radio waves. Thus, wireless channels are usually modeled as *multipath channels*. Such a model is convenient to use in analyzing the behavior of the receivers in general and the synchronization functions in particular.

Another important property of most wireless channels is that they are time varying. This is especially true for mobile wireless communication systems, in which transmitters and/or receivers could be inside fast-moving vehicles. Over such channels, the signals received often experience different types of fading. Thus, mobile wireless channels are often called *fading channels*.

Fading in mobile channels can be divided into two categories: slow fading and fast fading. Slow fading, also called *shadowing*, occurs when mobiles are traveling through areas with different geographical characteristics and experience different radio wave propagation loss. Shadowing usually has a time constant of more than tens of seconds. Such a time constant affects the overall mobile system performance but has less special effect in receiver synchronization.

The second type of channel fading, which usually has a fading frequency from a few Hertz to hundreds of Hertz, is due to Doppler effects[11]. When a receiver is moving relative to the transmitter, the frequency of the transmitted waves observed by the receiver is different from that of the transmitted carrier. If the receiver moves toward the transmitter, the carrier signal frequency observed by the receiver becomes higher than the transmitted carrier frequency. Conversely, if they are moving away from each other, the observed carrier frequency becomes lower. The change in frequency is called *Doppler frequency shift* or simply *Doppler frequency*.

The value of Doppler frequency is equal to the carrier frequency times the speed of the relative movement between the transmitter and receiver divided by the radio wave propagation speed, i.e., the speed of light. Let us assume that a mobile is moving at 100 km/hour relative to the transmitter. If the carrier frequency is 2 GHz (gigahertz) and since the light speed is equal to 3×10^8m/sec, i.e., 1.08×10^9km/hour, the Doppler frequency shift that the mobile observes is equal to

$$f_d = \frac{100}{1.08 \times 10^9} \times 2 \times 10^9 \simeq 185 \text{Hz} \qquad (1.5)$$

If a vehicle is moving in open space, e.g., on the highway, the observed Doppler frequency shift behaves as a single tone. Namely, the carrier has a simple frequency change, which is equivalent to an offset of the carrier frequency that the receiver can correct by carrier synchronization. However, when a vehicle is in complex areas, e.g., in urban environments, the received signal's carrier frequency, phase, and magnitude will all be timing varying. In this case, the Doppler frequency can be modeled as a random variable. Its power spectrum, called *Doppler spectrum*, is distributed between $-f_{d,max}$ and $f_{d,max}$, where $f_{d,max}$ is computed by (1.5) at the vehicle speed. The most popular Doppler spectrum model, called the Jakes model, shown in Section 1.2 of [11], is given by

$$P(f) = \begin{cases} \dfrac{C}{\pi f_{d,\,max}} \left[\sqrt{1 - \left(\dfrac{f}{f_{d,\,max}} \right)} \right]^{-1} & f \le f_{d,\,max} \\ 0 & f > f_{d,\,max} \end{cases} \tag{1.6}$$

where $f_{d,max}$ is the maximum Doppler frequency and C is a constant to normalize the expression.

Above, we discussed some characteristics of the wireless channels, characteristics that are useful when considering the channels' impact on synchronization functions. More information regarding the general characteristics of wireless and mobile channels can be found in the literature, including the classic book [11] on this topic and other textbooks such as [5, 12, 13].

1.3.3   Receivers and Their Operations

The modulated signal at a carrier frequency transmitted over the channel is received by the receiver on the other end of the channel. The function of the receiver is to recover the transmitted information bits from the signal received from the channel. Its operations essentially undo the transmitter operations in the reverse order. The functional blocks of the receiver are frequency down-conversion, receiver filtering, digital sample generation, generation of the estimates of data symbols and/or decoding bit metrics, channel deinterleaving and decoding. After these operations, the transmitted information bits are regenerated with no or few errors. A block diagram of a typical receiver is shown in Figure 1.6.

The functions and operations of these receiver blocks are described below.

1.3.3.1   Frequency Down-Conversion

The transmitted passband signal at carrier frequency $f_c(t)$ is given by (1.4) with its spectrum shown in Figure 1.5(b). After passing through the channel, the received passband signal can be expressed as

$$r_c(t) = r_r(t) \cos 2\pi f_c t - r_i(t) \sin 2\pi f_c t = \text{Re}\left[r(t) e^{j2\pi f_c t} \right] \tag{1.7}$$

where $r(t) = r(t) + jr_i(t)$ is the complex baseband received signal, which is the transmitted baseband waveform convolved with the time response of the equivalent baseband channel [5]. It can be expressed as

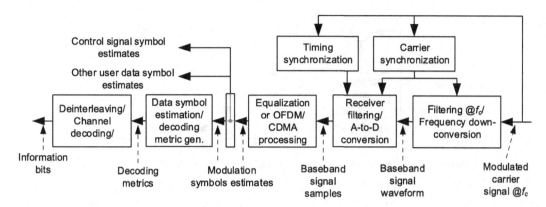

Figure 1.6 Block diagram of a typical receiver

$$r(t) = \sum_{k=-\infty}^{\infty} a_k g_c(t - kT - \tau) + z_c(t) \tag{1.8}$$

where $g_c(t)$ is the impulse response of the *composite channel* including the transmitter filter and the channel, τ is an unknown but constant delay, a_k's are the channel modulation symbols, and $z_c(t)$ is an additive white Gaussian noise (AWGN) process introduced by the channel.

The real and imaginary parts of $r(t)$, i.e., $r_r(t)$ and $r_i(t)$, also called the I and Q components, can be generated as follows.

Multiplying $r_c(t)$ by $\cos 2\pi \hat{f}_c t$, where $\hat{f}_c$ is the locally generated *down-conversion frequency*, sometimes called *demodulation frequency*, which is nominally equal to f_c, we obtain

$$r_c(t) \cos 2\pi \hat{f}_c t = r_r(t) \cos 2\pi f_c t \cos 2\pi \hat{f}_c t - r_i(t) \sin 2\pi f_c t \cos 2\pi \hat{f}_c t$$
$$= 0.5 r_r(t) \cos 2\pi (f_c - \hat{f}_c) - 0.5 r_i(t) \sin 2\pi (f_c - \hat{f}_c) t$$
$$+ \text{terms with frequency} f_c + \hat{f}_c \tag{1.9}$$

The last terms on the right side of (1.9) are at about twice the carrier frequency, which can be removed by low-pass filtering of $r_c(t) \cos 2\pi \hat{f}_c t$. The first two terms can be expressed as $0.5\mathrm{Re}\left[r(t)e^{j2\pi\Delta ft}\right]$, where $\Delta f \cong f_c - \hat{f}_c$ is called the carrier frequency offset of the baseband signal at the frequency down-converter output. Similarly,

$$r_c(t) \sin 2\pi \hat{f}_c t = -0.5\mathrm{Im}\left[r(t)e^{j2\pi\Delta ft}\right] + \text{terms with frequency} f_c + \hat{f}_c \tag{1.10}$$

Thus, $\mathrm{Im}\left[r(t)e^{j2\pi\Delta ft}\right]$ can be generated by low-pass filtering of $r_c(t) \sin 2\pi \hat{f}_c t$ with a sign change. The frequency down-converter implementation, which generates the baseband signal, based on (1.9) and (1.10) with carrier frequency offset Δf, is shown in Figure 1.7.

The sinusoids for frequency down-conversion are generated by a local oscillator (LO) with down-conversion frequency $\hat{f}_c$, which may be slightly different from the modulation frequency f_c due to the inaccuracy of the LO. As a result, there will be a nonzero frequency offset Δf. The offset can be corrected either by adjusting the LO frequency or

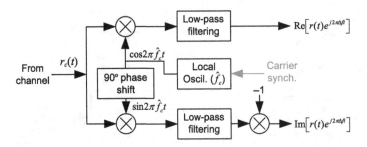

Figure 1.7 Frequency down-converter implementation

digitally later in the received signal samples. In both cases, the information from the carrier synchronization block is used, as will be shown in Chapter 5.

The low-pass filters only need to reject the image signals at twice the carrier frequency. Thus, their design and implementation are not very demanding. In practice, they also serve the purpose of antialiasing filtering for the analog-to-digital converters (ADCs) and need to meet the necessary requirements.

Thus, the frequency down-conversion process described above converts the modulated waveform at the carrier frequency to the equivalent baseband signal.

1.3.3.2 Receiver Filtering and Channel Signal Sample Generation

The baseband waveforms generated from frequency down-conversion block are converted to digital samples after optional additional filtering. The characteristics of the overall receiver front-end filtering functions, which we will call collectively the *receiver filter*, directly affect the receiver's performance. It has been shown in the literature that the optimal receiver filter is a matched filter (MF) of the received signal. The MF is optimal for receivers that employ equalizers for data symbol recovery in single-carrier communication systems [14]. It is also the optimal receiver front-end for various synchronization functions including initial acquisition, carrier synchronization, and timing synchronization as will be discussed in Chapters 2, 3, 5, and 6 of this book.

The implementation of the receiver filter as an MF requires the exact knowledge of the composite channel, which includes the transmitter filter and the channel. Practically, as this is not likely to be the case, the receiver filter is usually designed only to match the transmitter filter or the average composite channel characteristics. The receiver performance with such an approximate MF would be suboptimum.

The receiver filter can be implemented by using analog, digital, or mixed signal processing. With analog signal processing, the MF or approximate matched filtering is performed on the analog baseband signal. The output is sampled by ADCs at $1/T$ at the right time instant that is required by the MF filter, or sampled at m/T, where m is an integer. When using digital processing, the baseband signal is antialiasing filtered and sampled at a rate higher than its Nyquist sampling rate [8, 15], such as $2/T$ or higher.

The samples are filtered in the digital domain to implement the MF or approximate MF. Such a receiver filter can also be implemented by a combination of analog and digital filtering.

Without getting into the details of implementation, we can model the analog baseband signal at the receiver filter output as

$$y(t) = e^{j2\pi\Delta ft} \sum_{k=-\infty}^{\infty} a_k g(t - kT - \tau) + z(t) \tag{1.11}$$

where $g(t)$ is the impulse response of the *overall channel* including the transmitter filter, channel, and the receiver filter; τ is an unknown but constant delay; Δf is the frequency offset described in the last subsection; and $z(t)$ is an AWGN process. If the sampling delay is equal to τ and with the sampling rate of $1/T$, the sample y_n can be expressed as

$$y_n \cong y(nT + \tau) = e^{j2\pi\Delta f(nT+\tau)} \sum_{k=-\infty}^{\infty} a_k g(nT - kT) + z_n \tag{1.12}$$

where z_n is AWGN. If $g(lT) = 0$ for $l \neq 0$, such a channel satisfies the Nyquist (ISI-free) criterion [15, 16] and is referred to as the *Nyquist channel* and we have

$$y_n = a_n |g(0)| e^{j[2\pi\Delta f(nT+\tau)t+\theta_0]} + z_n \tag{1.13}$$

For example, if the transmitter filter is an SRCOS filter, the channel has a single path, i.e., no time dispersion, and the receiver filter matches the transmitter filter, then the overall channel has a raised-cosine (RCOS) frequency response. The CIR of such Nyquist channels has a form similar to a sinc function with its peak at $g(0)$ [5]. Thus, the sample given by (1.12) is at the peak of the CIR and y_n is ISI free. Otherwise, there will be ISI terms $e^{j2\pi\Delta f(nT+\tau)} \sum_{k=-\infty, k \neq n}^{\infty} a_k g(nT - kT)$ in y_n.

The digital samples are often generated at a rate of $1/T$ or $2/T$ depending on the type of processing that follows. The sampling rate should be synchronous to the remote transmitter's channel symbol rate, which is detected and controlled by the receiver timing synchronization block as will be discussed in Section 1.4.2 and Chapter 6.

There are two ways to generate digital samples that are synchronous to the remote transmitter symbol timing. Traditionally, they are generated by sampling the baseband waveform, and the sampling is done by using ADCs with a sampling clock that is synchronous to the remote transmitter timing. In recent years, it has become more common to generate digital samples first by using ADCs with sampling clocks that are free running or asynchronous to the remote transmitter timing. Samples with the desired timing are generated from the asynchronous digital samples through digital resampling. Digital resampling, also called digital rate conversion, is the subject of Chapter 7 of this book.

1.3.3.3 Data Symbol Detection/Estimation

The next step of the receiver operation is to estimate the transmitted data symbols from the digital samples of the baseband signal. This process varies depending on the type of the digital communications under consideration, namely, whether it is single-carrier, multicarrier (e.g., OFDM), or spread-spectrum (e.g., DS-CDMA) communications. In this section, we consider the first case.

In single-carrier communications, data symbols are directly used as the channel modulation symbols. If the overall channel satisfies the Nyquist criterion, as shown by (1.13) the digital samples at the receiver filter output are the transmitted data symbols a_n's scaled by the channel gain $|g(0)|e^{j\theta_0}$ with the additional phase rotation due to frequency offset Δf. They are corrupted by additive noise but are ISI free. Figure 1.8(a) shows such received signal samples with a carrier phase error $e^{j\theta(t)}$, where $\theta(t) = 2\pi\Delta f(t+\tau) + \theta_0$, introduced by the channel and the down-conversion frequency error, and corrupted by AWGN. After the phase error is corrected by the carrier synchronization block and with proper scaling, the samples are unbiased data symbol estimates as shown in Figure 1.8(b). On the other hand, when transmitted over a channel with ISI, in addition to AWGN, the received signal samples are also corrupted by ISI. In the latter case, an equalizer can be used to remove ISI for more reliable data symbol detection.

The most widely used equalizers are the linear equalizer (LE), the decision feedback equalizer (DFE), and the maximum likelihood sequence estimator (MLSE). The outputs of the first two types of equalizers are the estimates of the data symbols, which have the same form as the estimates of the symbols received from the ISI-free channels shown in Figure 1.8(b). In both cases, the symbol estimates are used to generate the decoding metrics for channel decoding. If MLSE is employed, decoding metrics can be directly generated from the received signal samples without recovering the data symbols.

Equalization is another important technical area in digital communications. Again, detailed descriptions of various equalization techniques are beyond the scope of this

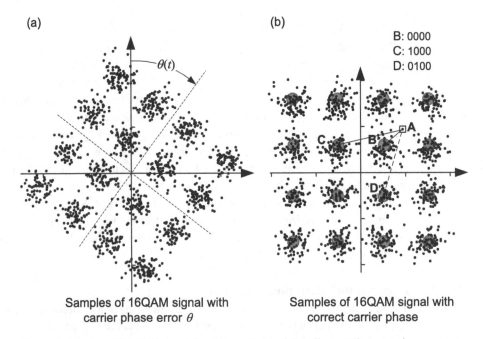

(a)   (b)

B: 0000
C: 1000
D: 0100

$\theta(t)$

Samples of 16QAM signal with
carrier phase error θ

Samples of 16QAM signal with
correct carrier phase

Figure 1.8 Samples of 16QAM signal for data symbol and decoding metric generation

book. However, they are well documented in textbooks on digital communications, including [3, 5, 17, 18].

The generation of data symbol estimates and the decoding metrics in CDMA and OFDM communication systems are discussed in Sections 1.5 and 1.6.

1.3.3.4 Generation of Decoding Metrics

The output samples generated from the front-end processing of many types of receivers with the carrier phase error removed and with proper scaling are the unbiased estimates of the transmitted data symbols. Such unbiased estimates can be modeled as the transmitted data symbols, which are the corresponding constellation points, with AWGN and other distortions.

Receivers that remove phase errors between the transmitted data symbols and their estimates before further processing are called *coherent receivers*. Communication systems with coherent receivers are called coherent communication systems. There are also communication systems that employ noncoherent coding/modulation techniques, which can operate without correction of carrier phase errors. However, the performance of noncoherent receivers is usually inferior to that of coherent receivers. Thus, coherent communications are preferred in system designs.

Data symbol constellations are usually in the form of BPSK, QPSK, MPSK, and QAMs. As an example, the estimates of transmitted 16QAM symbols with carrier phase errors are shown in Figure 1.8(a). Figure 1.8(b) shows the symbol estimates after carrier phase error correction and the expected constellation points.

To perform coherent data symbol detection, it is necessary to correct the carrier phase error so that the expectations of the symbol estimates align with the transmitted symbols' constellation points. As shown in Figure 1.8(b), square A represents the digital sample, which is an estimate of the transmitted data symbol, generated from the receiver filter. The gray dots in the background represent the constellation points, one of which is the transmitted data symbol. When there is noise and distortion introduced by the channel, the estimate will not be exactly on top of the transmitted symbol.

It can be shown that if the error in the estimate is solely due to AWGN, the maximum likelihood (ML) estimate of the symbol is the constellation point closest to the estimate, e.g., B in Figure 1.8(b). However, in systems with channel coding, it will be necessary to compute the *likelihood* or *log-likelihood* values of the coded bits that are mapped to the transmitted symbol for performing channel decoding.

The most commonly used decoding metrics of coded bits are the *log-likelihood ratios (LLRs)*.[6] The LLR values for BPSK and QPSK symbol constellations are relatively easy to compute. For the BPSK constellation on the real axis shown in Figure 1.3(a), the log-likelihood value of the transmitted bit is simply the real part of the estimate divided by the noise variance as will be shown by the example in Section 2.1.2. For the QPSK constellation shown in Figure 1.3(b), the real and imaginary parts of the estimate

---

[6] The concepts of the likelihood, log-likelihood, the maximum likelihood, and the log-likelihood ratio will be introduced in Chapter 2.

divided by the noise variance are the LLRs of the first and the second bits that determine the QPSK data symbol.

It is quite complex to compute the exact LLR of a coded bit for high-order data symbol constellations. In general, it is necessary to evaluate the likelihood values of the received signal sample with respect to all of the possible transmitted data symbols, i.e., all of the constellation points [19]. Therefore, in practical implementations, approximate LLRs are often used instead of the exact ones with little loss in performance.

One of such approaches to computing approximate LLRs is that, instead of evaluating all of the likelihood values, only the two largest ones, one for the bit equal to one and the other for the bit equal to minus one, are computed. Moreover, it can be shown that maximizing the probability of the received sample with regard to the transmitted data symbol is equivalent to minimizing the distance between the received sample and the corresponding constellation point [5]. Therefore, only the two closest constellation points corresponding to the coded bit being one and minus one, need to be considered.

Let us use the received signal sample A shown in Figure 1.8(b) as an example. For the first bit in the transmitted symbol, the closest constellation points for the first bit being one and minus one are B and C, respectively. When the noise is stationary, the LLR of the first bit is simply equal to the difference of the distance between A and C minus the distance between A and B divided by the noise variance. Similarly, the LLR of the second bit is equal to the difference of the distance between A and D minus the distance between A and B divided by the noise variance.

There are a number of methods to compute, with QAM constellations, the LLR of the coded bits of the transmitted data symbol. The theory and the details of these approaches can be found in [19, 20].

1.3.3.5   Deinterleaving and Decoding

The LLR values generated in the previous processing step are deinterleaved to undo the interleaving performed in the transmitter. After deinterleaving, the correlated noise, interference and other distortions contained in the received signal are nearly uniformly distributed in the resulting LLRs that are used by the decoder as decoding metrics. Under normal channel conditions, the channel decoder recovers the original transmitted information bits with no or few errors. These recovered bits are sent to the information sink to regenerate the original information sent by the transmitter.

The theory and operations of deinterleaving and decoding are not essential to the understanding of synchronization and will not be explored further.

The transmitter and receiver operations continue until all of the available information is sent and recovered.

1.4   Overview of Synchronization Functions in Digital Communication Systems

In Section 1.3, we provided a general view of the major transmitter and receiver functions in digital communication systems. The descriptions were given in the context of their operations in a data transmission/receiving session. In addition to what has been discussed, there are other important transmitter and receiver functions, which are introduced in this section. These functions, including initial acquisition, carrier synchronization, and timing synchronization, are commonly classified as synchronization functions.

Synchronization functions play an important role in achieving reliable and robust communication system performance. They are the key means to establish a reliable communication link during the initial stage of establishing communication and to maintain the receiver operations with desirable performances during normal data sessions. Synchronization is especially important for communication over mobile wireless channels because such channels change rapidly due to fast fading. As mobile/wireless communications have become popular in recent years, a good understanding of synchronization in digital communication systems becomes even more important today than before.

Synchronization in a communication system involves both the transmitter and the receiver. While most of the synchronization tasks are performed by the receiver, the transmitter also plays an important role. Specifically, special control signals are usually included in the transmitted signals to facilitate achieving synchronization by the receiver.

The theory, characteristics, and implementation of synchronization in digital communication systems are the subjects of this book. The details of the synchronization functions will be presented and discussed in later chapters. In this section, we provide brief overviews of the three major synchronization functions and their roles during various stages of transmission and reception in digital communication systems.

1.4.1 Initial Acquisition

Initial acquisition is the first task a device executes when it tries to establish a communication link with the desired remote transmitter. The objectives of initial acquisition are to establish quickly and reliably a communication link and to be ready for the data sessions that follow. It significantly affects the overall link and system performance.

To facilitate the initial acquisition by device receivers in a multiple-access communication system, the transmitters that connect to a network constantly transmit signals that are known to all of the devices in the area covered by the network. Receivers that need to establish communication with the network continuously search for the known signal that they expect to detect. Once such a signal is detected by a device, the information in the detected signal is extracted. Such information includes various system parameters, in particular, the timing of the start of the data frames. Thus, initial acquisition usually also performs *frame synchronization*.

When starting operations for the first time, the receiver's parameters are not accurately known to the device itself. Thus, another task of initial acquisition is to calibrate and initialize various receiver parameters by using the information obtained from the detected signal. Such parameters include the accuracy of the LO frequency, the desired timing phase and frequency of the digitals samples, receiver's gain, etc. Once the system information is acquired and the parameters are calibrated, the receiver is ready to start normal data sessions.

For point-to-point communications, the roles of the two entities on the two ends of the channel are symmetric in most systems. However, the communication protocols usually assign one entity as the master and the other as the slave. The master first starts transmitting synchronization signals and the slave will listen until such signals are detected. Then, their operations will continue according to the communication protocols.

The analysis, details of the operations, and the implementation of initial acquisition in various communication systems, will be presented in Chapter 3.

1.4.2   Timing Synchronization

Another major receiver synchronization function is *timing synchronization*, also often called *timing recovery* in the literature.

As shown in Section 1.3.1.4, the digital transmission signal, before it is converted to the analog form, consists of the *transmitter channel symbols*. These channel symbols are multiplied to analog pulses spaced at a fixed time interval T. The reciprocal of T, i.e., $1/T$, is called the *transmitter channel symbol rate*. The analog baseband waveform, which is the summation of the pulses multiplied by the transmitter channel symbols, is modulated onto carrier frequency and sent over the communication channel.

After down-converted from the carrier frequency to baseband, the received baseband signal is converted to digital samples at the received signal sampling rate for further processing. This sampling rate is nominally equal to $1/T$ or m/T, where m is an integer. For normal receiver operations, the receiver's sampling clock should be synchronous to the transmitter channel symbol rate of $1/T$. Thus, the objective of the receiver timing synchronization is to ensure that the received signal is sampled synchronously to $1/T$.

In addition, the sampling time should have a fixed delay, called *timing phase*, relative to the time at which the transmitter symbols are transmitted.

As shown by (1.13), if the samples are generated every T, the sampling time should be at the peak of the channel impulse response to maximize the received signal sample energy. Hence, the sampling clock must have the proper delay to compensate for the delay introduced by the overall channel.[7] In other words, the sampling clock must be *synchronous to the transmitter symbol clock in timing phase*. However, if the samples are generated at the Nyquist sampling rate of the received signal or higher, e.g., at $2/T$, there will be no information loss regardless of the timing phase. If such samples are used by the receiver to recover the transmitted data symbols directly, an exact sampling phase is not critical. However, it is important that the sampling phase is stable relative to the transmitter symbol clock, so that the channel observed by the receiver after sampling does not change with time.

In both cases, it is important to maintain a fixed timing phase relative to the timing phase of the transmitter channel symbol clock. It is equivalent to maintaining a

---

[7] The delay introduced by the overall channel may be longer than the transmitter symbol interval T and can be expressed as $M_D T + \tau$. Only the fractional part of the delay τ is observable and can be determined from the received signal at the physical (PHY) layer and is thus of interest to receiver synchronization. The integer part $M_D T$ can be determined at higher layers during system acquisition if necessary.

sampling frequency synchronous to the remote transmitter channel symbol rate. In other words, timing synchronization needs to achieve both phase and frequency synchronization between the receiver sampling clock and the transmitter channel symbol clock.

During initial acquisition, the signal detector searches for the signal energy peak by correlating a known transmitted symbol sequence with the received signal samples at different sampling times. Thus, the signal peak found can serve as the initial timing phase estimate. Once entering the data mode, the timing phase and frequency information contained in signal samples for data recovery are extracted and used to maintain and fine-tune the timing phase and frequency synchronization. Timing synchronization is usually achieved by forming a feedback loop and thus it is often called timing locked loop (TLL) or timing control loop (TCL). Once the TLL has converged, its output is synchronous to the remote transmitter timing, i.e., the remote transmitter's channel symbol clock.

The timing synchronization functions in receivers of different communication systems, such as single-carrier systems with or without spreading and OFDM systems, have their commonalities and differences. The theory and the details of timing synchronization operations of various communication systems will be presented in Chapter 6.

1.4.3   Carrier Synchronization

Carrier synchronization is yet another important component of synchronization in digital communication systems. It directly affects the overall system performance including system capacity and the robustness under adverse conditions. Specifically, carrier synchronization concerns the estimation and compensation of the carrier frequency and phase differences between the transmitted signal and the corresponding received signal.

As discussed in Section 1.3.3, the signal received from the communication channel at the carrier frequency is down-converted in frequency to baseband by using a locally generated reference clock. The baseband signal is filtered by the Rx filter and converted to digital samples by ADCs. The generated digital samples are processed to produce the estimates of the transmitted channel symbols.

The receiver channel symbol estimates are used to recover the transmitted data symbols and to generate the decoding metrics of the original transmitted information bits. To reliably recover the original transmitted data, the phase of a transmitted data symbol and the phase of its estimate should be the same. However, practically, this is usually not the case. As shown by (1.13) and Figure 1.8(a), in addition to being corrupted by noise and interference, a complex symbol estimate may have a phase different from that of the transmitted data symbol. Such a phase difference is caused by the complex channel response and the offset between the transmitter carrier frequency and receiver down-conversion frequency.

All these impairments degrade the receiver performance. While it is impossible or difficult to remove the additive random noise and interference, it is possible to accurately estimate and correct the phase errors in the receiver. Such phase error correction is particularly important to coherent communication systems. In coherent communication systems, to reliably recover the transmitted data, the phase of the symbol estimate should align with that of the original transmitter data symbols.

Hence, the receiver carrier synchronization function, also called *carrier recovery*, needs to perform both carrier phase and carrier frequency synchronization. Various schemes for carrier synchronization have been developed and implemented in digital communication systems. Briefly, the carrier synchronization block obtains the carrier phase and frequency offset information from the analog or digital form of the received signal. The information is processed and used to compensate for these offsets in the corresponding receiver blocks.

Detailed descriptions of the various aspects of carrier synchronization and the realization are discussed in Chapter 5.

So far, we focused our discussions on single-carrier communications. Due to the widespread applications of DS-CDMA and OFDM technologies in recent years, a significant portion of this book is devoted to synchronization in communication systems that employ these technologies. To facilitate our further discussion, the basics of these technologies are presented in the next two sections.

1.5   Basics of DSSS/DS-CDMA Communications

For a long time, commercial digital communication systems almost exclusively used single-carrier technology without spreading to wider band. The first commercial wireless communication system using DSSS technology was the IS-95 cellular system proposed in the late 1980s and standardized in a TIA standard committee in the United States in the early 1990s. Later, IS-95 evolved into the cdma2000–1x standard, one of the worldwide 3G wireless communication standards. They are known as DS-CDMA, or simply CDMA, cellular communication systems [21, 22]. Another member of this family, cdma2000 Evolution-Data Optimized, or simply EV-DO, was developed in the mid to late 1990s specifically for wireless data communications. In the late 1990s, the DS-CDMA technology was adopted by the European Telecommunications Standards Institute (ETSI) in a new wireless communication standard called *wideband CDMA (WCDMA)* [23]. It was later adopted by the 3GPP partnership of standard organizations, which have standard organizations in many countries including ETSI as a member. WCDMA is now a member of the *Universal Mobile Telecommunications System (UMTS)* family. WCDMA and cdma2000 are radio interfaces of IMT-2000, the 3G standard adopted by the *International Telecommunication Union (ITU)*.

In a DS-CDMA system, the transmitter first generates a *spreading sequence*, which is complex-valued, with white noiselike flat spectrum, and at a rate of f_{ch}, called *chip rate*. Here the term *chip* refers to a symbol interval of the spreading sequence and *chip rate* refers to the rate at which the CDMA channel symbols are transmitted. In that sense, a chip is equivalent to a channel modulation symbol in the conventional single-carrier systems discussed above. The spreading sequence is divided into multiple segments, each of which has N elements. Every data symbol for transmission is multiplied to a segment of the sequence. The resulting sequence is modulated to the transmitter carrier frequency and is then transmitted. Thus, the chip rate is equal to N times the input data symbol rate. The spectrum of such generated transmission signal is N times wider than the spectrum of direct transmission of data symbols without spreading. In other words, the spectrum of the transmitter signal is *spread* by a factor of N. Hence, N is called the *spreading factor*. This is where the term *spread spectrum* comes from.

Since the DS-CDMA signal occupies a wider spectrum than the original signal, it is not spectrum efficient. However, the receiver can combine coherently N received signal samples to generate an estimate of the data symbol. The SNR of the data symbol estimate will be N times higher than the received signal sample. This process is called *despreading*, which yields a *processing gain* of N. Due to the processing gain, the system can operate at a much lower SNR to achieve the same performance of a nonspread spectrum system. Essentially, the DS-CDMA technology trades spectrum efficiency for improved SNR. DS-CDMA is particularly suitable for voice communications.

During a conversation, one person usually speaks for only about 30 percent of the time. In other words, the *speech activity* is about 30 percent on average. DS-CDMA can effectively take advantage of this speech activity factor. Below, we consider the forward link communication from a base station to mobile devices as an example.

In base stations of a DS-CDMA system, such as IS-95/cdma2000–1x and WCDMA, the data from multiple users are spread by different sequences that are uncorrelated or orthogonal to each other and added together before transmission. At the receiver, when despreading the received signal using one user's spreading sequence for the user's data, the interference from other users' signals, which are spread with uncorrelated or orthogonal sequences, are suppressed. Hence, the desired user's signal is enhanced and recovered. Due to the speech activity factor, on average, the power of the interference from other users is not as high as when the users transmit full power all the time. This is why the DS-CDMA can support more simultaneous users than the early generations of TDMA and FDMA systems could.

For wireless data communications, the advantage of DS-CDMA disappears relative to other communication technologies. For data communication it is most efficient to totally avoid inter-user interference. As a result, most wireless data communication systems employ transmission technologies that create no or little such interference. For example, Long Term Evolution (LTE) and recent generations of Wi-Fi systems use the OFDM technology. EV-DO essentially employs TDMA for handling user data signals, even though it uses direct sequence spreading for other purposes.

Below, we describe the basic operations of DS-CDMA forward-link transmitters and receivers to establish the foundation for the device synchronization functions described in the later chapters.

1.5.1   DS-CDMA Transmitter Operations

DS-CDMA, or simply CDMA, is a single-carrier communication technology. Its transmitter is essentially the same as the transmitters of generic single-carrier communication systems described in Section 1.3.1. The main difference between them is that in a

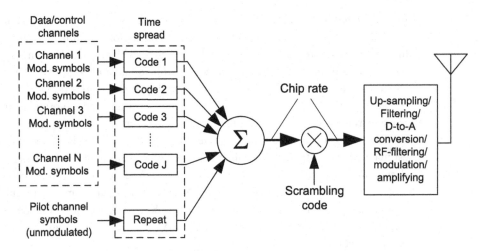

Figure 1.9 An exemplary DS-CDMA transmitter

CDMA system, multiple data and control channels are transmitted simultaneously in the same frequency band. To distinguish the data from different channels, each data symbol is multiplied by a channel-specific spreading code. The spread symbols from different channels are added together before being transmitted. A typical forward-link transmitter of a CDMA communication system is shown in Figure 1.9. Because many of its components are the same as that in generic single-carrier transmitters, only the parts unique to CDMA are shown.

Similar to what is described in Section 1.3.1.2, coded bits of CDMA traffic and control channels are mapped into data symbols at the symbol rate $1/T$. Each data symbol is multiplied to a channel-specific spreading code. In order for the receiver to distinguish the data symbols from different channels, the channel spreading codes are uncorrelated or orthogonal to each other.

The spread spectrum signals are generated in two steps. In the first step, a symbol from each channel is multiplied, also called *covered*, by a channel-specific orthogonal code. If the symbols of different channels have different rates, the length of the orthogonal code is inversely proportional to the data symbol rate of the channel. As a result, the spreader outputs of all of the channels will have the same chip rate f_{ch}. To simplify the discussion, in this section, we only consider the case that the symbol rates of all channels are the same. In addition, the transmitter also generates a pilot channel with unmodulated symbols, typically BPSK with value +1. The pilot symbols are repeated to be at the same rate as other channels. This step is called *time spreading*.

In the time-spreading step, the purpose of applying the orthogonal code covering is to eliminate, or at least to reduce, the interference between different channels. The most popular orthogonal spreading code is Walsh code [24]. Hence, we will use it as the example in the discussion below.

Walsh codes, also called Walsh functions or Hadamard codes, are a family of binary orthogonal codes with lengths of $N = 2^M$. There are N members in a group of length-N Walsh functions. Each of the length-N Walsh functions can be expressed as an

N-dimensional vector, denoted by $\mathbf{w}_N^{(i)}$ $i = 0, 1, \ldots, N-1$, with N elements, $w_{N,0}^{(i)}$, $w_{N,1}^{(i)}, \ldots, w_{N,N-1}^{(i)}$, each of which is equal to either -1 or $+1$, and

$$\left(\mathbf{w}_N^{(i)}\right)^t \mathbf{w}_N^{(j)} = \sum_{k=0}^{N-1} w_{N,k}^{(i)} w_{N,k}^{(j)} = \begin{cases} N & \text{for } i=j \\ 0 & \text{for } i \neq j \end{cases} \tag{1.14}$$

where the superscript t denotes the transpose operation of a vector or a matrix.

The Walsh functions look like variable-width square waves that do not have white spectrum. Because different Walsh codes are orthogonal to each other, the data symbols spread by different Walsh codes will not interfere with each other when transmitted and received over a single-path channel. For multipath channels, there is interference among the signals passing through the paths with different delays. However, the interference is reduced nonetheless. Details of Walsh codes and related topics have been well discussed in the literature, e.g., in [25].

Although, the chip rate of the time-spread sequence is higher than the symbol rate, the spectrum of such time-spread symbol sequences are not white as desired for transmission. To make it white noise like, the combiner output symbols are scrambled by pseudo-noise (PN) sequences, such as m-sequences [26]. This is the second step, which performs *frequency-spreading*, in generating the spread spectrum signals. In such a way, the data symbol $S_i(n)$ of the ith channel is expanded to a vector $\mathbf{w}_N^{(i)} S_i(n)$ with N elements and N is called the *spreading factor*. The chip sequences transmitted at nNT_c, denoted by vector $c(n)$, can be expressed as

$$\mathbf{c}(n) \cong \left(c_{nN} \quad c_{nN+1} \quad \cdots c_{(n+1)N-1}\right) = \mathbf{P}_{nN} \sum_{i \in \text{ active channels}} \mathbf{w}_N^{(i)} S_i(n) \tag{1.15}$$

where $\mathbf{P}_{nN}$ is a diagonal matrix with elements of sp_{nN}, sp_{nN+1}, $\cdots sp_{(n+1)N-1}$, which are the elements of the scrambling code. With no loss of generality, we will assume that these elements have unit magnitude. The pilot sequence after scrambling is the same as the scrambling sequence.

After the CDMA chips are up-sampled and low-pass filtered, they are converted to analog signals by the D-to-A conversion and further filtered by analog filters. Finally, the baseband analog signal is up-converted in frequency, i.e., modulated, to carrier frequency f_c and is then transmitted. These final steps are the same as those performed by the single-carrier transmitters discussed in Sections 1.3.1.4 and 1.3.1.5. The nominal transmission bandwidth is equal to f_{ch}. Commonly, the transmitter filter is designed so that the transmitted signal will have a baseband spectrum close to an SRCOS shape.

1.5.2 DS-CDMA Receiver Operations

A block diagram of a typical DS-CDMA receiver in a mobile device is shown in Figure 1.10. Similar to the generic single-carrier receiver described in Section 1.3.3, the DS-CDMA receiver's radio frequency (RF) front-end down-converts the received signal from carrier frequency f_c to baseband based on a locally generated reference clock. The generated baseband signal is low-pass filtered and converted to digital

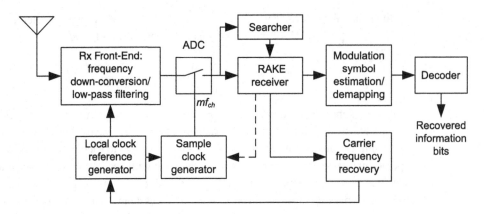

Figure 1.10 An exemplary DS-CDMA receiver in mobile devices

samples by an ADC at a rate of mf_{ch}, where m is an integer. The low-pass filter performs the antialiasing function. It is usually designed to match, at least approximately, the transmitter filter characteristics. The signal spectrum after receiver filtering is close to RCOS shape and approximately satisfies the Nyquist criterion for a single-path transmission channel.

For DS-CDMA systems, the communication channel is usually modeled as a multipath channel due to radio wave reflections. The impulse response of such a channel is a combination of multiple impulses with different delays and complex gains. Namely,

$$h_{ch}(t) = \sum_{l=0}^{L-1} g_l \delta(t - \tau_l) \qquad (1.16)$$

where $\delta(t)$ is the Dirac delta function, and g_l and τ_l are the complex gain and delay of the lth path of the channel, respectively. These paths are assumed resolvable, i.e., the spacing between two adjacent paths is greater than the chip duration T_c.

The digital samples generated from the ADCs are processed by a *RAKE receiver* [27]. The function of the RAKE receiver, which consists of multiple "fingers," is to generate the estimates of the transmitted data symbols. The RAKE receiver is a key component of a DS-CDMA receiver and will be discussed in detail in the next section.

To estimate data symbols effectively, the RAKE receiver must have the precise knowledge of the complex gain g_l and time delay τ_l of the paths. Thus, determining the existence of the paths with sufficient signal energy and their delay and phases is the main synchronization task of a DS-CDMA receiver.

The receiver operations in initialization and in data traffic states can be summarized as follows.

During initial acquisition, a searcher in the receiver looks for the pilot channel signal or special synchronization signals that are defined by system designs. Once such a desired signal is found, the receiver also acquires the carrier phase and time delay of at least one of the paths with sufficient energy. The searcher passes the information to at least one RAKE finger for it to demodulate the signal. At the same time, the carrier frequency offset value is estimated. It is then used to initialize the frequency value of the local clock generator to reduce the offset between the local reference and the remote transmitter carrier frequencies. The details of the DS-CDMA receiver operations in the initial acquisition stage are discussed in Section 3.5.

After the system parameters are acquired during initial acquisition, the communication link between the base station and the device can be established, and then the receiver enters the data mode.

In the data mode, the receiver continues improving the accuracy of the estimated carrier and timing parameters. It also searches for new paths with sufficient energy from the acquired base station and from other base stations. When a new path is found, the searcher passes its information to an available RAKE finger to demodulate its signal. If the new path's signal is stronger than one of the existing ones, the new path may be assigned to the weaker RAKE finger and the older path may be deassigned. These operations are called *finger management*, which is one of the most important functions of the DS-CDMA timing synchronization. It is probably also the most difficult and challenging operation of DS-CDMA receivers. The examples of the timing synchronization operations of DS-CDMA systems are discussed in Section 6.5.

After initial correction of a frequency offset, the RAKE fingers will track the carrier phase changes of the associated paths. The residual frequency offset can be computed from consistent phase changes. Carrier synchronization is usually less difficult to achieve than timing synchronization in DS-CDMA systems, except in some special cases.

In the receiver, the LO inaccuracy will cause a much larger carrier frequency offset than the timing frequency offset. Therefore, it is easier to correct the LO frequency from the estimated carrier frequency offset than from the estimated timing frequency offset. The estimated carrier frequency offset can then be used to correct the timing frequency error, since the carrier and timing frequencies are most likely derived from the same reference in the transmitter. However, it is also possible that these two frequencies are not locked. For example, additional frequency up and down-conversion may be involved during transmission. In such cases, it may be desirable to implement a second-order loop to correct the residual timing frequency error.

Once the initial finger timing phases are estimated accurately, the estimation of data symbols and combining of the outputs of multiple fingers are performed in a RAKE receiver, which generates the estimates of transmitted data symbols. Decoding metrics are derived from the symbol estimates and used by the decoder to recover the transmitted information bits. These operations are relatively less dynamic than the timing and carrier synchronizations. The details of recovering DS-CDMA symbols by RAKE receivers are described below.

1.5.3   RAKE Receiver: The DS-CDMA Demodulator

Recovering the transmitted data symbols in a DS-CDMA receiver is usually performed by the RAKE receiver, which was originally proposed in [27] and [28]. A form of the RAKE receiver appropriate for practical DS-CDMA systems was also proposed there. In this section, we describe the implementation of such a practical RAKE receiver and show that it is an approximate realization of a matched filter in a discrete multipath channel often encountered in a wireless communication environment.

Channel Model and RAKE Receiver Structure

Figure 1.11(a) shows an example of a multipath channel with four paths. The receiver filter output generated by a transmitted chip is shown in Figure 1.11(b). If the combined frequency response of the transmitter and receiver filters has an RCOS shape as desired, the impulse response of each path is close to a sync function. When the paths are resolvable as defined above, each of the peaks can be determined at the receiver. The signal sampled at a peak generates the highest energy received from the associated path.

Figure 1.12 shows a high-level RAKE receiver and an example of a RAKE finger. The operations of the RAKE receiver and how its fingers operate are described below.

Overview of RAKE Receiver Operations

As shown in Figure 1.12(a), the receiver front-end processes the received signal by converting it to baseband and generates samples at an integer multiple of the chip rate, i.e., mf_{ch}. Depending on the implementation, the value of integer m is commonly selected to be 2 or 8, as will be discussed in Chapters 6 and 7. The samples are sent to a number of RAKE fingers of the receiver for recovering the data symbols sent to this device. In the multipath channel shown in Figure 1.11, each RAKE finger demodulates the signal from a particular path. The samples sent to the finger are delayed according to the path delay.

The output of each of the RAKE fingers is an estimate of the transmitted symbol. The estimates of the same transmitted symbol from multiple RAKE fingers are added together to generate a combined symbol estimate. Bit LLRs are derived from the symbol estimates as shown in Section 1.3.3.4 to be used by a decoder for recovering the transmitted information bits.

(a)

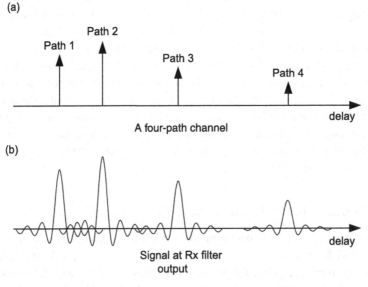

(b)

Figure 1.11 An example of a multipath channel at the receiver filter output

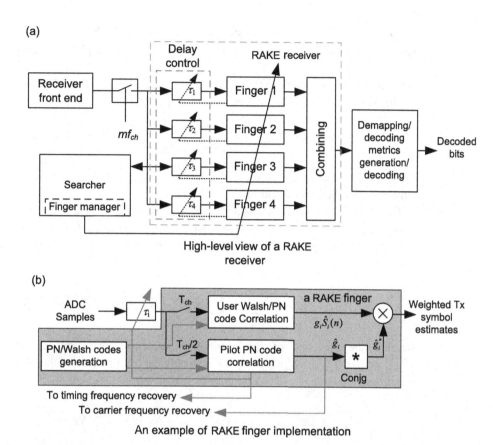

Figure 1.12 RAKE receiver block diagram

In addition to recovering the transmitted data symbols, the RAKE fingers also generate information for adjusting the delay of the samples as needed to achieve the best receiver performance. If the sampling rate is equal to $8f_{ch}$ or higher, the delay adjustment is simply to select the sample at the path delay. If the sampling rate is equal to $2f_{ch}$, the digital samples need to be resampled to obtain a higher timing resolution by digital interpolation, a process that will be described in Chapter 7. Moreover, the fingers also detect the existing carrier frequency offset, which is sent to the carrier-frequency synchronization block to reduce the offset. These operations will be described in detail in Chapters 5 and 6.

RAKE Finger Operations

A RAKE finger has two major blocks: a synchronization block and a demodulation block. The synchronization block performs two functions: to estimate the complex path gain and to optimize the sample timing relative to the $T_c/2$-spaced input samples. These samples are partitioned into two T_c spaced sample sequences. Each sample of the first sequence has the delay aligned with the peak of the combined transmitter and receiver filters shown in Figure 1.11(b) for the path to which the finger is assigned. Thus, each sample of the sequence is an estimate of a transmitted chip. The second sample sequence is used for the detection and correction of timing phase offset as will be discussed below.

Let us denote the sample, which is a scaled estimate of c_{nN+k} in (1.15) received by the lth finger $y_{nN+k}^{(l)}$. It can be expressed as

$$y_{nN+k}^{(l)} = g_l c_{nN+k} + z_{nN+k}^{(l)} = g_l \sum_{i\in \text{ active channels}} sp_{nN+k} w_{N,k}^{(i)} S_i(n) + z_{nN+k}^{(l)} \qquad (1.17)$$

where g_l is the complex gain of the lth path in the sample and $z_{nN+k}^{(l)}$ is the interference term including both noise and ISI.

To estimate the complex channel gain g_l, the synchronization block collects N samples, $y_{nN}^{(l)}, y_{nN+1}^{(l)}, \ldots, y_{(n+1)N-1}^{(l)}$, which are generated from the same data symbol $S_i(n)$. The channel gain estimate is obtained by summing the N unscrambled samples, i.e.,

$$\sum_{k=0}^{N-1} sp_{nN+k}^* y_{nN+k}^{(l)} = g_l \sum_{i\in \text{ active channels}} \sum_{k=0}^{N-1} \left|sp_{nN+k}\right|^2 w_{N,k}^{(i)} S_i(n) + \sum_{k=0}^{N-1} sp_{nN+k}^* z_{nN+k}^{(l)} \quad (1.18)$$

Because the elements of the scrambling sequence have unit magnitude and $\sum_{k=0}^{N-1} w_{N,k}^{(i)} = \sum_{k=0}^{N-1} w_{N,k}^{(0)} w_{N,k}^{(i)} = 0$ if $i \neq 0$, only the contributions of the pilot symbols remain and we have

$$\sum_{k=0}^{N-1} sp_{nN+k}^* y_{nN+k}^{(l)} = g_l N + z_{nN}' = N\hat{g}_l(nNT_c) \cong N\hat{g}_l(n) \qquad (1.19)$$

which is a scaled estimate of the complex channel gain at time nNT_c.

The estimate $\hat{g}_l(n)$ is used to correct the carrier phase of the estimated data symbols as described below. It can also be used to detect and correct the carrier frequency offset as will be discussed in Chapter 5.

The second function of the synchronization block is to adjust the sampling delay, i.e., the timing phase, for achieving the best possible receiver performance. This adjustment is necessary because the estimate of the sampling time from initial acquisition is usually coarse with an accuracy of only about $T_c/4$. Moreover, the sample timing may drift during the data mode and needs to be corrected. The second T_c spaced sample sequence with an offset of $T_c/2$ from the first sequence is used for this purpose. This function will be described in Chapter 6 when we discuss the timing synchronization of CDMA receivers.

Based on the information provided by the synchronization block, the demodulation block of the RAKE finger generates estimates of the transmitted symbols for the device. The demodulation block of the lth finger receives the same T_c-spaced sample sequences that are used for generating the path gain estimates in the synchronization block. These samples are unscrambled and correlated with the proper Walsh function for generating the estimates of the transmitted data symbols. Let us assume that the device needs to recover the transmitted symbol $S_1(n)$ from the samples $y_{nN+k}^{(l)}$,

$l = 0, \ldots, L-1$ and $k = 0, \ldots, N-1$ given by (1.17). Similar to (1.18), the correlation is performed by

$$\sum_{k=0}^{N-1} w_{N,k}^{(1)} sp_{nN+k}^* y_{nN+k}^{(i)} = g_l \sum_{i \in \text{active channels}} \sum_{k=0}^{N-1} \left[|sp_{nN+k}|^2 w_{N,k}^{(1)} w_{N,k}^{(i)} S_i(n) \right] + z_{nN,1}' \quad (1.20)$$

Note that $\sum_{k=0}^{N-1} w_{N,k}^{(1)} w_{N,k}^{(i)} = 0$, if $i \neq 1$ and $|sp_{nN+k}|^2 = 1$. Only the components in the sample due to $S_1(n)$ remain, i.e.,

$$\sum_{k=0}^{N-1} w_{N,k}^{(1)} sp_{nN+k}^* y_{nN+k}^{(l)} = g_l N S_1(n) + z_{nN,1}' \quad (1.21)$$

The carrier phase in g_l is removed by multiplying the conjugate of the estimated path gain $\hat{g}_l(n)$ obtained from the finger's synchronization block. The output is the estimate of the transmitted symbol weighted by the magnitude squared gain of the lth channel path as

$$\frac{1}{N} \hat{g}_l^*(n) \sum_{k=0}^{N-1} w_{N,k}^{(1)} sp_{nN_c+k}^* y_{nN+k}^{(l)} = |g_l(n)|^2 \hat{S}_1(n) \quad (1.22)$$

The weighted symbol estimates from the active fingers are added together to generate a combined estimate of the transmitted symbols. In CDMA systems under low SNR conditions, the noise variances in these estimates are approximately equal to each other. The squared path gain is proportional to the SNR of the path. Thus, the summation approximately implements *maximum ratio* combining and is optimal if these noises and interferences are spatially white.

Optimality of RAKE Receiver

Let us assume that the overall transmitter impulse response, $h_T(t)$, has a finite time duration of $[-T_p, T_p]$.[8] The composite time response of the transmitter and the multipath channel with CIR given by (1.16) can be shown to be

$$g(t) = \sum_{l=0}^{L-1} g_l h_T(t - \tau_l) \quad (1.23)$$

In the analysis below, we consider the chip c_{Nn}, sent at $t = nNT_c$. After passing through the channel, the signal containing c_{Nn} that arrives at the receiver is

$$y_{c_{nN}}(t) = c_{nN} g(t - nNT_c) = c_{nN} \sum_{l=0}^{L-1} g_l h_T(t - nNT_c - \tau_l) \quad (1.24)$$

---

[8] In this book, we adopt the notations $[a,b]$, $[a,b)$, $(a, b]$, and (a, b) of an interval (region) of the real numbers between a and b, as commonly used in mathematics. A square bracket, [or], means the associated endpoint is included in the interval. A parenthesis, (or), means that the associated endpoint is excluded from the interval.

The matched filter output at the optimal sampling time for the received signal $y_{c_{nN}}(t)$ is equal to

$$c_{nN} \int_{T_0} \left[\sum_{j=0}^{L-1} g_j^* h_T^* \left(u - nNT_c - \tau_j \right) \sum_{l=0}^{L-1} g_l h_T (u - nNT_c - \tau_l) \right] du \qquad (1.25)$$

where T_0 is the support region of $g(t)$, i.e., $g(t)$ is equal to zero outside this region. If the overlap between $h_T(t - \tau_i)$ and $h_T \left(t - \tau_j \right)$, for $i \neq j$, can be ignored, the matched filter output of c_{Nn} can be approximately expressed as

$$c_{Nn} \sum_{l=0}^{L-1} |g_l|^2 \int_{nNT_c - T_p + \tau_l}^{nNT_c + T_p + \tau_l} |h_T (u - nNT_c - \tau_l)|^2 du \qquad (1.26)$$

Now let us consider the RAKE receiver. If the receiver filter matches the transmitter filter, the output of the lth rake finger sampled at $nNT_c + T_p + \tau_l$, i.e., $y_{nN,l}$, can be expressed as

$$y_{nN,l} = c_{nN} g_l \int_{nNT_c - T_p + \tau_l}^{nNT_c + T_p + \tau_l} |h_T (u - nNT_c - \tau_l)|^2 du \qquad (1.27)$$

It is the peak value at the receiver filter output due to the signal from the lth path. Weighting $y_{nN,l}$ by the conjugate of the lth path gain estimated and adding the weighted RAKE finger outputs together, we obtain the RAKE receiver output

$$\sum_{l=0}^{L-1} g_l^* y_{nN,l} = c_{nN} \sum_{l=0}^{L-1} |g_l|^2 \int_{nNT_c + \tau_l}^{nNT_c + T_p + \tau_l} |h_T (u - nNT_c - \tau_l)|^2 du \qquad (1.28)$$

This is the same as that given by (1.26).

Thus, we have shown that the RAKE receiver output generated from the peaks of the channel paths is an approximation of the output of the MF. The main difference between the RAKE receiver and the matched filter is due to the interference between the signals received over different paths. When DS-CDMA systems operate under low SNR conditions, such interpath interferences are much weaker than the additive noise and other interference. As a result, the difference between the RAKE and the optimal MF receiver front end will be insignificant, and the RAKE receiver can be viewed as nearly optimal under the ML criterion.

In general, the performance of the MF is not optimal according to the minimum-mean-square-error (MMSE) criterion when ISI exists. However, in the case that the other noise and interference are significantly higher than the ISI, the effect of the ISI may be ignored. Thus, the RAKE receiver can also be viewed as nearly optimal under the MMSE criterion for DS-CDMA receivers. Similar arguments were originally introduced in [27] to show the optimality of the RAKE receiver.

1.6   Introduction to OFDM Communication

Earlier we described various aspects of the communication systems that are based on single-carrier communication technologies with and without spread spectrum. In recent years, OFDM has become the technology of choice in the latest generation of digital, especially wireless, communications.

The first major digital communication standard employing OFDM was DVB-T, the European digital terrestrial television standard, which was developed around the late 1990s [29]. In the United States, a variation of OFDM, called *discrete multitone* (*DMT*) *modulation*, was adopted for the digital subscriber line (DSL) standards [30]. For wireless data communications, OFDM was first standardized in the HyperLAN, ETSI's wireless local area network (LAN), also called Wi-Fi, standard. In 1999, OFDM technology was adopted by IEEE 802.11 standard committee as the 802.11a Wi-Fi standard for the 5 GHz band. OFDM technology was extended to the 2.4G MHz band in the 802.11g standard ratified in 2003. Later on, new 802.11 standards have emerged that further improved wireless LAN's performance [31].

Another important OFDM wireless communication standard is the *Long-Term Evolution* (*LTE*) [32], also commonly known as 4G LTE. It is for wireless high-speed data communications targeted to mobile phones and other devices in wide area network (WAN) environments. LTE is the evolution of 3G wireless standards including cdma2000 and WCDMA. It was first proposed by NTT DoCoMo in Japan, then developed and finalized at 3GPP. The main advantage of LTE is that it can effectively use a wideband wireless spectrum, e.g., up to 20 MHz or even wider, in each direction to achieve very high data throughput, e.g., 300 Mbps and higher.

The popularity of OFDM is the result of a number of its salient features as summarized below.

First, OFDM is a type of multicarrier communication technology. The OFDM signal transmitted from a transmitter consists of a plurality of carriers, called *subcarriers*. Multiple data symbols, each of which is modulated onto a separate subcarrier, are grouped in an OFDM symbol and transmitted. The subcarriers are orthogonal to each other. Hence, there will be no intersubcarrier interference (ICI) and no loss of efficiency even with overlaps in spectra between them. Its theoretical channel capacity is the same as that of a single-carrier system.

Second, when communicating over multipath channels, with proper design, OFDM communication can more effectively and efficiently combat ISI than single-carrier communication. As ISI is one of the major limiting factors affecting the transmission efficiency, OFDM has the potential to provide higher spectrum efficiency than the single-carrier technology with a reduced implementation complexity at the same time.

Third, an OFDM receiver eliminates ISI without performing channel inversion as equalizers in single-carrier receivers. In theory, OFDM communication can achieve the Shannon channel capacity more efficiently than single-carrier communications.

Finally, OFDM signaling is a natural fit for multiple input, multiple output (MIMO) technology, which has been used to further improve the spectrum efficiency of digital communication systems.

However, OFDM also has its shortcomings. The main deficiency of a communication system using OFDM signaling is its higher overhead than that of the single-carrier signaling. To eliminate ISI, the first portion of an OFDM symbol, called a cyclic prefix (CP), duplicates the last portion of the same symbol. Therefore, the CP does not carry information that should be transmitted, i.e., it constitutes an overhead. In addition, it is necessary for OFDM signaling to include *reference symbols*, or *pilots*, to facilitate data symbol recovery and synchronization functions. In contrast, single-carrier communications do not have anything equivalent to CP and may be able to dedicate less energy on reference/pilot signals when they are necessary. Nevertheless, after making trade-offs, the advantages of OFDM outweigh its disadvantages. As a result, OFDM has become attractive and important in today's wideband high-speed data communications systems.

Another potential issue with OFDM is its sensitivity to frequency and timing errors. To combat ISI, it is desirable to design the subcarriers to have narrow bandwidths. However, such a design results in long OFDM symbol lengths and OFDM becomes susceptible to carrier frequency error, Doppler effects, and, albeit to a lesser degree, timing error. Therefore, proper design and implementation of synchronization functions in OFDM receivers are crucial.

The theory, characteristics, and implementations of OFDM communication systems have been thoroughly studied and presented in the literature, including [33, 34, 35]. In this section, we consider the basics of OFDM system implementation and signaling design to facilitate the discussion on synchronization in OFDM communication systems later in this book.

Because of the need to reduce the complexity of implementation, today's OFDM systems are based, almost with no exception, on a structure that was first proposed in [36]. In such OFDM system implementations, the modulation and demodulation of data symbols are accomplished by using inverse discrete Fourier transforms (iDFT) and discrete Fourier transforms (DFT) that are usually implemented by using the computationally efficient inverse fast Fourier transform (iFFT) and fast Fourier transform (FFT) algorithms. To begin, we provide an overview of the operations of such OFDM transmitters and receivers, including their synchronization functions.

1.6.1   OFDM Transmitter Operations

The block diagram of such an OFDM transmitter is shown in Figure 1.13.

In a typical OFDM system, the binary input bits are organized into N_D parallel sequences, where N_D is the number of the active OFDM data subcarriers and is smaller than the number of the DFT/iDFT size N_{DFT}. These input bits may be from single or multiple sources, and may have been encoded by error correction codes and interleaved. To take advantage of the efficient FFT/iFFT implementation, N_{DFT} is usually chosen to be equal to 2^M, where M is an integer although DFT/iDFT with any size can be used. The bits in each of the N_D sequences are mapped to data modulation symbols, which are represented by complex numbers. These modulation symbols may be mapped to different forms of symbol constellations, such as PSK and QAM. In the nth OFDM

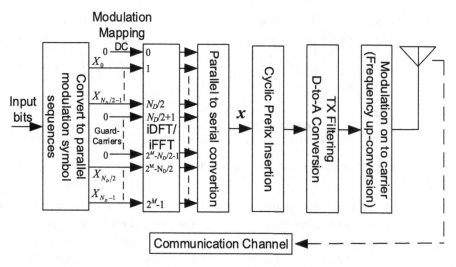

Figure 1.13 A typical iDFT/iFFT-based OFDM transmitter

symbol, N_D of such modulation symbols, one per sequence, together with $(N_{DFT} - N_D)$ zeros form an input data vector $\mathbf{X}_{N_{DFT}}(n)$ of the iDFT block.

The iDFT block performs the iDFT of the input vector $\mathbf{X}_{N_{DFT}}(n)$ to generate an output vector $x(n)$, which consists of N_{DFT} output samples, $x(n,l)$, $l = 0, \ldots, N_{DFT} - 1$. The OFDM transmitter treats the N_{DFT} output samples as a time sample sequence. This step is usually called *parallel-to-serial conversion*. These samples are transmitted as *OFDM channel samples* at a rate of f_s. The sample time interval T_s is equal to $1/f_s$. As shown in [36], iDFT effectively modulates the modulation data symbols, i.e., the elements of the input vector, onto the corresponding subcarriers.

In the frequency domain, the spectrum of such an OFDM signal is periodic with a period of f_s. It has a total of N_{DFT} subcarriers that span the entire bandwidth of f_s, which is called the *DFT bandwidth*. Thus, the spacing between adjacent subcarriers, which is called the *subcarrier spacing* and denoted by f_{sc}, is equal to f_s/N_{DFT}. The kth element of the DFT input vector is modulated onto the kth subcarrier, which is a windowed complex sinusoid function at frequency kf_{sc} in the form of

$$v_k(t) = \begin{cases} e^{j2\pi kf_{sc}t} & 0 \leq t < 1/f_{sc} \\ 0 & otherwise \end{cases} \tag{1.29}$$

where $1/f_{sc} = N_{DFT}/f_s = N_{DFT}T_s$ is equal to the time spanned by the N_{DFT} samples at the iDFT output. It can be shown that $\int_0^{1/f_{sc}} v_l^*(t)v_k(t)dt = 0$, for $l \neq k$, i.e., they are orthogonal to each other. Due to the rectangular time windowing, the spectrum of each $v_k(t)$ has the shape of the sinc function.

An example of an iDFT input data vector with N_{DFT} elements is shown in Figure 1.14. The data vector has N_D nonzero data elements including the frequency-domain multiplexed (FDM) pilots to be discussed later. The zero elements in the input data vector correspond to the so-called *guard subcarriers*.

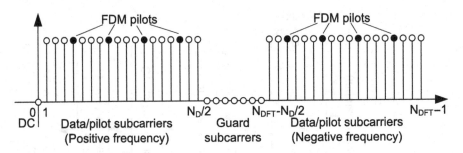

Figure 1.14 Structure of the iDFT input vector

Conventionally in digital signal processing, the indices of the iDFT and DFT input vectors run from 0 through $N_{DFT}-1$ as shown in Figure 1.13 and Figure 1.14. Following this convention, the 0th element in the input vector corresponds to the DC subcarrier in the frequency domain. The data symbols assigned to the 1st through $(N_D/2)$th elements are modulated onto the positive frequency subcarriers at frequencies from f_{sc} to $N_D f_{sc}/2$. The data symbols assigned to the $(N_{DFT}-N_D/2)$th through $(N_{DFT}-1)$th elements are on the subcarriers with negative frequency from $-N_D f_{sc}/2$ to $-f_{sc}$. The guard subcarriers are located at most positive and most negative frequencies. It is usually desirable to allocate one null symbol to the position corresponding to the DC subcarrier in order to simplify the receiver implementation. Alternatively, the iDFT can also be performed with index from $-N_{DFT}/2$ to $N_{DFT}/2 -1$.

The spectrum of the iDFT output is shown in Figure 1.15. It can be seen that the subcarriers in frequency domain have the shape of the sinc function. At the peak position of a nonzero subcarrier, the contributions from all other subcarriers are equal to zero. In other words, the subcarriers do not interfere with each other. This fact also demonstrates the orthogonality among them.

In addition to the N_{DFT} samples from the iDFT output, which is called the *main portion of an OFDM symbol* in this book, each OFDM symbol contains additional parts to facilitate the implementation and for the performance improvement of OFDM communications.

In order to eliminate or, at least, to reduce the inter-symbol interference (ISI) between the adjacent OFDM symbols, a guard-interval (GI) is added before the main portion of the OFDM symbol. Commonly, the GI portion is the same as the last part of the main portion and can be viewed as its cyclic extension. Therefore, such a GI is also called the *cyclic prefix*, i.e., CP, of the OFDM symbol.

The CP is the key element of the OFDM signaling for effectively combating ISI. If the CP is longer than the span of the multipath channel, the interference of ISI from the multipaths can be completely eliminated. In addition, as will be shown later, the CP can also play a useful role in achieving synchronization in OFDM systems. However, the CP also constitutes an overhead of transmitted signal energy. Thus, it is important to choose the CP length carefully in order to achieve the best possible system performance.

Two complete OFDM symbols with their CPs are depicted in Figure 1.16. The portions with the same shade indicate that they have the same data. T_M is the length

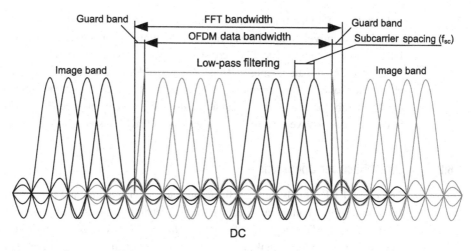

Figure 1.15 Spectrum of iDFT output

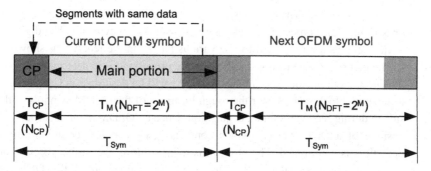

Figure 1.16 OFDM symbols with CP

of the main portion of the OFDM symbol with N_{DFT} samples, T_{CP} is the length of CP with N_{CP} samples, and T_{Sym} is the duration of the entire OFDM symbol with $(N_{DFT}+N_{CP})$ samples.

Since the time domain OFDM channel samples are generated at every T_s, the time duration T_M of the main portion of an OFDM symbol is equal to $N_{DFT}T_s$. The length of the entire OFDM symbol is $T_{Sym} = (N_{DFT} + N_{CP})T_s$.

The baseband OFDM sample sequence with the CP extensions is converted to analog waveforms by D-to-A conversion as shown in Figure 1.13. The generated analog signals are modulated onto the carrier frequency, amplified to the desired power, filtered again if necessary, and radiated from the transmitter antenna. These operations are the same as the corresponding ones in any other form of digital transmission, e.g., single-carrier or CDMA transmission, described previously. The total bandwidth span of the N_{DFT} subcarriers at the DFT output is equal to $f_s = 1/T_s$. Since only N_D out of the N_{DFT} subcarriers are modulated by data symbols, the bandwidth occupied by the OFDM signal, called its data bandwidth, is only slightly wider than $(N_D/N_{DFT})f_s$.

During the OFDM transmitter operations, OFDM symbols are generated and transmitted consecutively. An OFDM symbol is the unit of transmission in the time domain. Furthermore, an OFDM symbol consists of N_D active subcarriers, each of which is modulated by a data symbol. Thus, the OFDM signals can be viewed as being transmitted over a time-frequency two-dimensional space with the time unit of an OFDM symbol and with the frequency unit of the subcarrier spacing f_{sc}. This concept will be used when we discuss the properties of OFDM systems.

1.6.2   OFDM Receiver Operations

The OFDM RF signals radiated from the transmitter antenna and passing through the communication channel are received by a receiver. As is shown in Figure 1.17, the received OFDM signal is frequency down-converted from carrier to baseband. The analog baseband signal is filtered by an anti-aliasing filter and then converted to the digital form by the A to D conversion. Due to the existence of guard subcarriers, a sampling rate of $1/T_s$ satisfies the Nyquist sampling theorem [15] and no sub-chip sampling is needed. These operations are also essentially the same as the corresponding ones in the other types of digital receivers discussed previously. The digital baseband signal samples are processed by the various blocks in the receiver as shown below.

The first step of the received signal sample processing is to determine the locations of the OFDM symbols in the sample stream. As will be described in Chapter 3, the initial acquisition block searches for the rough locations of the OFDM symbols in the sample stream during receiver's initial acquisition. Once an OFDM symbol is found, the initial position of a DFT window, which contains N_{DFT} samples for performing the DFT operation, is determined. The position is refined during the further receiver operations on the received signal. Determining the correct position of the DFT window is

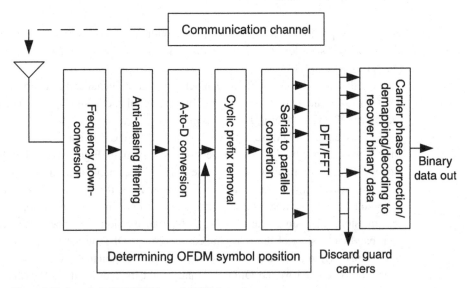

Figure 1.17 A typical DFT/FFT-based OFDM receiver

equivalent to achieving timing synchronization in OFDM receivers, as no sub-chip adjustment is needed. More details of this procedure will be provided in Chapter 6.

The initial acquisition block also estimates the frequency offset generated during frequency down-conversion. The estimated offset is refined by the frequency synchronization block as will be discussed in Chapter 5.

For each OFDM symbol, the N_{DFT} received signal samples inside the DFT window are grouped together as an input data vector and sent to the DFT block. This operation is often called *serial-to-parallel conversion*. The N_{DFT} elements of the DFT block output are the estimates of the N_{DFT} modulation symbols at the transmitter's iDFT input with possibly rotated phase, scaled magnitude and corrupted by additive noise/interference due to channel impairments. The DFT block performs demodulation of OFDM symbols as it converts the data symbols modulated on the subcarriers to baseband. As shown below, the transmitted modulation symbols are recovered after the phase rotations and magnitude scaling are corrected. These recovered symbols are corrupted by the additive noise introduced in the transmission process. A decoder is used to regenerate the information bits sent with no or few errors.

1.6.3 Demodulation of OFDM Signal Over Single- and Multipath Channels

In Section 1.6.1, the main portion of the OFDM symbol is defined as the output of the iDFT operation with the data vector $\mathbf{X}_{N_{DFT}}(n)$ as its input. To facilitate analysis, we denote the input vector of the n-th OFDM symbol to the iDFT block by

$$\mathbf{X}_{N_{DFT}}(n) = [X(n,0), X(n,1), ..., X(n, N_{DFT} - 1)]^t \qquad (1.30)$$

where $X(n,k)$, $k = 0, \ldots, N_{DFT} - 1$, are the modulation symbols to be transmitted with the null guard subcarrier symbols added. Each element of $X(n,k)$ is a complex number representing a modulation symbol to be modulated onto the kth subcarrier of the nth OFDM symbol by the iDFT operation. The iDFT output vector denoted by $\mathbf{x}(n)$, i.e., the main portion of the OFDM symbol, is

$$\mathbf{x}(n) = [x(n,0), x(n,1), ..., x(n, N_{DFT} - 1)]^t \qquad (1.31)$$

To simplify the description below, we adopt matrix expressions of the DFT and iDFT operators. The length-N_{DFT} DFT operator can be expressed as

$$\mathbf{W}_{N_{DFT}} = \frac{1}{\sqrt{N_{DFT}}} \begin{pmatrix} 1 & 1 & 1 & 1 & \cdots & 1 \\ 1 & W_{N_{DFT}} & W_{N_{DFT}}^2 & W_{N_{DFT}}^3 & \cdots & W_{N_{DFT}}^{N_{DFT}-1} \\ 1 & W_{N_{DFT}}^2 & W_{N_{DFT}}^4 & W_{N_{DFT}}^6 & \cdots & W_{N_{DFT}}^{2(N_{DFT}-1)} \\ 1 & W_{N_{DFT}}^3 & W_{N_{DFT}}^6 & W_{N_{DFT}}^9 & \cdots & W_{N_{DFT}}^{3(N_{DFT}-1)} \\ \vdots & \vdots & \vdots & \vdots & \ddots & \vdots \\ 1 & W_{N_{DFT}}^{N_{DFT}-1} & W_{N_{DFT}}^{2(N_{DFT}-1)} & W_{N_{DFT}}^{3(N_{DFT}-1)} & \cdots & W_{N_{DFT}}^{(N_{DFT}-1)^2} \end{pmatrix}$$

$$(1.32)$$

where $W_{N_{DFT}} = e^{-j2\pi/N_{DFT}}$ [8].

The iDFT operator is $\mathbf{W}_{N_{DFT}}^H$, where the superscript ^{H} represents the complex conjugate and transpose, or Hermitian, operation. It is easy to verify that $\mathbf{W}_{N_{DFT}}$ satisfies

$$\mathbf{W}_{N_{DFT}}^H \mathbf{W}_{N_{DFT}} = \mathbf{W}_{N_{DFT}} \mathbf{W}_{N_{DFT}}^H = \mathbf{I}_{N_{DFT} \times N_{DFT}} \tag{1.33}$$

which is an identity matrix. Therefore, $\mathbf{W}_{N_{DFT}}$ is an *orthonormal matrix*.

Using the matrix notations, the time samples generated by the iDFT operation in the transmitter can be expressed as

$$\mathbf{x}(n) = \mathbf{W}_{N_{DFT}}^H \mathbf{X}(n) \tag{1.34}$$

The demodulation of OFDM symbols, i.e., recovering the elements of the data vector $\mathbf{X}(n)$ is performed by the DFT operation.

Demodulation of Signals Transmitted Over a Single-Path Channel

Let us consider that the OFDM symbols are transmitted over a single-path channel with channel coefficient $h = e^{j\theta_h}|h|$. Assume that with proper DFT window placement, each DFT input vector consists of the received signal samples corresponding the main portion of an OFDM symbol to be demodulated. It can be expressed as

$$\mathbf{y}(n) = [y(n,0), y(n,1), ..., y(n, N_{DFT} - 1)]^t = h\mathbf{x}(n) + \mathbf{z}(n) = h\mathbf{W}_{N_{DFT}}^H \mathbf{X}(n) + \mathbf{z}(n) \tag{1.35}$$

where $\mathbf{z}(n)$ is an N_{DFT} dimensional noise vector, whose elements are AWGN with variance of σ_z^2. The vector $\mathbf{y}(n)$ is processed by the receiver DFT block. The output vector resulting from the DFT operation, $\mathbf{Y}(n) = [Y(n,0), Y(n,1), ..., Y(n, N_{DFT} - 1)]^t$, can be shown as

$$\mathbf{Y}(n) = \mathbf{W}_{N_{DFT}}[h\mathbf{x}(n) + \mathbf{z}(n)] = h\mathbf{W}_{N_{DFT}}\mathbf{W}_{N_{DFT}}^H \mathbf{X}(n) + \mathbf{W}_{N_{DFT}}\mathbf{z}(n) = h\mathbf{X}(n) + \mathbf{z}'(n) \tag{1.36}$$

where $\mathbf{z}'(n)$ is a transformed noise vector, whose elements are also AWGN with the variance of σ_z^2 because the operator $\mathbf{W}_{N_{DFT}}$ is orthonormal.

From (1.36), we observe that the elements $Y(n,k)$ of $\mathbf{Y}(n)$ are the estimates of $X(n,k)$ of $\mathbf{X}(n)$ scaled by the complex channel gain h. If h is known to the receiver, $Y(n,k)/h$ are unbiased estimates of $X(n,k)$. Moreover, it can be shown that the elements $h^*Y(n,k)$ are the log-likelihood estimates of $X(n,k)$. For communication over single-path channels, only the main portions of the OFDM symbols are used and the CPs are not needed. The CPs play an essential role in combating the ISI generated from multipath channels as shown below.

Demodulation of Signals Transmitted Over a Multipath Channel

When passing through a multipath, i.e., frequency selective, channel, the received signal sample vector $\mathbf{y}(n)$ contains multiple copies of the transmitted signal vector $\mathbf{x}(n)$, each of which is associated with a delayed path as shown in Figure 1.18.

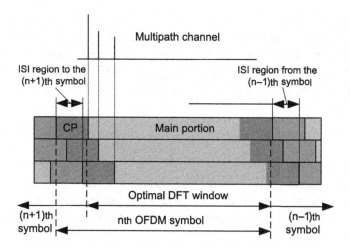

Figure 1.18 Received samples and DFT window placement

From the figure, we observe that the $(n-1)$th OFDM symbol generates ISI in the nth symbol and the nth symbol generates ISI in the $(n+1)$th symbol, as they overlap with each other. However, if the CP length is longer than the multipath span, it is possible to place the DFT window in such a way that there is no ISI from the OFDM symbol prior to the one to be demodulated. Moreover, the samples inside the DFT window contain one complete circularly shifted $\mathbf{x}(n)$ from each path. Thus, no information of the transmitted data gets lost. Note that the optimal DFT window location is not unique if the span of the multipath channel is shorter than the CP. The receiver performance for an OFDM signal transmitted over multipath channel is analyzed below.

The discrete-time impulse response of a multipath channel, excluding the transmitter and receiver filters, can be expressed as

$$h(t) = \sum_{m=0}^{M-1} h_m \delta(t - d_m T_s) \tag{1.37}$$

where h_m is the complex value of the mth channel path coefficient, d_m is its delay in units of the channel sample intervals T_s, and $\delta(t)$ is the Dirac delta function.

With a sampling frequency equal to $1/T_s$, the continuous time frequency response of this channel is the Fourier transform of $h(t)$, which can be expressed as

$$H(f) = \sum_{m=0}^{M-1} h_m e^{-j2\pi d_m(f/f_s)} \tag{1.38}$$

This is a periodic function with a period of $f_s = 1/T_s$.

Now, let us consider the nth OFDM symbol, which consists of the transmitter channel samples $\{x(n, k)\}$, $k = 0, 1, \ldots, N_{\text{DFT}}-1$, in the form given by (1.31) with the CP extension. As can be seen from Figure 1.18, after passing through the discrete multipath channel with frequency response $H(n,f)$ given by (1.38), the received samples contain M segments of the transmitted OFDM symbol with delays $\{d_m T_s\}$ and weighted by path coefficients h_m's. Each of such segments can be considered as a segment of samples taken from a periodic sample sequence of the repeated $x(n)$. Moreover, the N_{DFT} samples inside the DFT window to be sent to the DFT processing block contain M periods of the circularly shifted $x(n)$, each of which is associated with one of the paths. We can express the received signal samples in the DFT window, $y(n,k)$, $k = 0, \ldots, N_{DFT} - 1$, as

$$y(n, k) = \sum_{m=0}^{M-1} x\left(n, (k - d_m)_{\text{mod } N_{DFT}}\right) h_m + z(n, k) \tag{1.39}$$

where $(\bullet)_{\text{mod } N_{DFT}}$ represents the modulo N_{DFT} operation.

The data vector $\mathbf{y}(n)$, which consists of the samples $y(n,k)$ given by (1.35), is processed by the DFT block to generate the output vector $\mathbf{Y}(n)$. The kth element, $Y(n, k)$, of $\mathbf{Y}(n)$ can be expressed as

$$Y(n, k) = \sum_{m=0}^{M-1} h_m \sum_{l=0}^{N_{FFT}-1} x\left(n, (l - d_m)_{\text{mod } N_{DFT}}\right) W_{N_{DFT}}^{lk} + z'(n, k) \tag{1.40}$$

where $z'(n, k)$ is the noise term in the kth element of the DFT output. By using the property of the DFT of circularly shifted input sequences [37], we have,

$$Y(n, k) = X(n, k) \sum_{m=0}^{M-1} h_m W_{N_{DFT}}^{d_m k} + z'(n, k) \tag{1.41}$$

From (1.38) and because $W_{N_{DFT}}^{d_m k} = e^{-j2\pi d_m k/N_{DFT}}$, we can rewrite (1.41) as

$$Y(n, k) = X(n, k)H(n, k) + z'(n, k) \tag{1.42}$$

Represented in continuous time, $H(n, k) \cong H(n, kf_{sc})$ is simply the value of $H(n,f)$ sampled at kf_{sc}. In other words, the frequency-domain channel coefficient of subcarrier k is equal to the channel frequency response sampled at the frequency of this subcarrier.

Equation (1.42) shows that, when receiving the nth OFDM symbol, the kth element of the DFT output is an estimate of the kth modulation symbol, $X(n,k)$, times the channel frequency response at the kth subcarrier. The modulation symbol can be recovered if the frequency response of the channel is known or can be estimated. Denoting the estimate of $H(n,k)$ by $\hat{H}(n, k)$, the estimate of the transmitted symbol $X(n,k)$ can be computed as

$$\hat{X}(n, k) = \hat{H}^*(n, k)Y(n, k)/|H(n, k)|^2 \tag{1.43}$$

Thus, when the cyclic prefix length and the DFT window position are selected properly, the OFDM transmission channel can be viewed as consisting of N_D independent subchannels, each of which is carried by a subcarrier. If the transmitter carrier frequency and receiver down-conversion frequency are synchronized, there will be no or little inter-(sub)carrier interference, or ICI, between them. The impact of the frequency offset on OFDM receiver performance and how to achieve OFDM carrier synchronization will be discussed in Chapter 5.

In order to achieve reliable detection of the transmitted symbols, it is necessary to estimate as accurately as possible the frequency-domain channel coefficient of each subcarrier. For transmission over fading channels, the phases and amplitudes of the subcarriers change with time, but the coefficients of adjacent subcarriers are correlated. The subcarrier phase estimation is done as a part of overall frequency-domain channel estimation. The channel frequency response can be estimated either directly in the frequency domain or computed indirectly from the estimated time-domain CIR according to (1.38). Due to the importance of the accuracy of this estimate, considerations must be taken when designing the OFDM system. One of the approaches to facilitating this task is to send known data symbols called pilot or reference symbols together with information bearing data symbols, as described in the next section.

1.6.4   TDM and FDM Pilot/Reference Signals in OFDM Communication

Different types of pilot, or reference, signals are usually inserted in the signal streams of OFDM transmission. Even though the pilot signals are known or partially known to the receiver and thus carry no or little information for data transmission, they play a crucial role in achieving reliable synchronization. Below we describe two main types of pilot signals: the *time-division multiplexed* (*TDM*) pilots and the *frequency-division multiplexed* (*FDM*) pilots.

1.6.4.1   TDM Pilot Signals

TDM pilot signals, or TDM pilots, are transmitted as entirely known OFDM symbols with all known symbol modulated subcarriers. They are mostly used to assist the receivers during initial acquisition to detect the existence of the desired signals, to initialize the carrier frequency and/or phase estimates, and to set the initial receiver sample timing. To accomplish these functions, the TDM pilots are often sent at the beginning of the transmission for establishing the communication link. They may also be sent periodically for new users to acquire the link in multiple-access systems. We will discuss their roles in detail later in this book. Below we provide brief descriptions of these types of pilots with a few examples.

802.11a/g Wi-Fi Communications

Defined in 802.11a/g Wi-Fi standards, two types of TDM pilot symbols, also called *preambles*, are transmitted at the beginning of each data frame before the data transmission begins. The preambles consist of 10 *short preamble symbols* and 2 *long preamble symbols*. Each of the short preamble symbols is 16 OFDM transmitter channel samples long and repeats 10 times to form a short preamble. Following the short preamble is a long preamble. The long preamble has 2 periods of a long preamble symbol. Each period is 64 samples long with a CP of 32 samples. The lengths of the preambles are different from the length of the regular data OFDM symbols, each of which is 80 samples long.

As the first step to establish the communication link with the access points (APs), the short preamble is mainly used by the Wi-Fi device to detect the existence of the systems and their frame boundaries. They are also used for determining and correcting the rough frequency offset and for establishing approximate timing. The long preamble symbols are then used to perform channel estimation. In addition to being used for demodulation of data symbols transmitted in the frame, the generated channel estimate is used to determine the more accurate DFT window position, i.e., the more precise timing. They are also used to further refine the estimate of the frequency offset. After these steps, the receiver is ready to communicate with the Wi-Fi network.

More details of the Wi-Fi initial acquisition procedures based on the preamble TDM pilots will be provided in Section 3.6.2.

LTE

The Primary Synchronization Signal (PSS) and the Secondary Synchronization Signal (SSS) defined in the LTE standard can also be viewed as TDM pilots. Given that LTE is a wideband system, in order to efficiently use the available resources and for systems' robustness, the PSS and SSS occupy only parts of the total bandwidth. The PSS and SSS are transmitted every half-frame (5 ms). They occupy a 1.4 MHz frequency band at the center of the entire LTE channel bandwidth, which can be as wide as 20 MHz.

The PSS and SSS are used in LTE's two-step system acquisition procedure. In the process, the receiver also determines other system parameters such as slot and frame boundaries, the cell's FDM pilot positions, and it performs initial frequency offset correction. The detailed operations of the initial acquisition using the PSS and SSS are presented in Section 3.6.1.

TDM Pilot-Assisted Channel Estimation and Data Symbol Recovery

In some systems, TDM pilots are sent periodically during data transmission after the link is established. Their function is to assist carrier and time synchronization of the receivers. For example, in the Chinese digital terrestrial television broadcasting (DTMB) standard, TDM pilots are sent in the form of PN sequences as GIs instead of the CPs [38]. It is also used for channel estimation for data symbol recovery instead of using FDM pilots. Because the TDM pilots serve a dual purpose, the system overhead is reduced. However, it appears that such a design may cause other system deficiencies. As a result, the overall system performance may not be better than the conventional OFDM system designs discussed in this book.

1.6.4.2   FDM Pilot Signals

FDM pilot signals, or FDM pilots, generally refer to the known modulation symbols on a subset of subcarriers in every one or every few OFDM symbols. Using the time-frequency grid concept, an example of OFDM transmission signal with embedded FDM pilots is shown in Figure 1.19. The OFDM symbol index is shown along the x, or time, axis and the subcarrier index in an OFDM symbol is shown along the y, or frequency, axis. The white squares are the data symbol modulated subcarriers, and the gray squares denote the pilot symbol modulated subcarriers.

Two types of FDM pilots are shown in the figure. The first type of pilots, shown as the light gray boxes, is in every OFDM symbol on the same subcarrier. This type of

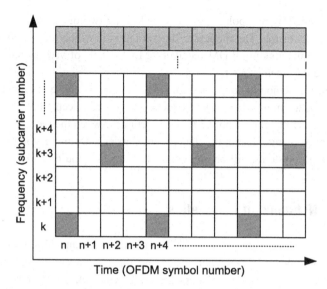

Figure 1.19 FDM pilots in OFDM signal

pilots is called *continuous pilots*, which are mainly used for carrier frequency estimation. The second type of pilots, shown as dark gray boxes, is evenly distributed over the entire two-dimensional grid and called *scattered pilots*. They are mainly used for channel estimation for performing data symbol recovery as explained below.

The main purpose of sending FDM pilots is to perform channel carrier phase estimation for coherent detection of the data symbols. As shown by (1.43), to perform coherent detection of a data symbol, we must have the estimate of the frequency-domain channel coefficient of the subcarrier, on which the data symbol is modulated.

From (1.42), we observe that the estimate of the frequency-domain subcarrier coefficient $H(n,k)$ can be computed if $X(n,k)$ is known to the receiver, i.e., if it is an FDM pilot. For example, in Figure 1.19 the modulation symbol on OFDM symbol n, the subcarrier k is a pilot, denoted by $X_p(n,k)$. The estimate of $H(n,k)$ can be computed as

$$\hat{H}(n,k) = X_p^*(n,k)Y(n,k)/\left|X_p(n,k)\right|^2 \qquad (1.44)$$

If the channel coherence time is longer than the OFDM symbol time interval, the channel frequency response for the OFDM symbols that are close to each other in time is correlated. The estimate $\hat{H}(n,k)$ can be used as the estimate for recovering data symbols on subcarrier k of these OFDM symbols, e.g., with time index $n-1$ and $n+1$. Similarly, if the coherence frequency bandwidth of the channel is wider than the subcarrier spacing, channel estimate $\hat{H}(n,k)$ can be used to recover the data symbols modulated on subcarriers near the kth subcarrier of the same OFDM symbol.

Moreover, because the modulation symbol on subcarrier $k+6$ is also a pilot, the channel coefficients of subcarriers $k+1$ through $k+5$ can be computed by interpolating $\hat{H}(n,k)$ and $\hat{H}(n,k+6)$ to obtain more accurate estimates. Theoretically, the best results can be obtained by interpolations performed in the two-dimensional plane over the frequency-time domains. The method of interpolation is widely used in frequency-domain channel estimations in OFDM receivers. More details of various interpolation techniques will be considered in Chapter 5.

In this section, for demodulation of the OFDM symbols, we have assumed that the carrier frequency and timing phase and frequency have been accurately acquired. In practice, some or all of them may not be known accurately when the receiver operation starts, and they may change during the data mode. Carrier and timing synchronization techniques in OFDM systems will be presented in Chapters 5 and 6.

1.7   Summary and Notations of Parameters

In this chapter, we first provided an overview of digital communication systems. The characteristics, components, and operations of the three main constituent blocks of digital communication systems, i.e., the transmitter, the channel, and the receiver, were then described. While the descriptions focused on the data mode operations of single-carrier systems, many of their components are shared by other types of communication systems, such as OFDM and CDMA systems. Moreover, they are also pertinent to performing the synchronization functions in these communication systems. Thus, what was presented in this first chapter forms the foundation for understanding and discussing synchronization functions in various communication systems, which are the subjects of this book.

Since the late 1980s, mobile wireless communications have been widely deployed and penetrated into the everyday life of our society. As a consequence, the technologies employed in mobile communication systems, in particular CDMA and OFDM, have attracted a great deal of attention in the digital communication technical community. In this chapter, two sections were devoted to the overviews of these two technologies to establish the foundation for the discussion of synchronization techniques in systems that use these technologies.

It should be noted that what was presented in this chapter is only an overview of communication systems and technologies. The purpose of this chapter is to facilitate the understanding and discussion of synchronization functions to be treated later in this book. The examples given are only for illustrating the related transmitter and receiver operations. There are many alternative approaches to performing these tasks. In other words, the examples should not be taken as a complete treatment of these subjects. Readers interested in gaining more complete knowledge and understanding on these topics are referred to the references provided, especially the many excellent textbooks available, such as [3, 5, 18].

In this book, we will cover the synchronization methods used in a number of different communication systems. To facilitate the discussion, notations of the parameters involved are summarized below. They will be used throughout this book as consistently as possible.

Single-Carrier Communication

T: Data symbol (data modulation symbol) interval/duration

$1/T$: Data symbol (data modulation symbol) frequency/rate

(*Note*: For such systems, data symbols are the same as the channel modulations symbols.)

DS-CDMA Communication

T: Data symbol (data modulation symbol) interval/duration

f_{Sym}: Data symbol (data modulation symbol) frequency, $f_{Sym} = 1/T$

T_c: Chip interval/duration, $T_c = T/N$, where N is the spreading factor

f_{ch}: Chip frequency/rate, $f_{ch} = 1/T_c = Nf_{Sym}$

(*Note*: A chip corresponds to a channel modulation symbol in single-carrier communications with no spreading.)

OFDM Communication

T_s: OFDM channel sample duration/interval, called *standard time unit* in LTE

f_s: OFDM channel sample rate, $f_s = 1/T_s$

W_{DFT}: OFDM DFT bandwidth, $W_{DFT} = 1/T_s = f_s$

f_{sc}: OFDM subcarrier spacing, $f_{sc} = W_{DFT}/N_{DFT} = f_s/N_{DFT}$

T_{Sym}: OFDM symbol duration/interval, $T_{Sym} = (N_{DFT} + N_{CP})T_c$

f_{Sym}: OFDM symbol rate, $f_{Sym} = 1/T_{Sym} = f_s/(N_{DFT} + N_{CP})$

N_D: Number of data (active) subcarriers

W_D: OFDM data bandwidth, $W_D = N_D f_{sc} = N_D f_s/N_{DFT} = N_D/(N_{DFT}T_s)$

(*Note*: An OFDM channel sample corresponds to a channel modulation symbol in single-carrier communication.)

References

[1] C. E. Shannon, "Communication in the Presence of Noise," *Proceedings of the IRE*, vol. 37, no. 1, pp. 10–21, 1949.

[2] W. E. Ryan and S. Lin, *Channel Codes – Classical and Modern*, Cambridge University Press, 2009.

[3] B. Sklar, *Digital Communications – Fundamentals and Applications*, 2nd edn, Upper Saddle River, NJ: Prentice Hall PTR, 2011.

[4] J. George, C. Clark, and J. B. Cain, *Error Correction Coding for Digital Communications*, New York: Plenum Press, 1981.

[5] J. G. Proakis and M. Salehi, *Digital Communications*, 5th edn, New York: McGraw-Hill, 2008.

[6] J. L. Ramsey, "Realization of Optimum Interleavers," *IEEE Transactions on Information Theory*, vol. IT-16, no. 3, pp. 338–45, 1970.

[7] G. D. Forney Jr., "Burst Correction Codes for the Classic Bursty Channel," *IEEE Transaction on Comm.*, vol. COM-19, no. 10, pp. 772–81, 1971.

[8] A. V. Oppenheim and R. W. Schafer, *Digital Signal Processing*, Englewood Cliffs, NJ: Prentice Hall, 1975.

[9] W. Kester, "Oversampling Interpolating DACs, Analog Devices Tutorial MT-107," October 2008. [Online]. Available: www.analog.com/media/en/training-seminars/tutorials/MT-017 .pdf. [Accessed October 16, 2015].

[10] National Instruments, "Interpolation and Filtering to Improve Spectral Purity, National Instruments White Paper 5515," May 29, 2014. [Online]. Available: www.ni.com/white-paper/5515/en/pdf. [Accessed October 16, 2015].

[11] W. C. Jakes, et al, *Microwave Mobile Communications*, W. C. Jakes, Ed., Piscataway, NJ: IEEE Press, 1974, reissued 1993.

[12] D. Tse and P. Viswanath, *Fundamentals of Wireless Communication*, Cambridge University Press, 2005.

[13] G. L. Stüber, *Principles of Mobile Communication*, New York: Springer-Verlag, 2012.

[14] G. D. Forney Jr., "Maximum-Likelihood Sequence Estimation of Digital Sequences in the Presence of Intersymbol Interference," *IEEE Transactions on Information Theory*, vol. 18, no. 3, pp. 363–78, May 1972.

[15] H. Nyquist, "Certain Topics in Telegraph Transmission Theory," *AIEE Transactions*, vol. 47, pp. 617–44, 1928.

[16] J. G. Proakis, *Digital Communications*, 3rd edn, New York: McGraw-Hill, 1995.

[17] R. F. H. Fischer, *Precoding and Signal Shaping for Digital Transmission*, New York: John Wiley & Sons, 2002.

[18] R. D. Gitlin, J. F. Hayes, and S. B. Weinstein, *Data Communications Principles*, New York: Plenum Press, 1992.

[19] S. Allpress, C. Luschi, and S. Felix, "Exact and Approximated Expressions of the Log-Likelihood Ratio for 16-QAM Signals," in *Conference Record of the Thirty-Eighth Asilomar Conference on Signals, Systems and Computers*, Pacific Grove, CA, 2004.

[20] J. D. Ellis and M. B. Pursley, "Comparison of Soft-Decision Decoding Metrics in a QAM System with Phase and Amplitude Errors," in *IEEE Military Communications Conference*, MILCOM 2009, Boston, MA, 2009.

[21] 3rd Generation Partnership Project 2, "Physical Layer Standard for cdma2000 Spread," 2004.

[22] A. J. Viterbi, *CDMA Principle of Spread Spectrum Communications*, Reading, MA: Addison-Wesley, 1995.

[23] A. Toskala, H. Holma, and P. Muszynski, "ETSI WCDMA for UMTS," in IEEE 5th International Symposium on Spread Spectrum Techniques and Applications, Sun City, South Africa, 1998.

[24] J. L. Walsh, "A Closed Set of Normal Orthogonal Functions," *American Journal of Mathematics*, vol. 45, no. 1, pp. 5–24, 1923.

[25] J. S. Lee and L. E. Miller, *CDMA System Engineering Handbook*, Norwood, MA: Artech House, 1998.

[26] M. K. Simon, J. M. Omura, R. A. Scholtz, and B. K. Levitt, *Spread Spectrum Communications Handbook*, rev. edn, Boston, MA: McGraw-Hill, 1985, 1994.

[27] R. Price, "Optimal Detection of Random Signals in Noise with Application to Scatter-Multipath Communication, I," *IRE Transactions*, vol. 6, no. 4, pp. 125–35, December 1956.

[28] R. Price and P. E. Green Jr., "A Communication Technique for Multipath Channels," *Proceedings of the IRE*, vol. 46, pp. 555–70, March 1958.

[29] U. Ladebusch and C. A. Liss, "Terrestrial DVB (DVB-T): A Broadcast Technology for Stationary Portable and Mobile Use," *Proceedings of the IEEE*, vol. 94, no. 1, pp. 183–93, January 2006.

[30] T. Star, M. Sorbara, J. M. Cioffi, and P. J. Silverman, *DSL Advances*, Upper Saddle River, NJ: Prentice Hall, 2003.

[31] IEEE LAN/MAN Standards Committee, "Part 11: Wireless LAN Medium Access Control (MAC) and Physical Layer (PHY) Specifications," New York, 2012.

[32] 3GPP TS 25.213, "Technical Specification Group Radio Access Network; Spreading and Modulation (FDD)," Valbonne, France, 2010.

[33] Y. G. Li and G. Stuber, Eds., *Orthogonal Frequency Division Multiplexing for Wireless Communications*, New York: Springer Science+Business Media, 2006.

[34] R. Prasad, *OFDM for Wireless Communications Systems*, Norwood, MA: Artech House, 2004.

[35] J. Heiskala and J. Terry, *OFDM Wireless LANs: A Theoretical and Practical Guide*, Indianapolis, IN: SAMS, 2002.

[36] S. B. Weinstein and P. M. Ebert, "Data Transmission by Frequency Division Multiplexing Using Discrete Fourier Transform," *IEEE Transactions on Communications*, vol. COM-19, no. 10, pp. 628–34, Oct. 1971.

[37] J. G. Proakis and D. G. Manolakis, *Introduction to Digital Signal Processing*, New York: Macmillan, 1988.

[38] M. Liu, M. Crussiere, J. Helard, and O. P. Pasquero, "Analysis and Performance Comparison of DVB-T and DTMB Systems for Terrestrial Digital TV," in 11th IEEE Singapore International Conference on Communication Systems, 2008. *ICCS 2008*, Guangzhou, China, 2008.

2 Selected Topics in Detection and Estimation Theory

Detection and estimation theory is an important component of the theoretical foundation of digital communications. Many components of a receiver in a digital communication system are judged by the receiver's optimality from a detection and estimation point of view.

Detectors or estimators can be designed based on different criteria. However, some of these criteria may not be easily applied in practice, because either some of the conditions cannot be met or it is not possible to have all of the information needed. The most common optimal estimators are based on the *maximum likelihood* criterion and the *maximum a posteriori (MAP)* criterion, which will be presented in this chapter.

In practice, it is desirable to compare the performance of a designed detector/estimator with that of a theoretically optimum one. A useful benchmark for such a comparison is the *Cramer–Rao bound (C-R bound)*, which will be presented and discussed later in this chapter. We will also describe the *Neyman–Pearson lemma*, which is essential to establishing the optimality of the thresholding decision rule used in the detectors for initial acquisition and frame synchronization.

What is presented in this chapter is primarily intended for facilitating the understanding of the subjects covered in this book, not for a general understanding of detection and estimation theory. Given the scope of this book, only the selected topics in detection and estimation theory that are pertinent to synchronization will be presented. For more information, readers are referred to the references on estimation and detection theory, e.g., [1, 2, 3, 4], and many other textbooks on statistics.

2.1 Likelihood Function and Maximum Likelihood Estimator

The *likelihood function* [4, 5] is widely used in the optimization of detection and/or estimation in digital communication applications. Its popularity is due mainly to the fact that it requires less information about the random variable involved in the given application than other optimization criteria. However, for the same reason, better estimators can be derived if additional information is available. In this section, we introduce the likelihood function, its logarithmic form, i.e., the *log-likelihood function*, and the ML estimate with the likelihood function as the metric.

2.1.1   Likelihood Functions of Random Variables

For a random variable X with a parameter θ, its *probability density function (pdf)*, $p_X(x|\theta)$,[1] is a function of x depending on the parameter θ. When the pdf is viewed as a function of the unknown parameter θ with the fixed observation x, it is termed the *likelihood function of θ given the observation x* denoted by $L(\theta)$. Namely,

$$L(\theta) = p_X(x|\theta) \tag{2.1}$$

Let us first consider a simple example, where X is a scalar random variable, that is equal to θ plus random noise sample n. The observation of X is simply

$$x = \theta + n \tag{2.2}$$

Its likelihood function is given by (2.1).

Second, we consider a series of samples $\{x_k\}$, $k = 0, 1, \ldots N-1$, which are observations of the random variable X. The sample x_k can be expressed as

$$x_k = \theta + n_k \tag{2.3}$$

where θ is the parameter to be estimated and n_k is the noise sample in x_k. The likelihood function of θ for the given observations $\{x_k\}$, denoted by a vector $\mathbf{x}$, is simply their joint pdf, $p_X(\mathbf{x}|\theta)$. In this case, the likelihood function of θ is

$$L(\theta) = p_{\mathbf{X}}(\mathbf{x}|\theta) \tag{2.4}$$

In many cases, it is more convenient to use the logarithmic form of the likelihood function instead of the function itself, i.e.,

$$l(\theta) \cong \ln[L(\theta)] = \ln[p_{\mathbf{X}}(\mathbf{x}|\theta)]^2 \tag{2.5}$$

Because the logarithm is a monotonic function and a pdf is nonnegative, the logarithm of a likelihood function preserves its key characteristics. It is commonly called the *log-likelihood function*.

Likelihood and log-likelihood functions are widely used for detection and estimation in digital communications.

2.1.2   Detection Based on Likelihood Functions

In a detection problem, the parameter θ can take multiple possible known values $\{\theta_k\}$. The objective is to determine how likely each value is based on the set of observations obtained. For a set of observations $\mathbf{x}$, the likelihood value for a given value θ_k indicates how likely θ takes the value θ_k. For example, if $L(\theta_1) > L(\theta_2)$, θ is more likely to take the value θ_1 than the value θ_2. Thus, we select θ_1 as the more likely outcome. Conversely, if $L(\theta_2) > L(\theta_1)$, we would select θ_2 as the more likely outcome.

---

[1] In the literature, the pdf $p_X(x|\theta)$ is often written as $p_X(x;\theta)$ to emphasize that θ is a parameter. The likelihood function $L(\theta)$ can be written as $L(\theta;x)$ to indicate that it is based on the observation x.

[2] In this book, "$\cong$" is used as the definition sign, and thus "$l(\theta) \cong \ln[L(\theta)]$" means $l(\theta)$ is defined as $\ln[L(\theta)]$.

In other cases, we are more interested in *how much* more likely one hypothesis is than the other. In these cases, the quantity of how much more likely can be characterized by $L(\theta_1)/L(\theta_2)$, called *likelihood ratio*, or $l(\theta_1) - l(\theta_2)$, called *log-likelihood ratio*. The decision-making criterion discussed above, i.e., if θ is more likely to have the value θ_1 or θ_2, is equivalent to determining if $L(\theta_1)/L(\theta_2)$ is greater than or less than 1, or $l(\theta_1) - l(\theta_2)$ is positive or negative.

Example 2.1 In (2.2), assume that θ equals to 1 or -1. If $L(\theta = 1) > L(\theta = -1)$, given the observation of sample x, the value of θ in x is more likely to be 1 instead of -1. Thus, the value of $\theta = 1$ is detected. Otherwise, if $L(\theta = 1) < L(\theta = -1)$, the value of $\theta = -1$ is detected. The log-likelihood function $l(\theta)$ can be used instead of $L(\theta)$ and yields the same result.

Moreover, if n is Gaussian, or normal, distributed with zero mean and a variance of σ^2, i.e., with distribution $\mathcal{N}(0, \sigma^2)$, the pdf of x is

$$p_X(x|\theta) = \frac{1}{\sqrt{2\pi\sigma^2}} e^{-\frac{1}{2\sigma^2}(x-\theta)^2} \tag{2.6}$$

It is more convenient to make the decision based on the log-likelihood value. Because

$$l(\pm 1) = \ln\left[\frac{1}{\sqrt{2\pi\sigma^2}} e^{-\frac{1}{2\sigma^2}(x\mp 1)^2}\right] = -\frac{\ln(2\pi\sigma^2)}{2} - \frac{x^2+1}{2\sigma^2} \pm \frac{x}{\sigma^2} \tag{2.7}$$

it is easy to verify that if $x > 0$, $l(1) > l(-1)$ and a decision $\theta = 1$ can be made. Conversely, if $x < 0$, it is likely that $\theta = -1$. In addition, the log-likelihood ratio is equal to

$$LLR(\theta = 1) = \log(L(1)/L(-1)) = l(1) - l(-1) = 2x/\sigma^2 \tag{2.8}$$

Thus, the decision can be made based on the sign of x. If θ is a coded bit from a channel encoder of the transmitter in a digital communication system, the LLR values are the optimal decoding metrics for performing channel decoding in the receiver.

2.1.3   Maximum Likelihood Estimator

For estimation, the objective is usually to determine the likely value of the parameter θ, which is unknown, based on the set of (discrete-time) observations $\mathbf{x}$ received. One intuitively good estimate of θ, denoted by $\hat{\theta}_{ML}$, for the given $\mathbf{x}$ is that it maximizes the pdf of X, $p_{\mathbf{X}}(\mathbf{x}|\theta)$, i.e.,

$$\hat{\theta}_{ML}(\mathbf{x}) = \arg\max_{\theta} p_{\mathbf{X}}(\mathbf{x}|\theta) \tag{2.9}$$

Given the pdf of X is the likelihood function of θ, $\hat{\theta}_{ML}$ is called the ML estimate of θ. If the likelihood function is well behaved, i.e., it is differentiable with respect to θ and has only one global peak with no local peaks, it can be obtained by taking the derivative of the pdf with respect to θ, setting the derivative equal to zero, and solving the equation. Because the logarithmic function is monotonic, finding the maximum of $p_{\mathbf{X}}(\mathbf{x}|\theta)$ is equivalent to finding the maximum of its logarithm. In many cases, using the latter definition is more convenient. The estimate obtained from either the likelihood function or its logarithmic form is called an ML estimate [6].

Example 2.2 Now we consider the case given by (2.4). Assuming that n_k's are independent and identically distributed (i.i.d.) with marginal distribution $\mathcal{N}(0,\sigma^2)$, we can express the joint pdf as

$$p_{\mathbf{X}}(\mathbf{x}|\theta) = \prod_{k=0}^{N-1} \frac{1}{\sqrt{2\pi}\sigma} e^{-\frac{(x_k-\theta)^2}{2\sigma^2}} = \frac{1}{\left(\sqrt{2\pi}\sigma\right)^N} e^{-\frac{1}{2\sigma^2}\sum_{k=0}^{N-1}(x_k-\theta)^2} \qquad (2.10)$$

The log-likelihood function of θ given $\{x_k\}$ is

$$l(\theta) = \ln\left(p_{\mathbf{X}}(\mathbf{x}|\theta)\right) = -\left(\sqrt{2\pi}\sigma\right)^N - \frac{1}{2\sigma^2}\sum_{k=0}^{N-1}(x_k-\theta)^2 \qquad (2.11)$$

By taking the derivative of (2.11) and setting it equal to zero, we have

$$\frac{d}{d\theta}\ln\left(p_X(\mathbf{x}|\theta)\right) = \frac{1}{2\sigma^2}\sum_{k=0}^{N-1} 2(x_k-\theta) = 0 \qquad (2.12)$$

By solving (2.12), we obtain the ML estimate of θ as

$$\sum_{k=0}^{N-1}\left(x_k-\hat{\theta}\right) = 0, \text{i.e.,} \hat{\theta} = \hat{\theta}_{ML} = \frac{1}{N}\sum_{k=0}^{N-1} x_k \qquad (2.13)$$

Given $x_k = \theta + n_k$ we have

$$\hat{\theta}_{ML} = \frac{1}{N}\sum_{k=0}^{N-1}(\theta + n_k) = \theta + \frac{1}{N}\sum_{k=0}^{N-1} n_k \qquad (2.14)$$

It can be shown from (2.14) that $\hat{\theta}_{ML}$ is an unbiased estimate of θ. Furthermore, because n_k's are independent of each other, it can be shown that the variance of $\hat{\theta}_{ML}$ is

$$\sigma_{\hat{\theta}_{ML}}^2 = \frac{1}{N^2} E\left[\left|\sum_{k=1}^{N} n_k\right|^2\right] = \frac{1}{N^2} N\sigma^2 = \frac{1}{N}\sigma^2 \qquad (2.15)$$

Therefore, the variance of the ML estimate of θ is inversely proportional to the number of samples N.

2.2   Likelihood Function of Continuous-Time Signal and Applications

In the last section, we presented the basics of the likelihood function and ML estimation based on a set of discrete-time observations. Such a likelihood function model is widely used in statistics and other technical areas. However, in the field of digital communications, it is especially important to have a good understanding of the likelihood functions and ML estimation when the observations are continuous-time signals.

The likelihood function based on continuous-time signal was originally presented in [3]. In this section, we state only the basic results and their applications to synchronization. More details and rigorous derivations can be found in the literature, such as [3, 7, 8].

2.2.1   Basics of Likelihood Functions and ML Estimation Based on Continuous-Time Signal Observations

Consider the baseband received signal $r(t)$ for performing estimation that is expressed by

$$r(t) = x(t, \mathbf{\theta}) + z(t) \tag{2.16}$$

where $x(t, \mathbf{\theta})$ is the transmitted complex baseband signal convolved with the equivalent baseband CIR containing a set of parameters $\mathbf{\theta}$ to be estimated and where $z(t)$ is an AWGN process with power spectrum density N_0. The parameters in $\mathbf{\theta}$ or a subset of them are unknown to the receiver. Each of the parameters may take more than one value.

The received signal $r(t)$ is an observation of a random process with the parameter set $\mathbf{\theta}$ in the time interval $T_0 > 0$. The objective of ML estimation is to find a set of parameter values of $\mathbf{\theta}$ that maximize the likelihood function of $\mathbf{\theta}$ given the observation $r(t)$. It has been shown [7, 8] that the likelihood function is given by

$$L(\mathbf{\theta}) = \exp\left\{ -\frac{1}{N_0} \int_{T_0} |r(t) - x(t, \mathbf{\theta})|^2 dt \right\} \tag{2.17}$$

Expanding the square term in (2.17) and noting that $\int_{T_0} |r(t)|^2 dt$ is a constant, we obtain:

$$L(\mathbf{\theta}) = \exp\left\{ \frac{2}{N_0} \int_{T_0} \mathrm{Re}[r(t)x^*(t, \mathbf{\theta})] dt - \frac{1}{N_0} \int_{T_0} |x(t, \mathbf{\theta})|^2 dt \right\} \tag{2.18}$$

The log-likelihood function of $\mathbf{\theta}$ given the observation $r(t)$ can be expressed as

$$l(\mathbf{\theta}) = \frac{2}{N_0} \int_{T_0} \mathrm{Re}[r(t)x^*(t, \mathbf{\theta})] dt - \frac{1}{N_0} \int_{T_0} |x(t, \mathbf{\theta})|^2 dt \tag{2.19}$$

The likelihood function given by (2.18) or the log-likelihood function given by (2.19) can be used for ML parameter estimation. Namely,

$$\hat{\boldsymbol{\theta}}_{ML}(r) = \arg\max_{\boldsymbol{\theta}} L(\boldsymbol{\theta}) \quad \text{or} \quad \hat{\boldsymbol{\theta}}_{ML}(r) = \arg\max_{\boldsymbol{\theta}} l(\boldsymbol{\theta}). \tag{2.20}$$

2.2.2 ML Parameter Estimation of Baseband Received Signal

As shown by (1.8), a large class of baseband received signals $r(t)$ can be expressed by

$$r(t) = x(t, \boldsymbol{\theta}) + z(t) = \sum_{k=-\infty}^{\infty} e^{j\phi(t)} a_k g_c(t - kT - \tau) + z(t) \tag{2.21}$$

where $g_c(t)$ is the composite CIR of the transmitter filter and the channel.

The set of parameters $\boldsymbol{\theta}$ to be estimated includes the carrier phase $\phi(t)$, the transmission delay τ, and the channel modulation symbols $\{a_k\}$ in $r(t)$ during the observation period. Substituting $x(t, \boldsymbol{\theta})$ in (2.19) with what is given by (2.21) for performing hypothesis testing, we have

$$l(\boldsymbol{\theta}) = \frac{2}{N_0} \int_{T_0} \text{Re}\left[r(t) \sum_{k=-\infty}^{\infty} e^{-j\phi(t)} a_k^* g_c^*(t - kT - \tau) \right] dt - \frac{1}{N_0} \int_{T_0} \left| \sum_{k=-\infty}^{\infty} e^{j\phi(t)} a_k g_c(t - kT - \tau) \right|^2 dt$$

$$(2.22)$$

The objective of ML estimation is to determine a parameter set $\boldsymbol{\theta}$, which contains $\phi(t)$, τ, and $\{a_k\}$, that maximize $L(\boldsymbol{\theta})$ or $l(\boldsymbol{\theta})$. For its applications to synchronization to be discussed in this book, the conditions listed below are valid.

1. In the observation time interval T_0, we assume that $r(t)$ only contains the terms $a_k g_c(t - kT - \tau)$, $n - K + 1 \leq k \leq n$. The data symbols contained in $r(t)$, i.e., $\{a_k\}$, $n - K < k \leq n$, will be simply denoted as vector $\mathbf{a}_{n,K}$ below.
2. The distortion introduced at the edges of the interval T_0 can be ignored assuming that $g_c(t)$ has finite support and T_0 is sufficiently long.
3. The carrier phase $\phi(t)$ can be viewed as an unknown constant ϕ_n during the time interval T_0, i.e., from $(n-K+1)T + \tau$ to $nT + \tau$.
4. The second term on the right side of (2.22), which is independent of the received signal $r(t)$, is a function of the symbol vector $\mathbf{a}_{n,K}$. For most of the applications to synchronization, $\mathbf{a}_{n,K}$ is a known constant vector. Thus, this term can be ignored when evaluating the log-likelihood function.

Based on these conditions, (2.22) can be rewritten as

$$l(\boldsymbol{\theta}) = \frac{2}{N_0} \int_{T_0} \text{Re}\left[r(t) e^{-j\phi_n} \sum_{k=n-K+1}^{n} a_k^* g_c^*(t - kT - \tau) \right] dt \tag{2.23}$$

The expression given by (2.23) is used for ML estimation of $\boldsymbol{\theta}$ according to (2.20).

2.2.3 Receiver Filtering for ML and Approximate ML Estimation

The expression of the log-likelihood function given by (2.23) can be rewritten as

$$l(\boldsymbol{\theta}) = \frac{2}{N_0} \mathrm{Re} \left[e^{-j\phi_n} \sum_{k=n-K+1}^{n} a_k^* \int_{T_0} g_c^*(t - kT - \tau) r(t) dt \right] \tag{2.24}$$

It can be shown that the integral in (2.24) can be expressed as

$$\int_{T_0} g_c^*(t - kT - \tau) r(t) dt = g_c^*(-t) * r(t) \big|_{t=kT+\tau} \cong y(kT + \tau) \tag{2.25}$$

where $y(t) \cong g_c^*(-t) * r(t)$, and $*$ denotes the convolution operation. Note that $y(t)$ is the output of the receiver filter with the impulse response of $g_c(-t)$, which is the MF of $r(t)$.

Using the result of (2.25), the log-likelihood function (2.24) can be expressed as

$$l(\boldsymbol{\theta}) = \frac{2}{N_0} \mathrm{Re} \left[e^{-j\phi_n} \sum_{k=n-K+1}^{n} a_k^* y(kT + \tau) \right] \cong \frac{2}{N_0} \mathrm{Re} \left[e^{-j\phi_n} u_n(\tau) \right] \tag{2.26}$$

where the scalar $u_n(\tau)$ is the correlation between data symbols $\{a_k\}$ and the output of the receiver filter sampled at $kT+\tau$, $n - K + 1 \leq k \leq n$, i.e., $u_n(\tau) \cong \sum_{k=n-K+1}^{n} a_k^* y(kT + \tau)$. The value of $u_n(\tau)$ depends on the parameter set $\boldsymbol{\theta} = \{\phi_n, \tau, \mathbf{a}_{n,K}\}$. The real part of $e^{-j\phi_n} u_n(\tau)$ represents the likelihood value of $\boldsymbol{\theta}$ based on the observation $r(t)$. Thus, the ML estimation of the parameter set $\boldsymbol{\theta}$ can be performed based on the scalar $u_n(\tau)$ defined above. Below we consider its applications to receiver synchronization.

Because the data symbol vector $\mathbf{a}_{n,K}$ is treated as a known constant vector rather than as parameters, $l(\boldsymbol{\theta})$ in (2.26) is the log-likelihood function of τ and ϕ_n with the known $\mathbf{a}_{n,K}$, i.e.,

$$l(\phi_n, \tau) = \frac{2}{N_0} \mathrm{Re} \left[e^{-j\phi_n} u_n(\tau) \right] \tag{2.27}$$

It is generated from the output of the receiver filter matched to $r(t)$. With proper treatment of τ and ϕ_n, $l(\phi_n, \tau)$ can be used as the metric for the detection of the received signal containing the expected symbol vector $\mathbf{a}_{n,K}$ in initial acquisition.

When the sampling delay is equal to the actual transmission delay τ_0, $y(kT+\tau_0)$ is the output of the matched filter of $r(t)$ sampled at the optimum time instant. In this case, the signal power of a_k in the sample is maximized. Thus, if both $\mathbf{a}_{n,K}$ and τ_0 are known, $l(\phi_n, \tau)$ in (2.27) is the log-likelihood function of parameter ϕ_n given the observation $r(t)$. It is the basis for ML estimation of the carrier phase presented in Section 5.4.1.

If both the carrier phase ϕ_n and vector $\mathbf{a}_{n,K}$ are known, the ML estimate of τ is the sampling delay that maximizes $l(\phi_n, \tau)$ in (2.27). In the case that ϕ_n is unknown, it would be sensible to treat it as a random variable with a uniform distribution between 0 and 2π. The ML estimate can be obtained by averaging the likelihood function $L(\phi_n, \tau) = e^{l(\phi_n, \tau)}$

over this range. This method will be used in Chapters 3 and 6 for initial acquisition and timing synchronization.

Above, we have shown that the optimal receiver filter that generates ML parameter estimates of the received signal should match the composite channel response $g_c(t)$. However, in order to implement such an optimal receiver filter, the characteristics of the transmitter filter and the channel must be known. In practice, such information is usually not available. Thus, the receiver filter can only be suboptimal. A common practice in system and receiver implementations is to let the frequency responses of both the transmitter filter and receiver filter have the *square-root-raised-cosine* form. Such an implementation is optimal if the channel is not frequency selective, i.e., it is a single-path channel. Otherwise, it is suboptimum.

Another option often used in wireline receiver designs is to design the receiver filter to match the typical channel response including the transmitter filter. In both cases, the estimates of $\boldsymbol{\theta}$ that maximize (2.26) from such suboptimal receiver filters are called the *approximate ML estimates* of $\boldsymbol{\theta}$ for the observation $r(t)$. We will use $y(t)$ to denote the receiver filter output in response to the received signal $r(t)$ even if the receiver filter does not match $g_c(t)$. The applications of such ML and approximate ML estimations to synchronizations are discussed in Chapters 3, 5, and 6.

2.3 Maximum a Posteriori Probability Detectors/Estimators

The ML estimate of the parameter θ discussed in Section 2.1.3 is an intuitively good estimate if the distribution of θ is unknown. However, if θ is a random variable and its pdf $f(\theta)$ is known, a better estimate of θ is the MAP probability estimate [1], which satisfies

$$\hat{\theta}_{MAP}(x) = \arg \max_{\theta} p_X(x|\theta) f(\theta) \tag{2.28}$$

The equation (2.28) indicates that $\hat{\theta}_{MAP}(x)$ is weighted more in the region where θ is more likely and weighted less in the region where θ is less likely.

If θ is uniformly distributed, the MAP estimate reduces to the ML estimate.

2.4 Cramer–Rao Bound

In an estimation problem, it is often of interest and importance to know if the estimate obtained is optimal, namely, if the estimate obtained is the best estimate under the conditions given. One criterion of the optimality of an estimate is the *Cramer–Rao bound* [9, 10], which is a lower bound of the variance of any unbiased estimator.

2.4.1 The Cramer–Rao Bound of a Scalar Unbiased Estimate

Assume that θ is an unknown, but deterministic, parameter to be estimated from the observation vector $\mathbf{x}$, which has a joint pdf of $p_X(\mathbf{x}|\theta)$. The variance of any unbiased estimator $\hat{\theta}$ of θ is lower bounded by

$$\mathrm{var}(\hat{\theta}) \geq \frac{1}{-E\left[\dfrac{\partial^2}{\partial \theta^2} \ln\left(p_{\mathbf{X}}(\mathbf{x}|\theta)\right)\right]} \tag{2.29}$$

The denominator in the above inequality $-E\left[\frac{\partial^2}{\partial \theta^2} \ln\left(p_{\mathbf{X}}(\mathbf{x}|\theta)\right)\right]$ is called the *Fisher information* $I(\theta)$ [1].

An unbiased estimator that attains this lower bound is called *efficient*. It is optimal in the sense that it achieves the MMSE between $\hat{\theta}$ and the actual value of θ among all unbiased estimators.

Example 2.3 Consider the simple case

$$x = A + n \tag{2.30}$$

where x is the signal (observation), which is equal to a constant A plus noise sample n, which is a random variable with distribution $\mathcal{N}(0, \sigma^2)$. Our objective is to estimate the parameter A from the observation x.

The pdf of x can be expressed as

$$p_X(x|A) = \frac{1}{\sqrt{2\pi\sigma^2}} e^{-\frac{(x-A)^2}{2\sigma^2}} \tag{2.31}$$

For a fixed observation x and parameter A, (2.31) is the likelihood function of A, as discussed above.

As a special case of (2.14), the ML estimate is $\hat{A}_{ML} = x$, which is an unbiased estimate of A and the variance of the estimation error is simply σ^2.

By definition, the Cramer–Rao bound of such an estimate is equal to

$$\mathrm{var}(\hat{A}) \geq \frac{1}{-E\left[\dfrac{\partial^2}{\partial A^2} \ln\left(p(x|A)\right)\right]} \tag{2.32}$$

Given that $\ln\left(p(x; A)\right) = -\ln\left(\sqrt{2\pi\sigma^2}\right) - (x-A)^2/2\sigma^2$ it can be shown that $-E\left[\frac{\partial^2}{\partial A^2} \ln\left(p(x; A)\right)\right] = \frac{1}{\sigma^2}$. According to the Cramer–Rao bound, $\mathrm{var}(\hat{A}) \geq \sigma^2$. The above estimator attains σ^2. Thus, it satisfies the Cramer–Rao bound. Moreover, it is an unbiased MMSE (UMMSE) estimator and is efficient.

Example 2.4 For the case discussed in Example 2.2, there are N received signal samples, $x_0, x_1, \ldots, x_{N-1}$, such that

$$x_i = A + n_i \tag{2.33}$$

As it was shown in Example 2.2, the ML estimate of A is $\hat{A}_{ML} = \sum_{i=1}^{N} x_i/N$, which is unbiased with a variance of estimation error, $\mathrm{var}(\hat{A}) = \sigma^2/N$. From (2.11) we have

$$-E\left[\frac{\partial^2}{\partial A^2}\ln\left(p_{\mathbf{X}}(\mathbf{x}|A)\right)\right] = \frac{\partial}{\partial A}\frac{1}{2\sigma^2}\sum_{k=0}^{N-1}2(x_k - A) = \frac{N}{\sigma^2} \tag{2.34}$$

Thus, this ML estimator is unbiased and achieves the Cramer–Rao bound. Therefore, it is optimal according to the UMMSE criterion, or efficient.

2.5 Binary Hypothesis Test and the Neyman–Pearson Lemma

In this section, we introduce the binary hypothesis test, which is pertinent to a number of the topics discussed in this book, in particular to initial acquisition. We then present the basic results of the Neyman–Pearson lemma [11] in order to establish the optimality of the methods used for performing initial acquisition. We will not provide its proof, which is readily available in many textbooks such as [1, 2].

2.5.1 The Binary Hypothesis Test

Consider a hypothesis test with two possible outcomes H_0 and H_1, based on an N-dimensional random vector $\mathbf{y} = (y_0, y_1, \ldots, y_{N-1})^T$ and some predetermined decision rules. Such a test is called a *binary hypothesis test*. The random data vector $\mathbf{y}$ has a joint pdf of $p_{\mathbf{Y}}(\mathbf{y}|\theta)$, where the parameter θ may take one of the two possible values θ_0 and θ_1. The objective of the binary hypothesis test is to determine if the random vector has the pdf of $p_{\mathbf{Y}}(\mathbf{y}|\theta = \theta_1)$, i.e., if H_1 is valid, or if $p_{\mathbf{Y}}(\mathbf{y}|\theta = \theta_0)$, i.e., if H_0 is valid. The binary test may be characterized by its false alarm probability and its miss probability or, equivalently, the detection probability, which equals to 1 minus the miss probability.

2.5.2 False Alarm, Miss, and Detection Probabilities

If the binary test determines that H_1, i.e., $\theta = \theta_1$, is the valid outcome but actually $\theta = \theta_0$, it is called a false alarm event. Its probability is called the *false alarm probability*, denoted as P_F. On the other hand, if the binary test determines that H_0, i.e., $\theta = \theta_0$, is the valid outcome but actually $\theta = \theta_1$, it is called a miss event. The probability of the miss event is called the *miss probability*, denoted as P_{miss}. If the test determines that H_1, i.e., $\theta = \theta_1$, is the valid outcome and H_1 is indeed true, it is called a detection event. The probability of the detection event among all of the tests in which $\theta = \theta_1$ is called the *detection probability*, denoted as P_D. Note that $P_D = 1 - P_{miss}$.

2.5.3 The Neyman–Pearson Lemma

For the binary hypothesis test described above, we first form the likelihood ratio:

$$\Lambda(\mathbf{y}) \cong \frac{L(\theta_1)}{L(\theta_0)} = \frac{p_{\mathbf{Y}}(\mathbf{y}|\theta_1)}{p_{\mathbf{Y}}(\mathbf{y}|\theta_0)} \tag{2.35}$$

We then define a decision rule for the hypothesis test such that

$$\Lambda(\mathbf{y}) = \frac{L(\theta_1)}{L(\theta_0)} \begin{cases} > \eta & \Rightarrow & H_1 : \theta = \theta_1 \\ \leq \eta & \Rightarrow & H_0 : \theta = \theta_0 \end{cases} \tag{2.36}$$

where η is a predetermined threshold value. The subset of the data vectors, which satisfy the likelihood ratio decision rule $\Lambda(\mathbf{y}) > \eta$, spans a subspace, or region, C of the entire N-dimensional space spanned by the random vectors $\mathbf{y}$. C is also called the *critical region* of this test.

Assume that the false alarm probability is equal to α under the decision rule defined by (2.36) based on the likelihood ratio. It is defined then that the critical region C has size α. The Neyman–Pearson lemma states that the critical region as defined above based on the likelihood ratio decision rule is the *most powerful* region of size α. In other words, the likelihood ratio test (2.36) has the highest detection probability compared to any other decision rule that yields a false alarm probability of α.

2.5.4   Basic Cases of Detecting a Known Signal in Additive Gaussian Noise

The Neyman–Pearson lemma provides the theoretical basis for constructing and analyzing the optimal binary hypothesis test. In this section, we provide two simple examples of applying such tests for detecting the scalar signal sample in noise. Such cases are often encountered in radar applications [12]. To make these examples mathematically tractable, we assume that the noise considered in the examples has a Gaussian, i.e., normal, distribution. The results obtained from these examples will be used for initial acquisition, discussed in the next chapter.

Example 2.5: *Detection of Signal with a Known Data Symbol in Gaussian Noise*
The observed sample y is given by

$$y = y_r + jy_i = A\theta + z \tag{2.37}$$

where θ is the parameter, which may take one of the two possible values $\theta_0 = 0$ and $\theta_1 = 1$, $A = A_r + jA_i \cong |A|e^{j\phi}$ is a complex constant with its phase ϕ known to the receiver, and noise sample z is complex and Gaussian distributed. The real and imaginary parts of z are independent real random variables, each of which has zero mean and a variance σ^2. The variance of z, denoted as σ_z^2, is equal to $2\sigma^2$.

When a data symbol existing in the signal, i.e., $\theta = \theta_1 = 1$, y takes the form

$$y = y_r + jy_i = A_r + jA_i + z = |A|e^{j\phi} + z \tag{2.38}$$

If there is no data symbol in the signal, i.e., $\theta = \theta_0 = 0$, the sample y has the form

$$y = z \tag{2.39}$$

If $\theta = \theta_1$, y takes the form of (2.38) and the pdf of y is

$$p_1(y) = \frac{1}{2\pi\sigma^2} e^{-\frac{(y_r - A_r)^2 + (y_i - A_i)^2}{2\sigma^2}} = \frac{1}{2\pi\sigma^2} e^{-|y-A|^2/2\sigma^2} = \frac{1}{2\pi\sigma^2} e^{-|y-|A|e^{j\phi}|^2/2\sigma^2}, \qquad (2.40)$$

and if $\theta = \theta_0$, y takes the form of (2.39) and its pdf is

$$p_0(y) = \frac{1}{2\pi\sigma^2} e^{\frac{y_r^2 + y_i^2}{2\sigma^2}} = \frac{1}{2\pi\sigma^2} e^{-|y|^2/2\sigma^2} \qquad (2.41)$$

We perform the hypothesis test to determine between H_1: $\theta = \theta_1$, i.e., $y = A+z$ and H_0: $\theta = \theta_0$, i.e., $y = z$ for signal detection. Moreover, we define a *decision metric D* as

$$D = \mathrm{Re}\left[e^{-j\phi}y\right] = \begin{cases} |A| + z' \\ z' \end{cases} \qquad (2.42)$$

where $z' = \mathrm{Re}\left[e^{-j\phi}z\right]$ and its variance is $\sigma_{z'}^2 = \sigma^2$.

For $H_1 : y = A + z$, the pdf of D in (2.42) is

$$p_1(D) = \frac{1}{\sqrt{2\pi}\sigma_z} e^{-(D-|A|)^2/2\sigma_z^2} \qquad (2.43)$$

However, if $y = z$ as in (2.39), the pdf of D is

$$p_0(D) = \frac{1}{\sqrt{2\pi}\sigma_{z'}} e^{-D^2/2\sigma_{z'}^2} \qquad (2.44)$$

The likelihood functions of the parameter θ given the observation D are defined by $L(\theta = \theta_1) = p_1(D)$ and $L(\theta = \theta_0) = p_0(D)$.

The decision rule is defined as follows. We first determine a threshold Th based on a certain criterion according to the desired detector performance. If D is greater than Th, we declare that H_1 is valid, i.e., the signal sample y contains the data symbol A. Conversely, if D is less than or equal to the threshold Th, H_0 is valid, i.e., y contains noise only.

Intuitively, we can see that this decision rule makes sense. If H_1 is the correct hypothesis, the sample y contains both data and noise and thus has higher energy than if H_0 is true where y contains noise only. Hence, the magnitude of D should be larger in the first case (H_1) than that in the second case (H_0). Below, we show that this decision rule also satisfies Neyman–Pearson lemma and, thus, it is optimal.

With the decision rule based on the value of D, the false alarm, miss, and detection probabilities, P_F, P_{miss}, and P_D are given, respectively, by

$$P_F = \int_{Th}^{\infty} p_0(D)dD, \quad P_{miss} = \int_{-\infty}^{Th} p_1(D)dD \text{ and } P_D = 1 - P_{miss} = \int_{Th}^{\infty} p_1(D)dD. \qquad (2.45)$$

As can be seen in (2.45), P_F is a monotonically decreasing function of Th because $p_0(D)$ is a positive function. As a result, there is a one-to-one correspondence between P_F and Th. Therefore, we can select a threshold Th such that it results in a false alarm probability P_F that is uniquely defined.

On the other hand, the likelihood ratio given in (2.35) in terms of D can be written as

$$\Lambda(D) \cong \frac{L(\theta = \theta_1)}{L(\theta = \theta_0)} = \frac{p_1(D)}{p_0(D)} = \frac{\frac{1}{\sqrt{\pi}\sigma_{z'}} e^{-(D-|A|)^2/2\sigma^2}}{\frac{1}{\sqrt{\pi}\sigma_{z'}} e^{-D^2/2\sigma^2}} = \frac{e^{-(D-|A|)^2/2\sigma^2}}{e^{-D^2/2\sigma^2}} = \frac{e^{2|A|D/2\sigma^2}}{e^{|A|^2/2\sigma^2}}$$

$$(2.46)$$

We observe from (2.46) that $\Lambda(D)$ is a monotonically increasing function of D. Hence, there is also a one-to-one correspondence between η and Th. In other words, the event that D is greater or less than a given Th corresponds to the event that $\Lambda(D)$ is greater or less than a unique value of η. Hence, η is a function of Th and can be expressed as $\eta(Th)$. Consequently, comparing D with the threshold Th is equivalent to comparing $\Lambda(D)$ with the threshold $\eta(Th)$. Therefore, the decision rule based on D as described in this section is optimal according to the Neyman–Pearson lemma. Furthermore, substituting the metric D in (2.46) with Th, we obtain the threshold on $\Lambda(D)$ as

$$\eta(Th) = e^{2|A|Th/2\sigma^2} / e^{|A|^2/2\sigma^2}$$

$$(2.47)$$

Example 2.6: *Detection of a Data Symbol with Unknown Phase in AWGN*

The sample y has the same form as that given by (2.38) and (2.39) in Example 2.5. Its pdfs, or likelihood functions, are given by (2.40) and (2.41). However, in this example, we assume that the phase ϕ is unknown.

When there is no a priori information of the unknown phase ϕ of y, it is sensible to assume that ϕ is a random variable uniformly distributed in the region $[0, 2\pi)$. The average likelihood function of $\theta = \theta_1$ can be obtained by integrating (2.40) with respect to ϕ from 0 to 2π, and we have

$$\begin{aligned}\bar{p}_1(y) &\cong E_\phi[L(\theta = \theta_1)] = \frac{1}{2\pi} \int_0^{2\pi} \frac{1}{2\pi\sigma_z^2} e^{-|y-|A|e^{-j\phi}|^2/2\sigma^2} d\phi \\ &= \frac{1}{2\pi\sigma^2} e^{-(|y|^2+|A|^2)/2\sigma^2} I_0\left(|A|\sqrt{|y|^2}/\sigma^2\right)\end{aligned}$$

$$(2.48)$$

where $I_0(x) = \frac{1}{2\pi} \int_0^{2\pi} e^{x\cos\phi} d\phi$ is the *0th order modified Bessel function of the first kind*. The average likelihood function is also called the *unconditional likelihood function* [13].

The log-likelihood function of $\theta = \theta_0$ is the same as that given by (2.41). By observing that both (2.48) and (2.41) are functions of $|y|^2$, we define the decision metric in this case as

$$D = |y|^2$$

$$(2.49)$$

It can be shown that the pdfs of D under hypotheses H_1 and H_0 are

$$p_1(D) = \frac{1}{2\sigma^2} e^{-(D+|A|^2)/2\sigma^2} I_0\left(|A|\sqrt{D}/\sigma^2\right) \qquad (2.50)$$

which is the *noncentral* χ^2 *distribution* with two degrees of freedom, and

$$p_0(D) = \frac{1}{2\sigma^2} e^{-D/2\sigma^2} \qquad (2.51)$$

is the *central* χ^2 *distribution* with two degrees of freedom, which is also called the *exponential distribution* [8], respectively. In both (2.50) and (2.51), D is greater than zero according to the definition (2.49). It can be observed that (2.50) and the average likelihood function (2.48) only differ by a scaling factor of $1/\pi$.

The detection process is the same as that in Example 2.5. Namely, the decision metric D is compared to a threshold Th. If D is greater than Th, we declare H_1 as the valid hypothesis; otherwise, H_0 is the valid hypothesis.

Using the average likelihood function of $\theta = \theta_1$ given by (2.48) and from (2.41) and (2.49), we can form the likelihood ratio as

$$\Lambda(D) = \frac{E[L(\theta = \theta_1]}{L(\theta = \theta_0)} = \frac{\overline{p}_1(y)}{p_0(y)} = e^{-|A|^2/2\sigma^2} I_0\left(|A|\sqrt{|y|^2}/\sigma^2\right) \cong e^{-|A|^2/2\sigma^2} I_0\left(|A|\sqrt{D}/\sigma^2\right)$$
$$(2.52)$$

In the region $D > 0$, I_0 is a monotonically increasing function of D. This is evident from the power series expansion form of $I_0(x) = \sum_{m=0}^{\infty} (x/2)^{2m+L-1}/(m!)^2$ [14]. Thus, the decision rule based on D is equivalent to the decision rule based on $\Lambda(D)$ with the corresponding thresholds. Therefore, the decision rule based on D is optimal according to the Neyman–Pearson lemma.

2.6 Summary and Remarks

In this chapter, we presented a few topics in detection and estimation theory that are pertinent to the subjects covered in this book. Below are a few remarks.

- The most widely used metric for receiver optimization is the likelihood function as defined in Section 2.1.1. The logarithm of the likelihood function, called log-likelihood function, is often used in the place of the likelihood function itself.
- In estimation, the most likely value of θ based on the set of observations **x** can be determined by maximizing $p_X(x|\theta)$. Such an estimate is called the maximum likelihood estimate.
- The basics of the likelihood function and ML estimation of continuous–time signal are provided based on the results that are well documented in the literature.
- It is shown that the receiver filter that produces the likelihood function of the parameters of the received signal is the matched filter of the received signal. The approximations of such an optimal receiver filter are considered.

- The ML estimate of the parameter θ is a sensible estimate if the distribution of θ is unknown. If the distribution is known, a better estimate of θ is the MAP estimate. When θ is uniformly distributed, the MAP estimate reduces to the ML estimate.
- To determine the optimality of an estimate, one of the criteria is the Cramer–Rao bound. It is a lower bound of the variance of any unbiased estimator.
- Binary hypothesis tests are often implemented by first forming a decision metric based on the observed data. The decision metric is compared to a threshold, and the outcome is determined by the result of the comparison.
- The optimality of such a binary hypothesis test, which is pertinent to initial acquisition discussed in Chapter 3, can be judged based on the Neyman–Pearson lemma.
- A binary hypothesis test that satisfies the Neyman–Pearson lemma is optimal in the sense that when its threshold is selected based on the expected false alarm probability P_F, the resulting detection probability P_D is the largest for the given P_F among all decision rules.

References

[1] H. V. Poor, *An Introduction to Signal Detection and Estimation*, 2nd edn, New York: Springer, 1994.
[2] L. L. Scharf, *Statistical Signal Processing*, Reading, MA: Addison Wesley, 1991.
[3] H. L. Van Trees, *Detection, Estimation, and Modulation Theory, Part I*, New York: John Wiley & Sons, 1968.
[4] S. M. Kay, *Fundamentals of Statistical Signal Processing: Estimation Theory*, Upper Saddle River, NJ: Prentice Hall PTR, 1993.
[5] A. F. Edwards, *Likelihood*, Cambridge University Press, 1972.
[6] H. Akaike, "Maximum Likelihood Identification of Gaussian Autoregressive Moving Average Models," *Biometrika*, vol. 60, no. 2, 1973.
[7] U. Mengali and A. N. D'Andrea, *Synchronization Techniques for Digital Receivers*, New York: Plenum Press, 1997.
[8] J. G. Proakis and M. Salehi, *Digital Communications*, 5th edn, New York: McGraw-Hill, 2008.
[9] H. Cramer, *Mathematical Methods of Statistics*, Princeton University Press, 1946.
[10] C. R. Rao, "Information and the Accuracy Attainable in the Estimation of Statistical Parameters," *Bulletin of the Calcutta Mathematical Society*, no. 37, pp. 81–9, 1945.
[11] J. Neyman and E. S. Pearson, "On the Problem of the Most Efficient Tests of Statistical Hypotheses," Philosophical Transactions of the Royal Society A: Mathematical, *Physical and Engineering Sciences*, vol. 231, pp. 289–337, February 1933.
[12] M. Skolnik, *Introduction to Radar Systems*, Boston: McGraw-Hill, 1980.
[13] A. J. Viterbi, *CDMA Principle of Spread Spectrum Communications*, Reading, MA: Addison-Wesley, 1995.
[14] G. N. Watson, *A Treatise on the Theory of Bessel Functions*, Cambridge University Press, 1944.

3 Initial Acquisition and Frame Synchronization

3.1 Introduction

As the name indicates, initial acquisition, or simply acquisition, is the first step when a device tries to establish a communications link with a transmitter, which is connected to a network in most cases. The performance of acquisition determines how fast a communication link can be established, as well as how reliable the initial communication is. Hence, a well-designed initial acquisition mechanism is important to the overall receiver and system performance.

At the initial acquisition stage, the receiver in a device knows very little about the characteristics of the communication link. Moreover, some of the receiver parameters, such as the frequency of the local oscillator, are not calibrated. This increases the difficulty of initial acquisition. Thus, efforts must be devoted to the design, analysis, and implementation of initial acquisition.

The first step in initial acquisition is for the device to determine if the expected communication signal exists. In a multiple-access system, there could be more than one of such signals, and the receiver needs to select one of them based on certain predetermined criteria. Once such a signal is detected, the receiver needs to perform its second task, namely to extract various types of information from the detected signal so that the communication link can be established. Thus, the second important task of initial acquisition is to ensure that the receiver is ready, even if not perfectly, for the reception of data and control channels.

One important aspect of the second task is to calibrate the receiver parameters, in particular, to calibrate the frequency error of the LO. This calibration is essential because the oscillator is free running prior to signal acquisition. This task is particularly crucial for system and device designers nowadays because, for cost reduction, low-cost and low-accuracy crystal oscillators (XOs) are preferred to more accurate yet more expensive ones. At the same time, the signals from the network transmitters must meet the system requirements, which are usually quite stringent, and thus must be very accurate. Once the expected signal is acquired, the device's receiver extracts the information carried by the signal and uses it to improve the accuracy of the local frequency reference.

In addition to the estimate of the frequency offset of the local XOs, other initial estimates required by the receiver's normal operations, including the timing for generating received signal samples, the state of the automatic gain control (AGC) circuitry and channel estimates in frequency and/or time domain are usually also determined at this stage. These estimates are used for initializing the respective functional blocks of the receiver.

A related topic to initial acquisition is frame acquisition, or frame synchronization. The objective of frame synchronization is to determine the starting point of a frame, or a data block. Frame synchronization may be performed during initial acquisition or after the communication link has been established. In the latter case, some of the parameters, such as the frequency offset of the LO and/or sample timing, are already known to the receiver quite accurately, and the detection process becomes easier.

In this chapter, we examine various aspects of the operations performed by device receivers in the initial acquisition stage. In Section 3.2, the typical detection process and the basic detectors are described. The relationship between the theoretical foundations of signal detection in noise presented in Section 2.5 and the practical basic detectors is established. In addition, the most important parameters of detection, i.e., the detection, miss, and false-alarm probabilities, are derived analytically and discussed in detail.

In Section 3.3, we will show how the performance of the basic detector is related to, and can be improved by, pre- and post-detection integration. Their theoretical performance and optimality are presented and discussed.

In Section 3.4, we consider various important theoretical and practical aspects of initial acquisition from a system engineering point of view. Among these aspects, the most difficult task for initial acquisition, i.e., establishing a reliable initial communication link when the local oscillator has a large frequency error due to the inaccuracy of the local XO, is analyzed and the remedies are discussed. The procedures for acquiring system parameters to establish the communication link at higher layers are briefly discussed.

In Sections 3.5 and 3.6, we describe initial acquisition and frame synchronization processes for CDMA and OFDM communication systems with examples of practical implementations. Finally, Section 3.7 concludes this chapter with a summary and some key remarks.

3.2   Basic Detection Process in Initial Acquisition

In most practical digital communications systems, the detection process can be modeled as the detection of known signal waveforms in AWGN process. The detection processes for different types of systems have many aspects in common. Thus, we will first describe such a common process. There are also some systems, especially OFDM-based systems, for which the initial acquisition processes may be different in a number of aspects. We will deal with such cases later in this chapter.

During initial acquisition, the timing of the signal to be detected is unknown and the carrier phase is also likely to be unknown. The channel may be static or time variant, frequency selective or nonselective. These factors are also considered.

3.2.1   An Overview of the General Detection Process in Initial Acquisition

The general detection process for initial acquisition can be summarized as follows.

Transmitters periodically send signals with known channel modulation symbol sequences, which are aligned with the transmitted data packets, often called frames. In some types of communication systems, e.g., in certain CDMA systems, the known sequences are sent continuously. In other systems, they are sent in bursts. The sequences from different transmitters are usually uncorrelated or orthogonal to each other in order to facilitate the identification of the transmitter that sends the signal. We will call such known channel modulation symbols *reference symbols*, and the transmitter symbol sequences that consist of such symbols *reference symbol sequences*. The transmitted signals that contain such sequences are called *expected signals*.

Signal Model and ML Detection of Known Symbol Sequence in Received Signal

Let us consider that a reference symbol sequence $\{a(k)\}$, $k = n, n-1, \ldots, n-K+1$, which is embedded in the transmitter symbol sequence, is transmitted periodically with a period of NT, where T is the transmitter channel symbol interval. In the discussion below, we denote such a segment of the transmitted symbol sequence by the *reference symbol vector* $\mathbf{a}_{n,K}$, i.e.,

$$\mathbf{a}_{n,K} = [a(n),\ a(n-1), \ldots,\ a(n-K+1)]^t \tag{3.1}$$

The objectives of initial acquisition are to search for the segment of the received signal $r(t)$ that contains $\mathbf{a}_{n,K}$ and, once it is found, to determine its position, i.e., its *timing*. Assume that both the baseband transmitter and the receiver filters have the SRCOS frequency response. For transmission over a single-path, i.e., frequency nonselective, channel, the overall channel has the RCOS shape and satisfies the Nyquist criterion. In this section, we will use this channel model for our discussion of the basic detection procedures and the analysis of the detection process. General multipath channels will be considered later.

The signal model used below has been discussed in Section 2.2 with its log-likelihood function given by (2.26). Consider that a segment of the received signal $r(t)$ given by (2.21) contains the elements of $\mathbf{a}_{n,K}$ specified by (3.1). Its log-likelihood function of the channel phase ϕ_n and sampling delay τ can be expressed as

$$l(\phi_n, \tau) = \frac{2}{N_0} \mathrm{Re}\left[e^{-j\phi_n} \sum_{k=0}^{K-1} a_{n-k}^* y((m-k)T + \tau) \right] = \frac{2}{N_0} \mathrm{Re}\left[e^{-j\phi_n} \mathbf{a}_{n,K}^H \mathbf{y}_{m,K}(\tau) \right] \tag{3.2}$$

where we assume that the detection starts at $t = (m-k)T+\tau$, $y((m-k)T + \tau)$ is the receiver filter output $y(t)$ sampled at the time $t = (m-k)T + \tau$, N_0 is the power spectrum density of the AWGN process in the received signal $y(t)$, and $\mathbf{y}_{m,K}(\tau)$ is the data vector defined by

$$\mathbf{y}_{m,K}(\tau) = [y(mT + \tau)y((m-1)T + \tau) \cdots y((m-K+1)T + \tau)]^t \tag{3.3}$$

The detection of the known reference symbol vector $\mathbf{a}_{n,K}$ can be performed directly based on the log-likelihood function given by (3.2). Specifically, given that the reference symbol sequence is sent every NT, the receiver can compute and evaluate all of the values of $l(\phi_n, \tau)$ with m from any m_0 to m_0+N-1, ϕ_n from 0 to 2π, and τ from 0 to T. The pair of the values of $mT+\tau$ and ϕ_n that generate the maximum of $l(\phi_n, \tau)$ is the ML estimates of sampling time and carrier phase and an indication of the presence of the reference symbol vector. However, such a brute force method is neither efficient nor reliable for the following reasons.

First, such an ML method does not tell us how reliable the outcome of the detection is. To avoid this problem, the *hypothesis testing* approach is employed to make detection decisions by comparing the log-likelihood value to a predetermined threshold. This approach will be described below.

Second, it is unnecessary to compute the log-likelihood value for the delay $mT+\tau$ at a resolution higher than $T/2$. Because the bandwidth of the baseband signal is usually less than $2/T$, according to the Nyquist sampling theorem, there is no information lost by sampling the analog signal every $T/2$.

Finally, in most cases of initial acquisition, the carrier phase ϕ_n is not known but can be assumed to be a random variable uniformly distributed between 0 and 2π. An average, also called *unconditional*, ML estimate [1] of τ is derived as follows.

First, we form the likelihood function from the log-likelihood function given by (3.2) as

$$L(\phi_n, \tau) = \exp\left\{l(\phi_n, \tau)\right\} = \exp\left\{\frac{2}{N_0}\mathrm{Re}\left[e^{-j\phi_n}\mathbf{a}_{n,K}^H\mathbf{y}_{m,K}(\tau)\right]\right\} \tag{3.4}$$

Then, the likelihood function is averaged with respect to ϕ_n over the range $[0, 2\pi)$, and we obtain,

$$\frac{1}{2\pi}\int_0^{2\pi} \exp\left\{\frac{2}{N_0}\mathrm{Re}\left[e^{-j\phi_n}\mathbf{a}_{n,K}^H\mathbf{y}_{m,K}(\tau)\right]\right\}d\phi_n$$

$$= \frac{1}{2\pi}\int_0^{2\pi} \frac{2}{N_0}\exp\left\{\mathrm{Re}\left[e^{-j\phi_n+j\theta_{n,m}(\tau)}\left|\mathbf{a}_{n,K}^H\mathbf{y}_{m,K}(\tau)\right|\right]\right\}d\phi_n$$

$$= \frac{1}{2\pi}\int_0^{2\pi} \exp\left\{\frac{2}{N_0}\cos\left(\theta_{n,m}(\tau)-\phi_n\right)\left|\mathbf{a}_{n,K}^H\mathbf{y}_{m,K}(\tau)\right|\right\}d\phi_n$$

$$= I_0\left(\frac{2}{N_0}\left|\mathbf{a}_{n,K}^H\mathbf{y}_{m,K}(\tau)\right|\right) \tag{3.5}$$

where $\theta_{n,m}(\tau) = \mathrm{Arg}\left[\mathbf{a}_{n,K}^H\mathbf{y}_{m,K}(\tau)\right]$ and $I_0(x)$ is the 0th order modified Bessel function of the first kind as was used in Section 2.5.4. Finally, given that $I_0(x)$ is a monotonic increasing function of x [2], the sampling delay that maximizes $\left|\mathbf{a}_{n,K}^H\mathbf{y}_{m,K}(\tau)\right|$ also maximizes (3.5), so it is the *unconditional ML estimate* of τ. Therefore, when the carrier phase is unknown, $\left|\mathbf{a}_{n,K}^H\mathbf{y}_{m,K}(\tau)\right|$ or $\left|\mathbf{a}_{n,K}^H\mathbf{y}_{m,K}(\tau)\right|^2$ is commonly used as the decision metric in the hypothesis testing. Such hypothesis testing is optimal according to the Neyman–Pearson lemma as shown in Section 2.5.4.

A high-level description of the detection process based on hypothesis testing for initial acquisition is stated below.

Detection Process Based on Hypothesis Testing

The detection process commonly used in practice involves a series of tests consecutively conducted on the digital samples generated by sampling the receiver filter output, usually every $T/2$. It starts by correlating a known reference symbol sequence $\{a_k\}$, $k = n$, $n-1, \ldots, n-K+1$, with a sequence of K T-spaced signal samples ending at the sampling time $t = mT + \hat{\tau}$. In the discussion below, the symbol sequence is denoted by the vector $\mathbf{a}_{n,K}$ defined by (3.1). The sample sequence is represented by the vector $\mathbf{y}_{m,K}(\hat{\tau})$ defined by (3.3). Note that the sampling time offset $\hat{\tau}$ is arbitrarily selected, but it does not change during the detection process.

If the time span of the samples of $\mathbf{y}_{m,K}(\hat{\tau})$ is shorter than the coherence time of the signal, the carrier phase ϕ_n of these samples can be treated as a constant. As a result, the K terms involved in the correlation shown by (3.2) are combined *coherently*. Consequently, such a correlation operation is also called *coherent integration*, or *coherent combining*. If the time span of the samples in $\mathbf{y}_{m,K}(\hat{\tau})$ is longer than the coherence time of the signal, $\mathbf{y}_{m,K}(\hat{\tau})$ can be divided into multiple subvectors such that the time span of each subvector is shorter than the coherence time. Correspondingly, the symbol vector is also divided into multiple symbol subvectors. The symbol subvectors are correlated with the corresponding sample subvectors. The squares of these correlations are combined *noncoherently* as will be discussed in Section 3.3.2.

The system model and the operations of coherent integration and noncoherent integration in initial acquisition will be discussed further in Section 3.3.

To begin with, we consider the case that the coherence time is long enough that only one correlation is performed and noncoherent integration is not needed. The correlation generates a *detected symbol*, which, in turn, produces a *decision metric D*. The decision metric D is compared with a predetermined decision threshold Th. If D is greater than the threshold, the detection is declared successful. The receiver uses the sampling time of the sample vector that resulted in successful detection as the reference time of the received signal samples for further processing. However, if the correlation output is less than the threshold, the detection fails and the result is discarded. After a failed detection, the correlation is performed again at the next sampling time by correlating the same reference symbol sequence with another segment of T-spaced sample sequence.

If noncoherent combing should be performed due to a short channel coherence time, the decision metrics generated from the multiple correlations are combined to generate a *composite decision metric*. This composite metric is used in the detection process, and the other processing steps are the same as described above.

With the output of the receiver filter sampled at every $T/2$ and after the correlation using the sample vector $\mathbf{y}_{m,K}(\hat{\tau})$ is completed but the detection is not successful, correlations using sample vectors $\mathbf{y}_{m,K}(T/2+\hat{\tau})$, $\mathbf{y}_{m+1,K}(\hat{\tau})$, $\mathbf{y}_{m+1K}(T/2+\hat{\tau})$, $\ldots$, are carried out. The detection process continues until a successful detection is declared.

It is also possible that the detection process yields too many failures without success. In this case, based on certain criteria we may conclude that there is no expected signal to receive at that time and the particular geographical location. The receiver may go to sleep and wake up to perform acquisition later and/or at a different location.

The performance of the acquisition process can be characterized by the *false alarm probability* and *detection probability* as in the hypothesis testing described in Section 2.4. If a successful detection is declared but there is actually no expected reference signal, or the detected transmitter ID is incorrect, it is called a *false alarm event*. Conversely, if the expected signal exists but no detection success is declared, it is called a *missed detection event*, or *miss event*. The ratio of the number of false alarm events to the total number of tests using signal sample sequences with no expected signal is called the *false alarm probability*, or *false probability*, denoted by P_F. The ratio of the number of successful detections to the total number of tests on signal sample sequences with expected signal is called the detection probability, denoted by P_D. One minus the detection probability is called the *miss probability*, or P_{miss}, i.e.,

$$P_{miss} = 1 - P_D.$$

3.2.2   Detection Procedure and Hypothesis Testing in Initial Acquisition

The detection for initial acquisition is essentially a hypothesis-testing problem well studied in detection theory, which was first investigated in the area of radar systems [3]. It is also widely used in digital communications and digital signal processing (DSP) [1, 3, 4]. In this section, we will look into the detection procedure from this point of view. To begin, let us first establish the system model for the detection process.

A typical baseband digital communication link from a transmitter to a receiver is depicted in Figure 3.1.

As shown in the figure, a_l's are the Tx channel modulation symbols and $g(t)$ is the overall CIR, including the transmitter filter, channel, and receiver filter. In addition, T denotes the transmitter symbol interval, $1/T$ is the transmitter symbol frequency, and $z_c(t)$ is the noise process introduced by the channel, which is usually assumed to be an AWGN process. Moreover, an AGC block is present before the sampler so that the power of $y(t)$ and the variance of the samples at its output can be treated as a constant. Overall, Figure 3.1 shows a baseband model of the communication link. The modulator in the transmitter and the frequency down-converter in the receiver are not shown.

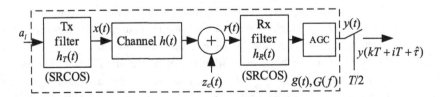

Figure 3.1 A typical baseband digital communication link

The baseband received signal $y(t)$ can be expressed as

$$y(t) = \sum_{l=-\infty}^{\infty} a_l g(t - lT - \tau) + z(t) \tag{3.6}$$

where τ is the delay introduced during transmission and $z(t)$ is a Gaussian noise process, which is the result of $z_c(t)$ filtered by the receiver filter. During initial acquisition, the a_l's are known. It is assumed that $g(t)$ is slowly time-varying, and the frequency offset between the transmitter carrier frequency and the receiver down-conversion frequency is small. Thus, in this section, the coherence time of the received signal is considered longer than the time span of the signal segment involved in performing correlation for the detection of the expected signal.

The output of the receiver filter $y(t)$ is converted to digital samples. If the receiver filter is the matched filter of the received signal $r(t)$, the samples of the received signal can be used to generate the log-likelihood values of the parameters to be estimated for the given $y(t)$, as discussed previously. Otherwise, the samples only produce approximate log-likelihood values. In the rest of this chapter, we will use values that are suitable for the detection of the reference symbols in both cases without concern to their optimality.

The sampling rate should be higher than or equal to the total transmitter signal bandwidth to ensure no loss of information. In single-carrier systems, since the transmitter signals are usually transmitted with an excess bandwidth that is less than $1/T$, the sampling rate is typically chosen to be $2/T$. The received signal samples are generated every $T/2$ with a fixed but unknown offset $\hat{\tau}$ from the timing of the transmitted symbols. Thus, the sampling time can be expressed as $t_{m,i} = mT + iT/2 + \hat{\tau}$, $i = 0$, 1 and $\hat{\tau} < T/2$. The samples generated are

$$y_{m,i} \cong y(mT + iT/2 + \hat{\tau}) = \sum_{l=-\infty}^{\infty} a_l g((m-l)T + iT/2 + \hat{\tau} - \tau_0) + z_{m,i} \tag{3.7}$$

where τ_0 is the actual channel delay.

Letting $m - l = k$ and assuming that $g(t)$ has a finite support, i.e., $g(t) = 0$ for $t < -K_1 T$ and $t > K_2 T$, we have

$$y_{m,i} = \sum_{k=-K_1}^{K_2} a_{m+k} g(kT + iT/2 + \hat{\tau} - \tau_0) + z_{m,i} \cong \sum_{k=-K_1}^{K_2} a_{m+k} g_{k,i}(\hat{\tau} - \tau_0) + z_{m,i} \tag{3.8}$$

where $g_{k,i}(\hat{\tau} - \tau_0) = g(kT + iT/2 + \hat{\tau} - \tau_0) \cong |g_{k,i}(\hat{\tau} - \tau_0)| e^{j\varphi_{k,i}(\hat{\tau} - \tau_0)}$, which is a complex coefficient of the CIR that is associated with a_{m+k} and sampled at $mT + iT/2 + \hat{\tau}$. The values of $g_{k,i}(\hat{\tau} - \tau_0)$ are unknown for initial acquisition and $g_{k,i}(\hat{\tau} - \tau_0) = 0$ for $k < -K_1$ and $k > K_2$. The CIR coefficients are time-varying in general but are assumed as constants during each hypothesis test described in this section. The transmitted reference symbols a_{m+k}'s are assumed to have a unit magnitude, e.g., to be BPSK or QPSK. Finally, $z_{m,i}$'s are the samples of the noise process. To simplify analysis, the noise samples are assumed to be i.i.d. and with the distribution $\mathcal{N}(0, \sigma_z^2)$.

Initial acquisition also determines approximately the timing of the signal samples. Here, we consider the problem to be testing the hypothesis whether a particular symbol a_n is transmitted as the last symbol of a known symbol sequence from a_{n-K+1} to a_n. If a successful test is declared and verified, the timing of the last sample in the sample sequence that generates the successful detection is an estimate of the timing at which a_n is received. When the signal is sampled every $T/2$, the timing obtained could have an error up to $T/4$. Additional enhancement of the timing accuracy may be needed and will be discussed later.

The objective of the detection is to determine if the sequence of T-spaced received signal samples is generated by the known reference symbol sequence $a_n, \ldots, a_{n-K+1}$. Thus, we perform a hypothesis test with two possible outcomes. Namely, either the sample sequence of the received signal is generated by the reference symbol sequence $a_n, \ldots, a_{n-K+1}$, i.e., hypothesis H_1, or it is not, i.e., hypothesis H_0. Hypothesis H_0 is true in two possible cases: Either there is no signal but only noise, or the transmitted signal exists but it is not generated by the expected reference symbol sequence.

As was discussed above, the detection process for initial acquisition involves a series of hypothesis tests at consecutive sampling time instances spaced by $T/2$ from each other. Each of such tests consists of three steps as listed below.

Step 1: *Detected Symbol Generation*
The reference symbol sequence used by the detector, i.e., $\{a_n, \ldots a_{n-K+1}\}$, is correlated with the sample sequence, $\{y_{m,i}, \ldots y_{m-K+1,i}\}$, $i = 0$ or 1, to generate a *detected symbol* $d_{m,i}$, such that

$$d_{m,i} = \frac{1}{\sqrt{K}} \sum_{k=0}^{K-1} a_{n-k}^* y_{m-k,i} = \frac{1}{\sqrt{K}} \sum_{k=0}^{K-1} \sum_{l=-L_1}^{L_2} a_{n-k}^* a_{m-k+l} g_{l,i}(\hat{\tau} - \tau_0) + \frac{1}{\sqrt{K}} \sum_{k=0}^{K-1} a_{n-k}^* z_{m-k,i}$$

$$= \sqrt{K} g_{n-m,i}(\hat{\tau} - \tau_0) + z'_{m,i} \tag{3.9}$$

To start with, we let $i = 0$ in (3.9). It is also assumed that the channel coefficient $g_{l,i}(\hat{\tau} - \tau_0)$ does not change in this segment of samples.

On the right side of (3.9), $g_{n-m,i}(\hat{\tau} - \tau_0)$ is the result of the summation of the terms with $n - k = m - k + l$, i.e., $l = n - m$, in $\frac{1}{\sqrt{K}} \sum_{k=0}^{K-1} \sum_{l=-L_1}^{L_2} a_{n-k}^* a_{m-k+l} g_{l,i}(\hat{\tau} - \tau_0)$. For $-K_1 \le n - m \le K_2$, $d_{m,i}$ is an estimate of $g_{n-m,i}(\hat{\tau} - \tau_0)$, which is equal to zero if $n - m$ is outside this range. The term $\tilde{z}_{m,i}$ includes both noise and interference, which can be expressed as

$$z'_{m,i} = \frac{1}{\sqrt{K}} \sum_{l=-L_1, l \ne n-m}^{L_2} g_{l,i}(\hat{\tau} - \tau_0) \sum_{k=0}^{K} a_{n-k}^* a_{m-k+l} + \frac{1}{\sqrt{K}} \sum_{k=0}^{K-1} a_{n-k}^* z_{m-k,i} \tag{3.10}$$

The last term on the right side of (3.10) is due to noise and is the sum of K independent terms, i.e., $a_{n-k}^* z_{m-k,i}$, $k = 0, \ldots, K - 1$, each of which has a variance of σ_z^2 with $|a_k| = 1$. The sum of $a_{n-k}^* a_{m-k+l}$, $l \ne n - m$ in $\sum_{k=0}^{K} a_{n-k}^* a_{m-k+l}$ on the right side of (3.10) is the *intersymbol interference* term. When the noise power is much higher than that of the interference, the variance of $z'_{m,i}$ is approximately equal to $K\sigma_z^2/\left(\sqrt{K}\right)^2 = \sigma_z^2$. This is also true if the sequence $\{a_m\}$ has an *impulse autocorrelation function*, so that $\sum_{k=0}^{K} a_{n-k}^* a_{m-k} = 0$, for $n \neq m$, and there is no ISI.

Moreover, if $n - m < -K_1$ or $n - m > K_2$, $g_{n-m,i}(\hat{\tau} - \tau_0) = 0$, we have

$$d_{m,i} = z'_{m,i} \tag{3.11}$$

In this case, $d_{m,i}$ is a zero-mean random variable. Its variance is approximately equal to σ_z^2 if ISI is ignored.

Step 2:   *Generating Decision Metric from the Detected Symbol*
A *noncoherent decision metric* $D_{m,i}$ is formed based on the detected symbol $d_{m,i}$. If the phase of the channel is unknown during initial acquisition as is usually the case, the detection can be based on the magnitude of $d_{m,i}$. Thus, the decision metric is formed by

$$D_{m,i} = |d_{m,i}|^2 = \left|\sqrt{K} g_{n-m,i}(\hat{\tau} - \tau_0) + z'_{m,i}\right|^2 \tag{3.12}$$

As discussed in Sections 2.5.4 and 3.2.1 such a noncoherent decision metric has been shown to be optimal under the assumption that carrier phase ϕ_m is unknown but uniformly distributed between 0 to 2π in the sense of the unconditional ML estimation. In the case that ϕ_m is known, a coherent decision metric can be generated according to (3.2) for hypothesis testing to achieve better performance.

Step 3:   *Comparing the Decision Metric with a Predetermined Threshold*
Finally, the decision metric is compared with a predefined threshold Th to make a decision if the sample sequence involved in the correlations is generated from the expected reference symbol sequence (Hypothesis H_1, or $\theta = \theta_1$) or not (Hypothesis H_0, or $\theta = \theta_0$), i.e.,

If $D_{m,i} < Th$: $H_0 \Rightarrow$ Not generated by the reference symbol sequence, detection failed;
If $D_{m,i} \geq Th$: $H_1 \Rightarrow$ Symbol sequence $a_n...a_{n-K+1}$ detected!

$$(3.13)$$

If the detection failed, the detection process repeats for the next sampling instance. A new sample sequence is used in the correlation process in step 1. As the samples are generated every $T/2$, this sequence has a $T/2$ timing offset from the sequence in the test at the previous instance, namely, the new sequence is $\{y_{m,i},\ldots y_{m-K+1,i}\}$, $i = 1$. If the detection failed again, the sample sequence should shift by an additional $T/2$. With the new sample sequence $\{y_{m+1,0},\ldots y_{m-K+2,0}\}$, the detection repeats again. Note that this sample sequence used in detection and the sample sequence used two time instances earlier have $K - 1$ elements in common.

The detection continues until it becomes successful, i.e., H_1 is determined valid, or until a time-out is reached. In the latter case, the receiver concludes that there is no expected signal at the given geographical location at the given time. The detection process will start again later and maybe at a different location.

Once a successful detection is declared, the receiver concludes that the expected reference symbols exist in the signal samples involved in the correlation. We then determine the timing of the signal according to the sample sequence $\{y_{m,i}, \cdots y_{m-K+1,i}\}$ and the sampling phase i. To further improve the accuracy of the detected signal timing, we should let the receiver go through the detection process a few more steps to check if there are other timing phases with even larger metric D. If there are, the timing associated with the largest D should be used.

If there are multiple transmitters in the area around the device, the received signal may contain reference symbol sequences from one or more of them. All of the possible symbol sequences from these transmitters should be tested in each of the detection steps. Once a success of detection is declared, the reference sequence contained in the received signal is determined. Consequently, the ID of the transmitter from which the signal is transmitted also becomes known to the device.

The detection based on the magnitude of the detected symbol $d_{m,i}$ is called *noncoherent* detection. For a time-varying channel, the number K, called the *coherent integration length*, is limited by the channel coherence time. While using a larger K can yield a higher SNR of the detected symbol if coherent integration is possible, the performance would degrade if K is too large, such that the integration becomes noncoherent. To improve detection performance, multiple decision metrics can be generated by coherent integration and then combined to form a composite decision metric to compare with a threshold. This situation will be analyzed in Section 3.3.

3.2.3   False Alarm, Miss, and Detection Probabilities

The performance of initial acquisition can be characterized by the false alarm and detection/miss probabilities [5]. To derive these probabilities, it is necessary to know the probability distribution functions of the decision metrics generated in the detection process discussed above.

As shown in (3.9), the detected symbol $d_{m,i}$ is modeled as a constant signal in AWGN. In a static channel when the signal exists, it has a normal distribution with a mean of $\sqrt{K}g_{n-m,i}(\hat{\tau} - \tau)$ and a variance of σ_z^2. Assuming that the intersymbol interference term is insignificant compared to the noise z, we can simplify (3.9) to

$$d_{m,i} = \frac{1}{\sqrt{K}}\sum_{k=0}^{K-1} a_{n-k}^* y_{m-k,i} \approx \frac{1}{\sqrt{K}}\sum_{k=0}^{K-1} g_{n-m,i}(\hat{\tau} - \tau_0) + \frac{1}{\sqrt{K}}\sum_{k=0}^{K-1} a_{n-k}^* z_{m-k,i}$$

$$= \sqrt{K}g_{n-m,i}(\hat{\tau} - \tau_0) + z_{m,i}' \tag{3.14}$$

The variance of $z_{m,i}'$, $\sigma_{z'}^2$, is equal to σ_z^2.

To begin, we consider the case of the ideal single-path channel with SRCOS transmitter and receiver filters as assumed previously. With no loss of generality, we let the main coefficient of the CIR, which is approximately a sinc function, be $g_{0,0}(0)$, i.e., we assume that $\hat{\tau} = \tau_0$. The coefficient $g_{0,0}(0)$ is also much larger than other channel coefficients, or *taps*, $g_{k,i}(0)$, for $i = 0$ and $k \neq 0$. We will discuss more complex yet practical cases later.

The SNR of $y_{m,0}$ can be shown to be $\gamma_y = |g_{0,0}(0)|^2/\sigma_z^2$, for $|a_k| = 1$. The detected symbol $d_{n,0}$ that could lead to the correct decision can be expressed as

$$d_{n,0} = \frac{1}{\sqrt{K}} \sum_{k=0}^{K-1} a_{n-k}^* a_{n-k} g_{0,0}(0) + \frac{1}{\sqrt{K}} \sum_{k=0}^{K-1} a_{n-k}^* z_{n-k,0}$$

$$= \frac{1}{\sqrt{K}} \sum_{k=0}^{K-1} g_{0,0}(0) + \frac{z_{n,0}'}{\sqrt{K}} \cong \sqrt{K} g_{0,0}(0) + z_{n,0} \qquad (3.15)$$

and the SNR of the detected symbol is equal to

$$\gamma_{d_{n,0}} = E\left[|d_{n,0}|^2\right]/\sigma_z^2 = \left|\sqrt{K} g_{0,0}(0)\right|^2/\sigma_z^2 = K|g_{0,0}(0)|^2/\sigma_z^2 = K\gamma_y \qquad (3.16)$$

Namely, the SNR of $d_{n,0}$ is K times the SNR of $y_{m,0}$. The pdf of $d_{n,0}$ is $p_1(d_{n,0}) = \frac{1}{\pi\sigma_z^2} e^{-|d_{n,0}-\sqrt{K}g_{0,0}(0)|^2/\sigma_z^2}$. In the case that the samples do not contain the reference symbols, the detected symbol $d_{m,i}$ has a zero-mean normal distribution with the same variance. Its pdf is $p_0(d_{m,i}) = \frac{1}{\pi\sigma_z^2} e^{-|d_{m,i}|^2/\sigma_z^2}$.

When the phase of $g_{0,0}$ is unknown, $D \cong D_{m,i} = |d_{m,i}|^2$ is used as the decision metric. If the samples do not contain the reference symbols, D has the central chi-square distribution with two degrees of freedom with the pdf of

$$p_0(D) = \frac{1}{\sigma_z^2} e^{-D/\sigma_z^2}, \quad D \geq 0 \qquad (3.17)$$

In the case where the reference symbols exist in the samples used in the detection and align with the reference symbols involved in the correlation, the distribution of D is the noncentral chi-square distribution with two degrees of freedom [1, 6]. Its pdf is

$$p_1(D) = \frac{1}{\sigma_z^2} e^{-(D+\lambda^2)/\sigma_z^2} I_0\left(\frac{2\lambda\sqrt{D}}{\sigma_z^2}\right), \quad D \geq 0 \qquad (3.18)$$

where λ^2 is equal to $|E[d_{n,0}]|^2$ and $\lambda = \sqrt{K}|g_{0,0}(0)|$.

By definition, the false alarm event occurs when the hypothesis H_1 is determined to be true, but the reference symbol sequence was not transmitted. Based on the decision rule given by (3.13), the false alarm event is equivalent to $D > Th$ while D has a pdf given by (3.17). Thus, the probability of false alarm events P_F, is equal to

$$P_F = \int_{Th}^{\infty} p_0(u)du = \int_{Th}^{\infty} \frac{1}{\sigma_z^2} e^{-\frac{D}{\sigma_z^2}} du = e^{-\frac{Th}{\sigma_z^2}} \qquad (3.19)$$

where $p_0(\cdot)$ is given by (3.17).

In contrast, a miss event occurs when the hypothesis H_0 is determined to be true, but the reference symbol sequence is actually transmitted in the sample sequence that performs correlation. According to (3.13), it is equivalent to that $D \leq Th$ while D has a pdf given by (3.18). Thus, the probability of miss events P_{miss}, is equal to

$$P_{miss} = \int_{-\infty}^{Th} p_1(u)du = \int_{-\infty}^{Th} \frac{1}{\sigma_z^2} e^{-(u+\lambda^2)/\sigma_z^2} I_0\left(\frac{2\lambda\sqrt{u}}{\sigma_z^2}\right) du \qquad (3.20)$$

where $p_1(\cdot)$ is given by (3.18).

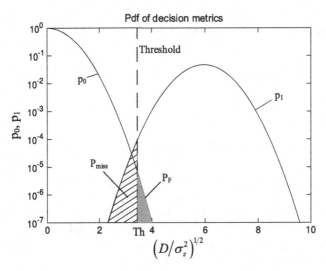

Figure 3.2 Distributions of the decision metric for hypothesis testing with graphic illustrations of P_F and P_{miss}

The detection probability defined as the success rate of detection in all events when H_1 is true is denoted by P_D and $P_D = 1 - P_{miss}$.

Figure 3.2 shows the pdfs $p_0(D)$ and $p_1(D)$ given by (3.17) and (3.18), respectively, with $\lambda/\sigma_z = 6$. It also provides the graphic illustrations of the false alarm and miss probabilities using the decision rule by comparing the decision metric D with the threshold Th.

The false alarm probability only depends on the noise variance and is independent of the energy and statistical properties, e.g., fading characteristics, of the received signal. Therefore, it is convenient to determine the detector's threshold value Th based on the distribution of the noise according to (3.19) for a given false alarm probability α. For a given noise distribution, the threshold is a function of the noise variance. Once Th is determined, the miss and detection probabilities are functions of the SNR of the detected symbol $d_{m,i}$. As shown in Section 2.5, such an approach is optimal according to the Neyman–Pearson lemma [7] in the following sense. *Based on this decision rule according to the given threshold yielding a predefined false alarm probability P_F, the detection probability, P_D, is the highest among all possible decision rules, for which the false alarm probability is no greater than P_F* [1, 5, 7].

Figure 3.3 shows the detection probabilities when the thresholds are set according to the predefined false alarm probabilities for hypothesis testing based on the decision rule given by (3.13). The three curves shown are the detection probabilities with respect to the SNRs of the detected symbol when the false alarm probabilities are chosen to be 0.01, 0.001, and 0.0001, respectively. It shows that to achieve the same detection probability, the required SNR of the detected symbol $d_{m,i}$ needs to be higher to ensure lower false alarm probability.

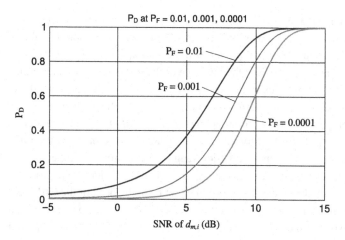

Figure 3.3 P_D's for prescribed P_F's

Note that the SNR of the received signal sample is K times lower than the SNR of the detected symbol, where K is the number of the samples involved in the correlation, called the length of coherent integration. For example, to achieve a detection probability of 0.98 at a false alarm rate of 0.001, the SNR of the detected symbol should be about 12 dB as shown in the figure. However, if the decision symbol is generated by integrating 16 received signal samples, i.e., $K = 16$, the required SNR of the samples is equal to 0 dB, which is usually practically achievable.

In the analysis given above, we assumed that the channel is a single-path channel, the signal is sampled at the peak of the CIR, and/or the ISI is much weaker than the additive noise. In practice, none of these three assumptions may be completely true. In addition, readers may also notice that we have only considered the sampling phase for $i = 0$ and have not dealt with the case for $i = 1$. These conditions are addressed below.

First of all, the channels are most likely multipath channels in practice. Moreover, even if a channel is a single-path channel, its observed CIR coefficients, also called *taps*, depend on sampling time relative to the continuous-time CIR. For example, even if the overall channel has an ideal RCOS frequency response, it behaves as a single-path channel only if the sampling time coincides with its peak, i.e., $n = m$, $i = 0$ and $\hat{\tau} - \tau_0 = 0$, as defined in the discussion above. If it does not, e.g., if the sampling time is $0.5T$ away from the peak, there would be two large equal-magnitude CIR coefficients. This is exactly the case for $i = 1$ in the above discussion. As a result, H_1 may become true when the detected symbol is at one of the larger CIR coefficients. In cases like this, the signal power in the detected symbol is equal to the power of the signal carried by the corresponding CIR coefficient and is usually lower than the total received signal power.

Second, the power of the ISI becomes negligible only if the power of the additive noise is very high and/or the transmitter symbol sequence has a perfect impulse autocorrelation. These two conditions are only true in some special cases. In most practical cases, they are not satisfied, i.e., ISI cannot be ignored. Taking these two factors into account, the signal power of the detected symbol is lower than K times the total received signal power and the variance of the total noise and interference is greater than the noise variance used in the P_F calculation. Hence, the detection probability would be smaller than that in the ideal case. However, the analysis given above is still valid, as long as the right signal power and noise/interference power are used in calculating the pdf $p_1(D)$. Moreover, the signal to noise/interference ratio is always proportional to the coherent integration length K, the same as in the noise-only case.

Another factor we need to consider is that when H_1 is determined to be true and the detection is valid, the detected symbol may not be at the peak of the CIR. As a result, the determined timing is not as good as it could be. In this case, it is advisable to perform a few more tests on time offsets near the sampling time that yields successful detection to find an even larger peak.

To improve the detection probability over multipath channels, it may be desirable to generate more than one of the detected symbols that have relatively large magnitudes and are close to each other. The sum of the squared magnitudes of these decision symbols is an estimate of the total channel energy and can be used as the decision metric for the detection to be more reliable.

As shown by (3.14), the detected symbols in successful hypothesis-tests are the estimates of the channel taps, which are useful for other receiver functions.

3.3   Detection Performance with Pre- and Post-Detection Integration

In the last section, we discussed the basic procedure of reference symbol detection for initial acquisition. Since initial acquisition is the first stage in establishing a communication link, very little information about the desirable signal and channel condition is available to the receiver. Thus, the detection could be difficult, especially under adverse channel conditions. In order for a communication session to be reliable, it is desirable to make the detection performance as robust as possible. In this section, we are looking into some of the techniques to improve the robustness of the receiver at the stage of initial acquisition.

The key step of the basic detection procedure is to generate a detected symbol by correlating the reference symbol sequence with the T-spaced sample sequence as given by (3.9). Note that each item in the double summation, i.e., $a_{n-k}^* a_{m-k+l} g_{l,i}(\hat{\tau} - \tau_0)$, for $n-k = m-k+l$ is an estimate of the channel coefficient $g_{n-m,i}(\hat{\tau} - \tau_0)$ with the same channel phase for transmission over the channel. These terms are combined coherently in the summation. We call this step *coherent integration* or *coherent combining*. It is also called *predetection integration* or combining. When the coherent integration is not appropriate due to channel phase changes, *noncoherent integration/combining*, or *post-detection integration/combining*, can be used to improve the performance of the detector. Analyses of the characteristics, optimality, and limitations of coherent integration and noncoherent integration, as well as their performance are presented in this section.

3.3.1 Predetection Coherent Integration: Optimality and Limitations

Using the single-path channel model at the optimal sampling time, we know from (3.15) that $d_{n,0}$ is the summation of K items each of which is equal to $g_{0,0}(0)$ divided by the square-root of K plus a noise term. When the noise is i.i.d. with normal distribution, such an estimate is the ML estimate of the channel coefficient $g_{0,0}(0)$ scaled by the square root of K, as can be shown by the example in Section 2.1.3. Moreover, since this estimate satisfies all of the conditions of Example 2 in Section 2.4.1, it achieves the Cramer–Rao bound. Thus, this estimate is optimal in the sense that it attains the minimum error variance among all unbiased estimators.

The SNR of the detected symbol is proportional to K. Therefore, the detector could achieve better performance by increasing the integration length K. However, this is only true if the channel coherence time is much longer than KT. If the channels are time-varying and/or if there is a nonnegligible offset between the transmitter carrier frequency and the receiver down-conversion frequency, the integration length is limited by the coherence time even if more reference symbols and signal samples are available for performing detection.

During initial acquisition, the receiver has not yet acquired the transmitter carrier frequency. Hence, the existing frequency offset may limit the length of coherent integration. The impact of the frequency offset on the allowable length of coherent integration can be analyzed as follows.

Equation (3.15) can be rewritten as

$$d_{n,0} = \frac{1}{\sqrt{K}} \sum_{k=0}^{K-1} |g_{0,0}(0)| e^{j\phi_{0,0}(n-k)} + z_{n,0} \tag{3.21}$$

When a frequency offset exists, the phase of the coefficient $g_{0,0}(n-k)$ is a linear function of $n-k$, i.e., $\phi_{0,0}(n-k) = \phi_0 + \Delta\theta \times (n-k) = \phi_0 + \Delta\theta \times n - \Delta\theta \times k$, where $\Delta\theta$ is the phase change per sample time interval due to frequency offset and ϕ_0 is a constant phase term. Denoting the frequency offset by f_{offset}, we have $\Delta\theta = 2\pi f_{offset} T$. Thus, (3.21) can be expressed as

$$d_{n,0} = \frac{1}{\sqrt{K}} \sum_{k=0}^{K-1} |g_{0,0}(0)| e^{-j(\phi_0 + 2\pi f_{offset} T(n-k))} + z_{n,0}$$

$$= e^{j(\phi_0 + 2\pi f_{offset} nT)} |g_{0,0}(0)| \frac{1}{\sqrt{K}} \sum_{k=0}^{K-1} e^{-j2\pi f_{offset} T \times k} + z_{n,0} \tag{3.22}$$

The summation in the first term on the right side of (3.22) can be viewed as the output of a complex sinusoidal signal with frequency f_{offset} passing through a rectangular FIR filter with K coefficients. The magnitude of such a filter's frequency response is

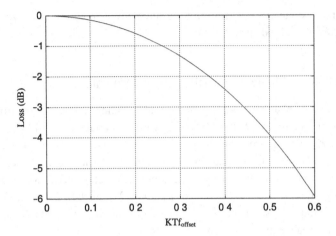

Figure 3.4 Coherent integration loss versus integration time and frequency offset

$$|H(f)| = \frac{\sin(\pi KTf)}{K \sin(\pi Tf)} \approx \mathrm{sinc}(KTf) \tag{3.23}$$

At f_{offset}, we have $E[d_{n,0}] \approx e^{j\phi_0 + 2\pi j \times f_{offset}T\left(n - \frac{K-1}{2}\right)} \mathrm{sinc}\left(KTf_{offset}\right)\sqrt{K}g_{0,0}(0)$. There is an energy reduction of approximately $20\log_{10}\mathrm{sinc}\left(KTf_{offset}\right)$ dB relative to the ideal case where there is no frequency offset [1]. This energy loss also results in an interference term with the power of $K \times |g_{0,0}|^2 \times \left(1 - \mathrm{sinc}^2\left(KTf_{offset}\right)\right)$.

The result given by (3.23) is plotted in Figure 3.4. It shows the loss of coherent integration in terms of the product of frequency offset and the integration length with the unit of the transmitter symbol interval T. If we want the loss to be less than 1 dB, 3 dB, and 6 dB, this product should be less than about 0.262, 0.36 and 0.6, respectively. These results are useful for receiver parameter selection as will be shown in Section 3.5 and 3.6.

Considering all of these factors, we can derive the equation that shows the relationship between the detector gain G, i.e., the ratio of detected symbol SNR (γ_d) to the input SNR (γ_y), and K at different Tf_{offset} with the input SNR (γ_y) as a parameter. Specifically, we have

$$G = \frac{\gamma_d}{\gamma_y} = \frac{1}{|g_{0,0}|^2/\sigma_z^2} \times \frac{\mathrm{sinc}^2\left(KTf_{offset}\right)K|g_{0,0}|^2}{\left(\sigma_z^2 + \left[1 - \mathrm{sinc}^2\left(KTf_{offset}\right)\right] \times K|g_{0,0}|^2\right)}$$

$$= \frac{K\mathrm{sinc}^2\left(KTf_{offset}\right)}{\left(1 + \left[1 - \mathrm{sinc}^2\left(KTf_{offset}\right)\right]K\gamma_y\right)} \tag{3.24}$$

In (3.24), we observe that when KT_{offset} is zero, i.e., there is no frequency offset, $\mathrm{sinc}^2\left(KTf_{offset}\right)$ is equal to 1 and $G = K$. If an offset exists, $\mathrm{sinc}^2\left(KTf_{offset}\right)$ is less than one and there is a loss. Namely, G is less than K due to scaling on the estimate and the additional noise term $\left[1 - \mathrm{sinc}^2\left(KTf_{offset}\right)\right]K\gamma_y$. Since the additional noise term is proportional to $K\gamma_y$, its relative impact is greater when the SNR of the detected symbol is higher.

Without frequency offset, the detector's output SNR is proportional to the integration length K. Thus, if K is increased by a factor of 2, the SNR is 3 dB higher. With nonzero frequency offset, the detector's output SNR will have a loss, which is at least equal to $\text{sinc}^2(KTf_{offset})$, relative to the case when $f_{offset} = 0$ as shown in Figure 3.4. An alternative is to use noncoherent, or post detection, integration to improve the detector performance, as discussed in the next section.

3.3.2 Post-Detection Noncoherent Integration

For initial acquisition of a signal transmitted over a time-variant channel and/or when there is a frequency offset, the length of coherent integration is limited by the coherence time of the signal. To improve the detection performance, multiple detected symbols are generated by correlating different segments of the received signal samples with corresponding multiple reference symbol sequences. The decision metrics generated from these detected symbols are combined to form a *composite decision metric*. It is used to perform hypothesis testing for initial acquisition instead of a decision metric based on a single detected symbol from coherent integration. The detection process is the same as that described in the previous section.

Let us consider the symbol sequence $\{a_k\}$, $k = n,\, n-1,\, \ldots,\, n-KL+1$ and the sample sequence $\{y_{j,i}\}$, $j = m,\, m-1,\, \ldots,\, m-KL+1$ and $i = 0$ or 1. They are divided into L pairs of subsequences, each of which has K symbols and K samples. A symbol subsequence convolves with a sample subsequence to form a detected symbol as follows.

We first define L symbol subsequences $\{a_n, a_{n-1}, \ldots, a_{n-K+1}\}$, $\{a_{n-K}, a_{n-K-1}, \ldots,$ $a_{n-2K+1}\}$, $\ldots$, $\{a_{n-(L-1)K}, a_{n-(L-1)K-1}, \ldots, a_{n-LK+1}\}$ and L sample subsequences $\{y_{m,i}, y_{m-1,i}, \ldots, y_{m-K+1,i}\}$, $\{y_{m-K,i}, y_{m-K-1,i}, \ldots, y_{m-2K+1,i}\}$, $\ldots$, $\{y_{m-(L-1)Ki},$ $y_{m-(L-1)K-1,i}, \ldots, y_{m-LK+1,i}\}$. The correlations of the L pairs of symbol and sample subsequences generate L detected symbols, $d_{m,i}, d_{m-K,i}, \ldots, d_{m-(L-1)K,i}$. The generation process of each of these detected symbols is the same as that in step 1 of the basic detection procedure described in Section 3.2.2 according to (3.9).

Second, the L detected symbols generate L noncoherent decision metrics that are added together to form a composite noncoherent decision metric, such that

$$D_{m,i}^{(L)} = \sum_{j=0}^{L-1} D_{m-jK,i} = \sum_{j=0}^{L-1} \left| d_{m-jK,i} \right|^2 \tag{3.25}$$

Finally, a hypothesis test is performed by comparing $D_{m,i}^{(L)}$ to a predetermined threshold similar to the test performed on $D_{m,i}$ according to (3.13) shown in Section 3.2.2.

Detection performance is improved by performing the hypothesis testing on the composite decision metric from *post-detection integration* given by (3.25), not on individual decision metrics. Because the integration is done on the squared magnitudes of the detected symbols and no phase information is involved, the post-detection integration is also called *noncoherent integration*. In the above description, the subsequences of symbols and samples can be adjacent to each other, have gaps between each other, or even overlap with each other. They can be used to form the detected symbols, as long as the data symbol subsequences and sample subsequences involved in the same correlation correspond to each other.

If the carrier phases of the detected symbols are known but different from each other, the detected symbols can be coherently combined after their phases are corrected. The resulting composite coherent decision metric can be used in the detection to achieve better performance than that obtained by using the composite noncoherent detection metric.

3.3.3   Performance Analysis of Detector with Post-Detection Integration

The composite decision metric $D_{m,i}^{(L)}$ is the sum of L squared magnitudes of detected symbols. In (3.25), each of $d_{m-jK,i}$ has a similar form as that given by (3.9). Specifically,

$$d_{m-jK,i} = \frac{1}{\sqrt{K}}\sum_{k=0}^{K-1} a_{n-jK-k}^* y_{m-jK-k,i} = \sqrt{K}g_{n-m,i}^{(j)}(\hat{\tau}-\tau_0) + z_{m_j,i}'\tag{3.26}$$

To simplify the discussion, we will again assume that the channel has a single main path, $g_{0,0}^{(j)}(0)$, which is a function of j, that the intersymbol interferences are much weaker than the additive noise, and that the coherence time of the channel is longer than the coherent integration duration KT. The noise terms are i.i.d. and with normal distribution $\mathcal{N}(0, \sigma_z^2)$.

If $m = n$, $i = 0$ and $\hat{\tau} - \tau_0 = 0$, then H_1 is the correct hypothesis. The detected symbol $d_{n-jK,0}$ has a normal distribution with a mean of $\sqrt{K}g_{0,0}^{(j)}(0)$ and a variance of σ_z^2. However, if H_0 is the correct hypothesis, $d_{m-jK,i}$ has a normal distribution with zero mean and a variance σ_z^2. Because the channel is time-varying, the channel coefficients $g_{0,0}^{(j)}(0)$ may be different for different j's.

It is known from the probability theory [1, 6] that, if H_0 is the correct hypothesis, the composite decision metric $D_{m,i}^{(L)}$ has a central chi-square distribution with $2L$ degrees of freedom with the pdf of

$$p_{0,L}(D) = \frac{1}{(L-1)!(\sigma_z^2)^L}D^{L-1}e^{-D/\sigma_z^2}, \quad D = D_{m,i}^{(L)} \geq 0, \ m \neq n\tag{3.27}$$

However, if H_1 is the correct hypothesis, i.e., $m = n$ and $i = 0$, $D_{n,0}^{(L)}$ has a noncentral chi-square distribution with $2L$ degrees of freedom [1, 6] with pdf

$$p_{1,L}(D) = \frac{1}{\sigma_z^2}\left(\frac{D}{s^2}\right)^{(L-1)/2} e^{-(D+s^2)/\sigma_z^2} I_{L-1}\left(\frac{2s\sqrt{D}}{\sigma_z^2}\right), \quad D = D_{n,0}^{(L)} \geq 0\tag{3.28}$$

where $s = \sqrt{\sum_{j=0}^{L-1}|\lambda_j|^2}$, $\lambda_j = E[d_{n-jK,0}] = \sqrt{K}g_{0,0}^{(j)}(0)$, and

$$I_{L-1}(x) = \sum_{m=0}^{\infty}\frac{1}{m!(m+L-1)!}\left(\frac{x}{2}\right)^{2m+L-1}\tag{3.29}$$

is the modified Bessel function of the first kind of an integer order $L-1$ [2]. Since $|\lambda_j|^2$ is the signal energy of the jth detected symbol $d_{n-jK,0}$, s^2 is the total signal energy of the

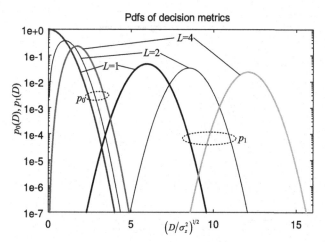

Figure 3.5 pdfs p_0 and p_1 of normalized decision metric $\left(D_{n,0}^{(L)}/\sigma_z^2\right)^{1/2}$ for L = 1, 2, and 4

decision metric $D_{n,0}^{(L)}$, in which each of the detected symbols may have different signal energy. The expressions of (3.17) and (3.18) are the degenerated forms of (3.27) and (3.28) with $L = 1$, respectively.

Figure 3.5 shows the pdfs of the normalized decision metrics given by (3.27) and (3.28), $L = 1$, 2, and 4 plotted against the decision metric $D \cong D_{n,0}^{(L)}$ normalized by the noise variance σ_z^2.

Recall that a false alarm event occurs when the decision metric is greater than the preset threshold, but the received signal samples do not contain the reference symbol sequence. The false alarm probability P_F is given by

$$P_F = \int_{Th}^{\infty} p_{0,L}(u)\,du \tag{3.30}$$

where $p_{0,L}(\cdot)$ has the form of (3.27). The *detection probability, P_D,* is given by

$$P_D = \int_{Th}^{\infty} p_{1,L}(u)\,du \tag{3.31}$$

where $p_{1,L}(\cdot)$ has the form of (3.28).

The complement of P_D is the *miss probability P_{miss},* which can be expressed as

$$P_{miss} = 1 - P_D = \int_{-\infty}^{Th} p_{1,L}(u)\,du \tag{3.32}$$

The threshold is most conveniently calculated based on the required P_F as it depends only on the noise/interference characteristics and is independent of the strength and other characteristics of the signal. Once the threshold is determined, the probabilities P_F, P_D, and P_{miss} are functions of the SNR of the signal received over the channel. The type of channel, e.g., static or time-varying, also has a major impact on these probabilities. Below we will examine the results for static channels. The results for Rayleigh fading channels can be found in [1].

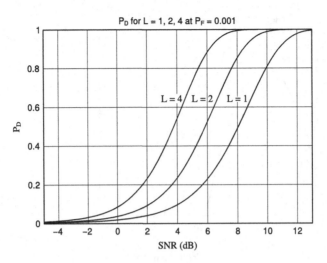

Figure 3.6 Detection probabilities for $P_F = 0.001$ and L = 1, 2, 4

From Figure 3.5, it can be seen that when L increases, the curves of p_0 move toward the left. This is because more terms with the same noise variance are added up to make the total noise variance larger. Hence, to achieve the same false probability P_F, the threshold Th should increase. However, the space between p_0 and p_1 increases even more for a larger L. As a result, for the same P_F, P_{miss} will be smaller, or equivalently, P_D becomes larger. Thus, the acquisition performance improves with a larger L at the same SNR.

Figure 3.6 shows the detection probabilities P_D versus SNR per detected symbol when the false alarm probability P_F is equal to 0.001.

The figure shows that when the number of the signal samples that generate a decision metric increases by a factor of two, the required SNR reduces by about 2 dB to yield the same P_D. Recall that for coherent integration, doubling the number of signal samples to form a decision metric yields a 3 dB improvement. Thus, in this case, there is a loss of about 1 dB when noncoherent integration, rather than coherent integration, is used.

Figure 3.7 shows the miss probabilities P_{miss} versus SNR per detected symbol when the false alarm probability P_F is equal to 0.0001. It can be observed that at higher SNRs the SNR gain due to doubling the number of samples for detection is a little higher than the SNR gain in the previous case. It can be up to 2.5 dB. In other words, in this case, the loss of noncoherent integration relative to coherent integration reduces to about 0.5 dB.

Based on the above results, we conclude that there is a gain of 1.8 to 2.5 dB, or roughly 2 dB, when the number of detected symbols that form a decision metric is doubled. It is less than the corresponding gain of 3 dB for coherent integration. In other words, there is a *noncoherent integration loss*. Even though this observation is heuristic rather than rigorous and the results vary under different conditions, it can be useful as a rule of thumb to get a quick and rough estimate for noncoherent integration methods with different parameters.

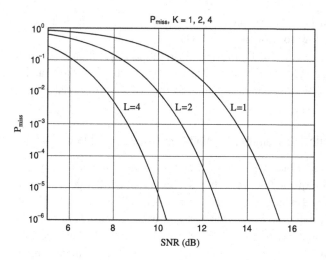

Figure 3.7 Miss probabilities for $P_F = 0.0001$ and L = 1, 2, 4

Finally, by using the power series expansion form of $I_{L-1}\left(2s\sqrt{D}/\sigma_z^2\right)$ in $p_{1,L}(D)$ given by (3.29) [2], we can see that the likelihood ratio of $p_{1,L}(D)/p_{0,L}(D)$ is also a monotonically increasing function. Thus, similar to the argument made in Section 2.4, the decision rule of comparing the composite decision metric $D_{m,i}^{(L)}$ to a given threshold Th is equivalent to the decision rule by comparing the likelihood ratio to a corresponding threshold $\eta(Th)$ to yield the same false alarm probability P_F. Therefore, the decision rule based on thresholding of $D_{m,i}^{(L)}$ is optimal according to the Neyman–Pearson lemma. Namely, it yields the highest detection probability P_D among all decision rules that result in the same P_F based on $D_{m,i}^{(L)}$.

3.4 Theoretical and Practical Aspects of Design and Implementation of Initial Acquisition

In the previous two sections, we described the basic detection procedure for initial acquisition and analyzed the acquisition performance when using predetection integration and/or post-detection integration. These basic procedures are widely used in initial acquisition of almost all modern digital communication systems. Thus, the understanding of their properties and related basic theories are of fundamental importance. However, other aspects of initial acquisition also need to be considered due to their theoretical and practical importance in the practical system design and implementation. In this section, we discussed a few of these aspects, including:

- Desired properties of the reference symbol sequences for initial acquisition and practical examples;
- Noise variance estimation and its impact on false alarm and detection probabilities;
- The impact on acquisition performance by false alarm, miss, and detection probabilities;

- The impacts of carrier frequency offset during initial acquisition and its remedy for using low-cost/low-accuracy XOs;
- The use of the estimation results from initial acquisition to facilitate frequency and timing synchronization discussed in the ensuing chapters; and finally
- An overview of system acquisition as the last stage of initial acquisition.

3.4.1   Ideal and Practical Transmitter Symbol Sequences Suitable for Initial Acquisition

In the previous sections, we described and analyzed the detection of the signal with known symbol sequence in noise. It has been assumed that the noise is AWGN, and there is no signal when H_0 is the correct hypothesis. In digital communication systems, this is only one of the two possible cases where H_0 is true. The other case of H_0 is that there exists a transmitted signal, but the transmitted symbols in the received signal samples are not the expected reference symbols. In this case, the reference symbol sequence is preferably orthogonal or uncorrelated to all other possible transmitted symbol sequences. In particular, we would like the reference symbol sequence for acquisition to be orthogonal to the shifted versions of itself. Specifically, for a periodic symbol sequence $\{a_k\}$ with a period of N and $|a_k| = A$, the length K correlation satisfies

$$\left| \sum_{k=0}^{K-1} a^*_{(m+k) \bmod N} a_{(m+l+k) \bmod N} \right| = \begin{cases} KA^2, & \text{for } l = 0 \\ 0, & \text{for } l \neq 0 \end{cases} \tag{3.33}$$

for any m.

In other words, the autocorrelation function of this sequence is an impulse. Sequences of this kind are particularly desirable for initial acquisition and channel estimation because they can eliminate ISI during detection. Practically, however, such an orthogonality requirement is difficult to satisfy. This requirement can be relaxed such that the sequence is uncorrelated, rather than orthogonal, to its own shifts, i.e., for any m

$$\sum_{l=1}^{N} \sum_{k=0}^{K-1} a^*_{(m+k) \bmod N} a_{(m+l+k) \bmod N} = 0. \tag{3.34}$$

Such a sequence reduces the output variance for $l \neq 0$ but does not eliminate it.

Below, we consider two of such sequences that are commonly used in practice for initial acquisition in digital communication systems.

3.4.1.1   Maximum-Length Shift Register Sequence

Maximum-length shift register sequences, or *m-sequences*, belong to the category of *pseudo-noise sequences* [8, 9]. As the name indicates, a PN sequence is a noise-like sequence that is deterministically generated. An m-sequence is generated by using an N-bit shift register. It is a periodic binary sequence with a period of 2^N-1 bits. M-sequences are widely used in direct-sequence spread-spectrum communication systems because they are convenient to generate and have good periodic and aperiodic correlation properties [9].

(a)

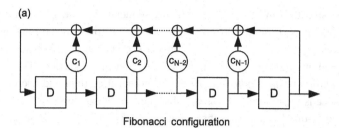

Fibonacci configuration

(b)

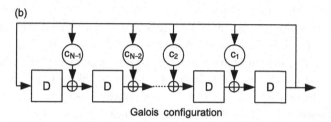

Galois configuration

Figure 3.8 LFSR m-sequence generators (Fibonacci/Galois configurations)

An m-sequence can be generated according to the following recursive relation:

$$a_n = c_1 a_{n-1} + c_2 a_{n-2} + \cdots + c_N a_{n-N} \tag{3.35}$$

where c_i's are the generator coefficients, c_i's and a_{n-i}'s are all binary, and "+" represents modulo-2 addition, i.e., exclusion-OR operation. The function $f(x) = 1 + c_1 x + c_2 x^2 + \cdots + c_{N-1} x^{N-1} + x^N$ is called the *characteristic polynomial* for the shift register sequence generator given by (3.35). This sequence can be generated by the linear feedback shift register (LFSR) structure [9] with the so-called *Fibonacci configuration* as shown in Figure 3.8(a). The values of the registers in the figure are called the *state of the m-sequence*. Alternatively, the m-sequence can also be generated by the shift register structure with the so-called *Galois configuration*, shown in Figure 3.8(b). With proper characteristic polynomials, the sequence generated has a period of 2^N-1 bits. It is the largest period of a sequence that can be generated by an N-bit shift register. This is why they are called *maximum length* shift register sequences.

M-sequences have many nice properties. Below we look at a few of them that are relevant to initial acquisition. Interested readers are referred to other references such as [1, 9] for more information.

When used as modulation symbols, a zero bit in an m-sequence is mapped to the real value one, and a one bit is mapped to the real value minus one. With such a mapping, there are $2^{N-1}-1$ plus ones and 2^{N-1} minus ones in the real sequence. Thus, the mean of the entire sequence is equal to -1. It is quite close to zero mean given that the entire m-sequence can be quite long when N is not too small. More importantly, the autocorrelation function of an m-sequence is

$$\sum_{k=0}^{2^N-2} a_k a_{(k+m)_{\mathrm{mod}\ (2^N-1)}} = \begin{cases} 2^N - 1 & m = 0 \\ -1 & m \neq 0 \end{cases} \tag{3.36}$$

which is almost an ideal impulse function.

However, even with a modest value of N, the sequence can be quite long. During initial acquisition, it is difficult and maybe impossible to perform correlation with a period of the entire sequence. As a result, only a segment, or a subsequence, of the sequence is used in the correlations that generate decision metrics. This is called *aperiodic correlation* or *aperiodic integration*.[1] Another important and useful property of the m-sequence is the randomness of its subsequences. It can be shown that there are no two identical N-bits (or longer) subsequences in a 2^N-1 bit long m-sequence. The subsequences of the m-sequence can be shown to be uncorrelated. By using the mapped m-sequence as the transmitter reference symbol sequence for initial acquisition, we can use any of its subsequences as the reference symbol sequence in the (aperiodic) correlation, as long as it is longer than N symbols.

If the received signal contains the samples that are generated by the reference symbol sequence used in detection, the correlator of the detector is likely to generate a large correlation peak. Otherwise, the correlator output would be smaller because the shifted subsequence is uncorrelated with the reference sequence. However, some pairs of the subsequences can generate correlation peaks larger than those generated by other pairs. If this is the case, a false alarm event could occur. Noncoherent integration or multi-dwell search, as described in [1, 9], can be used to reduce the probability of such false alarm events.

3.4.1.2   Chirp Sequences

Another type of useful pseudo-random sequences is the so-called *chirp sequence*. Recall that the desired property of a symbol sequence for initial acquisition is that the magnitude of the correlation between the reference sequence and its shifted versions should be ideally equal to zero. This is equivalent to that the reference sequence has an impulse autocorrelation function as given by (3.33).

Let us assume that the acquisition sequence is sent periodically. As is well known in DSP theory, a discrete periodic sequence has a periodic impulse autocorrelation function if and only if it has a flat spectrum, i.e., all of the elements of its DFT output have the same magnitude. Thus, to construct a time-domain periodic sequence with a period of N samples that has a periodic impulse autocorrelation function, we can first form a frequency-domain sequence with N elements, all of which have the same magnitude. Performing iDFT of such a frequency-domain sequence generates a period of the time-domain periodic sequence that has an impulse autocorrelation function.

However, such a construction method does not tell us anything about the time-domain characteristics, in particular, the peak-to-average power ratio (PAPR), of the resultant sequence. If we want to reduce the PAPR of the constructed sequence, additional constraints need to be applied to the phases of the elements of the frequency-domain sequence. It has been shown that letting the phases of the samples

---

[1] Aperiodic correlations also occur when one period of the m-sequence is sent but not continuously in a periodic fashion. As a result, the correlation of the sequence with itself could only partially overlap.

follow the square of its index, i.e., letting the phase of the kth sample proportional to k^2 or $k(k+1)$, leads to a low peak-to-average ratio of the time-domain sequence [10]. Such sequences are also called chirp sequences because, if audible, they would sound like a bird's chirp. Unlike the PN sequences, the elements of chirp sequences are not binary but are complex valued.

Such chirp sequences have been proposed for a number of radar and communication applications. One such sequence, in particular, the frequency-domain Zadoff–Chu sequence [10, 11], has been adopted in the 3GPP LTE standard as the first-stage (primary) synchronization signal with some modifications.

As described in [10] a family of chirp sequences can be expressed by its frequency-domain form as

$$x_u(k) = \exp\left(-j\pi uk(k+1)/N_{ZC}\right) \quad \text{for odd } N_{ZC} \tag{3.37}$$

and

$$x_u(k) = \exp\left(-j\pi uk^2/N_{ZC}\right) \qquad \text{for even } N_{ZC} \tag{3.38}$$

where N_{ZC} is the length of the sequence and $0 \leq k < N_{ZC}$, u is an integer parameter that defines a family of such sequences. The parameter u is relatively prime to N_{ZC} and satisfies $0 < u < N_{ZC}$.

The sequences defined by (3.37) and (3.38) are called the constant amplitude zero autocorrelation (CAZAC) sequences, i.e., their periodic autocorrelations are impulses. As can be seen from (3.37) or (3.38), $x_u(k)$ has a constant magnitude that is equal to 1. Therefore, the iDFT of the sequence also has an impulse autocorrelation function. Hence, it can be used as the sequence for initial acquisition. In addition, it has a few interesting properties.

The Zadoff–Chu sequences are periodic with a period of N_{ZC}. If N_{ZC} is prime, the DFT of the Zadoff–Chu sequence is another Zadoff–Chu sequence. In this case, both of the sequences have constant amplitudes. Moreover, the cross correlation between two Zadoff–Chu sequences with the same N_{ZC} but different u's is equal to $1/\sqrt{N_{ZC}}$.

It should be pointed out that, while Zadoff–Chu sequences have ideal periodic autocorrelation properties, their aperiodic correlations are not perfect even though still quite good. Given that aperiodic correlation is very important for initial acquisition, caution must be taken when practical systems that use Zadoff–Chu sequences are implemented for such applications.

3.4.2   Noise Variance Estimation

The implementation of the initial acquisition is relatively straightforward. One of the key steps is to determine the threshold based on the expected noise variance. The performance of the initial acquisition and its agreement with the theoretical analysis greatly depends on the accuracy of noise variance estimate. In this section, we discuss a few aspects of noise variance estimation useful in practical communication systems.

3.4.2.1   Noise/Interference Dominated System

If the device is operated in a low SNR environment, the received signal is dominated by the additive noise and external interference. Such a situation is typical, for example, in CDMA systems for voice communications. In this case, the total signal power is almost the same as the noise power. Thus, the threshold can be set based on the total signal power, which is relatively easy to measure.

As shown in Figure 3.1, the received signal is typically adjusted by AGC circuits before the signal is converted to digital samples by ADCs. The signal at the AGC output maintains approximately a constant power, as does its sampled version. Let us express the mean squared value of the signal samples by $E[|y_{m,i}|^2] = B^2$, which is a known constant from the design of the AGC or can be computed digitally by averaging multiple $|y_{m,i}|^2$ values. When the sample sequence involved in the correlation does not contain the reference symbol sequence, the samples $y_{m,i}$'s contain only noise and interference. Moreover, $y_{m,i}$'s and a_n's are uncorrelated. If $E[|a_n|^2] = 1$, the variance of the detected symbols given by (3.9) is

$$E\left[|d_{m,i}|^2\right] = E\left[\left|\frac{1}{\sqrt{K}}\sum_{j=0}^{K-1}a_{n-k}y_{m-k,i}\right|^2\right] = \frac{1}{K}\sum_{j=0}^{K-1}E\left[|a_{n-k}|^2|y_{m-k,i}|^2\right]$$

$$= \frac{1}{K}\sum_{j=0}^{K-1}E\left[|y_{m-k,i}|^2\right] = B^2 \tag{3.39}$$

Because a detected symbol is the sum of K independent noise and interference terms, it can be considered as Gaussian distributed. The detection threshold to achieve a given false alarm probability can be set based on the value of B^2, which is known from the design of the AGC circuit and is roughly a constant. Alternatively, B^2 can be estimated relatively accurately if multiple squared digital samples are averaged. In either case, the computation of the threshold is straightforward when the goal is to achieve a given false alarm probability.

When the expected symbol sequence exists, $y_{m,i}$ contains a signal component and a noise component. Given that the signal and noise are uncorrelated, from (3.7) we have

$$E\left[|y_{m,i}|^2\right] = P_s + P_n = B^2 \tag{3.40}$$

When the SNR is low, we have $P_n = \sigma_z^2 \approx B^2$. As shown by (3.14), the noise portion of the detected symbol $d_{m,i}$ has a variance $\sigma_{z'}^2$ equal to σ_z^2. In other words, the variance of the noise in the detected symbol is approximately equal to the variance of the signal samples. Thus, the detection probability calculated by using B^2 instead of $\sigma_{z'}^2$ is relatively accurate.

3.4.2.2   System Operated in Medium to High SNR Environments

Time division multiple access (TDMA), frequency division multiple access (FDMA), or OFDM communication systems or CDMA systems for data communications operate in medium to high SNR environments. For example, a communication link for data transmission is often implemented in the TDM mode to avoid intra-cell interference. This is true even if the overall system is operated in the code division multiplexing (CDM) or OFDM mode.

As shown by (3.40), the power at the output of a perfect AGC, or with digital normalization, is equal to the sum of the signal power and the noise power. The total power is approximately constant. When the received signal does not contain the reference symbol sequence, with the assumption that the signal plus noise are uncorrelated with the reference symbol sequence, the correlator output approximately behaves as a Gaussian random variable to the detector. Thus, to determine if the H_0 hypothesis is true, the threshold to achieve a given false alarm probability P_F can be set based on the variance of the received signal samples. This is the same as in the low SNR case discussed above.

However, as can also be seen from (3.40), if the reference symbol sequence exists, i.e., when the hypothesis H_1 is true, the noise power is equal to $P_n = B^2 - P_s < B^2$. In other words, when the reference symbol sequence exists and the SNR is high, the variance of the noise and interference in the detected symbol is smaller than the variance of the signal samples. To achieve the optimal detection, the detection threshold for testing the H_1 hypothesis should be set according to P_n. However, this is impractical because P_n is not known when setting the threshold.

A realistic approach, which is commonly used in practice, is to use the same threshold for both H_0 and H_1 tests, even though it is not optimal according to the Neyman–Pearson lemma. Thus, it is of interest to analyze the behavior of the detector under such conditions. A simple analysis for a single-path channel with coefficient $g_{0,0}(0)$, the same as that discussed in Section 3.2.3, is shown below.

When H_0 is true, the samples contain only noise and interference, denoted by n_1, with the variance $P_{n_1} = B^2$. On the other hand, when the signal sample sequence contains the reference symbol sequence, i.e., when H_1 is true, denoting the noise and interference in the signal samples by n_2 and the SNR by γ_y, we have

$$B^2 = P_s + P_{n_2} = \gamma_y P_{n_2} + P_{n_2} = (1 + \gamma_y) P_{n_2} \qquad (3.41)$$

i.e.,

$$P_s = |g_{0,0}(0)|^2 = \frac{\gamma_y B^2}{(1 + \gamma_y)} \text{ and } P_{n_2} = \sigma_z^2 = \frac{B^2}{(1 + \gamma_y)} \qquad (3.42)$$

In (3.41) and (3.42), the SNR $\gamma_y = P_s/P_n = |g_{0,0}(0)|^2/\sigma_z^2$ if $E[|a_m|^2] = 1$.

As shown by (3.15), the detected symbol $d_{n,0}$ can be expressed as $d_{n,0} = \sqrt{K}g_{0,0}(0) + z'_{n,0}$, where K is the integration length for generating the detected symbol. The variance of $z'_{n,0}$, σ_z^2, is equal to $B^2/(1 + \gamma_y)$ and $|E[d_{n,0}]|^2 = |\sqrt{K}g_{0,0}|^2 = KB^2\gamma_y/(1 + \gamma_y)$. The SNR of the detected symbol $d_{n,0}$ is equal to $K\gamma_y$ as expected.

When the threshold Th is set by using B^2 as the noise variance in the detected symbol based on a given false alarm probability, the detection probability can be computed by the use of (3.18) with the understanding that

$$\lambda^2 = |E[d_{n,0}]|^2 = KB^2\gamma_y/(1 + \gamma_y) = KB^2/(1 + 1/\gamma_y) \qquad (3.43)$$

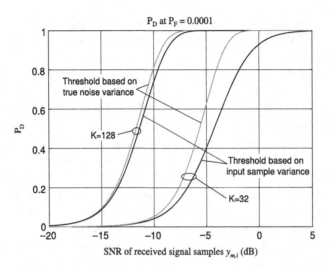

Figure 3.9 P_Ds versus SNR of received signal samples ($P_F' = 0.0001$, threshold determined based on received signal samples variance and true noise variance)

The results obtained above are useful for the evaluation of the expected detection probability for a given SNR of γ_y if the threshold Th is set based on the estimate of B^2. Figure 3.9 plots the detection error probabilities with $P_F = 0.0001$ as functions of the SNR of the input samples according to (3.43) for $K = 32$ and 128. As a comparison, a set of hypothetical curves of the P_D's with the threshold determined based on the true noise variance is also shown. The second set of curves corresponds to the result given in Figure 3.3 for $P_F = 0.0001$. However, given that the definitions of SNRs are different, the curves are shifted by $10\log_{10}K$, i.e., 15 and 21 dBs to the left from the original curve for $K = 32$ and 128, respectively. As can be observed from this figure, the detection probability with the threshold computed based on the variance of the received signal samples is lower than that computed from the true noise variance as expected. With a longer integration length, i.e., a larger processing gain, the loss due to the nonideal threshold setting gets lower.

Above, we have shown that the detection threshold for initial acquisition can be determined based on the variance of the signal samples involved in the detection. In most practical cases, when H_1 is true, the true noise variance in the detected symbols is unknown. The threshold setting based on the total received signal power is the only realistic approach. Moreover, this approach is also desirable from the detector's performance point of view. It is because this method makes it possible to perform longer averaging on the square values of signal samples. The more samples that are used in averaging, the more accurate the estimate of the noise variance can be. This is important for ensuring the performance of initial acquisition because the accuracy of the noise variance estimate plays the most important role in achieving the expected false alarm rate P_F. As will be shown in Section 3.4.3, a low P_F is very important to the overall receiver performance at the early stage of receiver operations. It is interesting to note that with a longer integration length, i.e., a larger processing gain, the loss due to the nonideal threshold setting gets lower.

3.4.3   The Impact of P_F and P_{miss} on Acquisition Performance and the Multi-Dwell Search

The most important parameter of initial acquisition is the total time needed to find the correct timing at which the reference sequence was sent. It is called the *acquisition time,* T_{Acq}. Due to the random nature of the acquisition process, T_{Acq} is a random variable and is usually characterized by its mean, or expected value, and its distributions.

In a typical acquisition process, the reference sequence for acquisition is sent periodically, either continuously or in bursts. The detector may need to search up to all of the possible timing offsets of the received signal sample sequences containing the reference sequence to find the correct timing, or a "*hit.*" If the transmission of the reference sequence repeats every N_p symbols, it would take $\tau_p = N_p T$ (seconds) to search the entire period. If $T/2$ sampling resolution is used, up to $2N_p$ hypothesis tests may be needed to get a hit. On average, the required number of hypothesis tests is equal to half of that in the worst case, i.e., N_p. If the reference sequence exists and is detectable, the average time of achieving a successful detection is equal to $0.5\tau_p$ seconds. Once a hit is found, additional time is needed for further processing the received signal and verifying if it is indeed a correct one. By denoting the verification time as τ_V and assuming that both P_F and P_{miss} are equal to zero, the average acquisition time is

$$\overline{T}_{Acq} \cong E[T_{Acq}] = 0.5\tau_p + \tau_V \tag{3.44}$$

Above we assumed that the processing time could be ignored by using very fast processing hardware and/or software to perform the acquisition.

If we have a hit but it is a false alarm found from the verification process, we need to redo the acquisition process. When P_F is not equal to zero but P_{miss} is still assumed equal to zero, the average acquisition time can be expressed as

$$\begin{aligned}
\overline{T}_{Acq} &= (1 - \overline{P_F})(0.5\tau_P + \tau_V) + 2\overline{P_F}(1 - \overline{P_F})(0.5\tau_P + \tau_V) \\
&\quad + 3\overline{P_F}^2(1 - \overline{P_F})(0.5\tau_P + \tau_V) + ... \\
&= (1 - \overline{P_F})(0.5\tau_P + \tau_V)\left(1 + 2\overline{P_F} + 3\overline{P_F}^2 + 4\overline{P_F}^3 + ...\right) \\
&= \frac{1}{(1 - \overline{P_F})}(0.5\tau_P + \tau_V)
\end{aligned} \tag{3.45}$$

where $\overline{P_F}$ is the effective false alarm probability, i.e., the false alarm probability over the total number of hypothesis tests. It can be shown that the effective false alarm probability is approximately equal to $N_p P_F$ when P_F is low.

The average number of tests to attain a hit within one period of the reference sequence transmission can be quite large. For example, in the cdma2000–1x system, it is equal to $2^{15} = 32768$. To achieve an effective false alarm probability of 0.1, the P_F per test should be about 3×10^{-6}. Thus, it is desirable to have a very low P_F per test.

When $\overline{P_F}$ is not too high, the average acquisition time does not increase much. However, the variance of the acquisition time could be large.

The analysis of the penalty due to P_{miss} is similar. Assume that $P_F = 0$ in the analysis and there is no miss event. A hit would be declared in $0.5\tau_P$ on average. When the first P_{miss} event occurs, the average detection time is equal to $1.5\tau_P$. When there are two miss events, the average detection time is equal to $2.5\tau_P$. Thus, the average acquisition time can be expressed as

$$
\begin{aligned}
\overline{T_{Acq}} &= (1 - P_{miss})0.5\tau_P + P_{miss}(1 - P_{miss})(0.5\tau_P + \tau_P) \\
&\quad + P_{miss}^2(1 - P_{miss})(0.5\tau_P + 2\tau_P) + \dots \\
&= (1 - P_{miss})\left[0.5\tau_P\left(1 + P_{miss} + P_{miss}^2 + P_{miss}^3 + \dots\right)\right. \\
&\quad \left. + \tau_P P_{miss}\left(1 + 2P_{miss} + 3P_{miss}^2 + \dots\right)\right] + \tau_V \\
&= (1 - P_{miss})\left[\frac{0.5\tau_P}{1 - P_{miss}} + \tau_P \frac{P_{miss}}{(1 - P_{miss})^2}\right] + \tau_V \\
&= 0.5\tau_P + \tau_P \frac{P_{miss}}{(1 - P_{miss})} + \tau_V
\end{aligned}
\tag{3.46}
$$

By comparing (3.46) and (3.45), we can conclude that the penalty due to P_F is higher than the penalty due to P_{miss}. This is because it is necessary to go through a verification process to rectify a P_F event, a process that can involve the full-blown decoding and demodulation of the traffic or other control channels. When a P_{miss} event occurs, the search process continues, and only one verification is needed in the end.

It is interesting to note that the penalty due to P_F and P_{miss} is not very large if the expected value of T_{Acq} is used as a metric in the case where these error probabilities are not too high. Thus, the expected value of acquisition time may not be a very good metric to characterize the initial acquisition process. A more useful metric is the probability of T_{Acq} that is beyond a certain threshold given that an event like this seriously affects the overall system performance.

Above, we performed some simple analysis of the penalty due to detection errors. That is, we did not consider the case that both miss and false events could occur in the same acquisition process. Such more complex cases can be described as multistate Markov chains. They are analyzed by generating a function flow graph for such a Markov chain that characterizes the acquisition process. We will not go into the details of such analyses, and the interested readers are referred to [9] for more information.

Another method to reduce the penalty due to a false alarm is that once an H_1 hypothesis is declared valid, additional independent detections are performed to verify that it is not a false alarm before demodulation and decoding are performed. The additional detection can be the same as the normal detection or can change the detection parameters, such as using a longer coherent integration length if possible or combining additional noncoherent detected symbols. Such detections are performed based on the detected timing and reference symbol sequences rather than by starting a new detection process. As a result, fewer hypothesis tests need to be done. Thus, the computation complexity for detection is less significant.

If the detection process involves additional detections using different parameters from that used in normal detection, it is called *multiple-dwell* technique [1, 9]. The objective of using such a technique for enhanced reliability is to provide significantly lower false alarm probability than regular detection and yet to have less overhead than full-blown verifications do.

3.4.4 Impact of Initial Frequency Offset and the Remedies

During initial acquisition, the receiver has little information about the transmitted signal that it is expected to receive, other than the nominal candidate carrier frequencies of the transmitted signal. In addition, it is likely that there is a frequency offset between the carrier frequency and the frequency generated by the local oscillator at the receiver for frequency down-conversion of the received signal. Such a frequency offset created by the receiver is usually the main impairments that degrade the initial acquisition performance.

The existence of the initial frequency offset imposes challenges to receiver/modem designers. It is especially true in recent years because in order to reduce the cost of devices, low-cost XOs are preferred as a local frequency reference. However, such XOs are usually less accurate than the more expensive ones such as *voltage-controlled temperature-compensated crystal oscillators* (VC-TCXOs).

At the acquisition stage, the XO is usually free running, and its frequency can be quite inaccurate. Thus, there would be a large frequency offset in the received signal after being converted to baseband. As we have analyzed in Section 3.3.1, the resulting frequency offset limits the coherent integration length of detection and degrades the acquisition performance.

3.4.4.1 Selection of Coherent Integration Length

If the integration time duration is equal to KT, the magnitude of the integration output is reduced by a factor of $\left(1 - \text{sinc}^2\left(KTf_{offset}\right)\right)$ relative to no frequency offset as shown in [1] and analyzed in 3.3.1. Even for a modest value of KTf_{offset}, the loss can be quite large. Thus, when f_{offset} is not negligible, it is preferable to limit the coherent integration length. If enough reference symbols are available, multiple detected symbols can be generated and noncoherently combined as shown in Section 3.3.2. However, that would result in a noncoherent integration loss relative to the ideal coherent integration as discussed in Section 3.3.3. Given these factors, it is important to choose the coherent integration length and the number of noncoherent integrations to achieve the best possible performance with a given number of samples involved in the detection.

The impact of a frequency offset to initial acquisition in practice is considered as follows.

The frequency accuracy of an XO is specified by the relative error from its nominal frequency in units of *parts per million* (*ppm*), i.e., 10^{-6}. A typical free-running XO may have an accuracy of 2 to 10 ppm, or larger. An XO with temperature compensation, i.e., TCXO, usually has an accuracy of less than 2 ppm, but it would cost more. A few examples are considered below.

In 2G and 3G cellular systems, the carrier frequencies are commonly in the 800 MHz, 900 MHz, 1.8 GHz, and 1.9 GHz frequency bands. For a performance loss of 1 dB in coherent integration to generate the detected symbol, as seen in (3.24) and Figure 3.4, KTf_{offset} should be less than 0.262.

Consider a cdma2000–1x cellular system operating in the 800 MHz band; the values of f_{offset} are 1600 Hz, 8000 Hz, and 80000 Hz for 2, 10, and 100 ppm XOs, respectively. The channel symbol (chip) duration is equal to 0.814 microsecond (μs). If the coherent integration's loss is about 1 dB relative to the ideal case, the coherent integration length K should be no longer than 201, 40, and 4 chips at these frequency offsets.

Next, let us consider a Universal Mobile Telecommunications Service wideband code division multiple access system operating at 1.9G band. The chip rate of the UMTS system is equal to 3.84 MHz, or $T_c = 0.26 \ \mu s$. The frequency offsets are equal to 3.8 kHz, 19 kHz, and 190 kHz for 2 ppm, 10 ppm, and 100 ppm XOs, respectively. We can compute that K should be less than 265, 54, and 5 chips for these frequency offsets.

From the above calculation, the coherent integration length cannot be too long when the frequency offset is relatively large due to the inaccurate XO. Additional detection gain could be achieved by using noncoherent combing. However, the coherent integration length should not be too short either. This is because aperiodic correlation properties are usually quite poor for commonly used reference sequences for initial acquisition. For example, for an m-sequence generated from an N-bit shift register, each segment of the sequence is uniquely defined only if its length is greater than or equal to N. Thus, in the case of cdma2000–1x, the coherent correlation length should never be less than 15 because its short PN sequences are m-sequences with $N = 15$. It is preferable to have the coherent integration length to be at least equal to $2N$. Moreover, we may allow a larger loss, e.g., 3 dB or even 6 dB instead of 1 dB, due to longer integration time, to have a better aperiodic correlation property.

When short coherent integration must be used, noncoherent integration of multiple decision metrics can reduce the adverse effect of the aperiodic correlations. Another method to reduce the false alarm probability is to perform multiple such hypothesis tests on these decision metrics. The final decision can be made by a majority vote.

3.4.4.2   Frequency Binning Method

A commonly used method to accommodate large frequency offsets is the so-called *frequency binning* method (see e.g. [12] and [13], Section 12.5). Its principal and implementation considerations are discussed below.

Assume that the magnitude of the frequency offset due to the inaccuracy of an XO is less than f_{max}, i.e., $-f_{max} < f_{offset} < f_{max}$. To implement initial acquisition using the binning method, the region $(-f_{max}, f_{max})$ is first divided into a few subregions, also called *bins*. For example, by dividing it into $2K_b + 1$ bins, each of the bins has a width of $f_b = 2f_{max}/(2K_b + 1)$. These bins are centered at the frequencies if_b, $i = -K_b, \ldots, -1$, $0, 1, \ldots, K_b$. The union of these bins covers the entire frequency offset region of the XO. A detection process is performed per subregion, i.e., it repeats $2K_b + 1$ times. The ith detection uses the samples generated from the ith frequency bin with the frequency offset shifted by if_b from its original value. As a result, the detection process only needs to handle a maximum frequency offset of $\pm 0.5f_b$.

Such frequency offset compensation can be performed by analog or digital means. In today's implementations, such compensations are most conveniently done by DSP. For example, to generate the samples in the ith frequency bin, a phase rotation of $-2\pi ikf_bT$ is applied to the kth sample $y_{m-k,i}$ for correlation as given by (3.9). Specifically, the detected symbol $d_{m,i}$ is generated according to

$$d_{m,i} = \frac{1}{K}\sum_{k=0}^{K-1} a_{n-k}^* \left(e^{-2\pi ikf_bT} y_{m-k,i}\right) = \frac{1}{K}\sum_{k=0}^{K-1} \left(e^{2\pi ikf_bT} a_{n-k}\right)^* y_{m-k,i} \qquad (3.47)$$

As can be seen from (3.47), the compensation can also be implemented by introducing a phase rotation of $2\pi ikf_bT$ to the expected data symbol a_{n-k} if preferred.

As the maximum frequency offset following the implementation of this method is equal only to $1/(2K_b+1)$ of the offset without frequency binning, the coherent integration length can be increased by a factor of $2K_b+1$ with the same correlation loss. This can greatly facilitate the selection of the length of the correlation. However, the cost is that the computational complexity of the initial acquisition could increase up to $2K_b + 1$ times.

Let us revisit the earlier example of initial acquisition for UMTS. If the maximum frequency error of the XO is 10 ppm and if $K_b = 1$, i.e., there are three bins, the maximum frequency offset of each bin is 19/3, or about 6.33 kHz. Then K can be 108 chips with a reasonable loss. This design would be much more practical and implementable.

The trade-off between using accurate XOs and using the binning method is the hardware cost of the XO and the cost of the increased computational complexity. Given the computation capability of today's DSPs or ASICs, which is usually not heavy loaded during initial acquisition, the binning method can be more attractive than using more accurate XOs.

3.4.4.3 Initial Frequency Offset Estimation

Once initial acquisition is successful, it is possible and desirable to perform preliminary estimation of the frequency offset to improve the performance of the receiver operations that follow.

If the frequency binning method is used in detection, the center frequency of bin i that generates successful detection can be used as the candidate for frequency correction. For example, a frequency correction of if_b can be introduced relative to the XO natural frequency. Such a correction may not be accurate enough for other operations, but, at least, it can be used as a good initial value for further fine-tuning.

In the case of using noncoherent integration of multiple detected symbols to generate the composite decision metric, when the detection is successful, each of the detected symbols is an estimate of the channel coefficient, as shown by (3.15). For two detected symbols generated from two sample sequences close to each other that yield the successful detections, there is a difference between their phases if the frequency offset is not equal to zero. From (3.22), it can be shown that the phase difference between $d_{n,i}$ and $d_{n+N,i}$ is equal to $2\pi f_{offset}NT$. An initial estimate of the frequency offset can be computed by

$$\hat{f}_{offset} = \frac{\text{Arg}\left(d_{n,i}^* d_{n+N,i}\right)}{2\pi NT} \approx \frac{\text{Im}\left(d_{n,i}^* d_{n+N,i}\right)}{2\pi NT |d_{n,i}|^2} \tag{3.48}$$

If multiple detected symbols are involved in generating the decision metric for successful detection, the frequency offset estimates computed between the adjacent $d_{k,i}$'s can be averaged to yield a more accurate one.

3.4.5   Initialization of Carrier and Timing Synchronization Parameters

System characteristics learned by the receiver from initial acquisition are essential for establishing the communication link between the transmitter and receiver. Moreover, they can also be used for initialization of the parameters of the timing and frequency synchronization blocks, which will be discussed in Chapters 5 and 6 of this book. As a preview, we provide some preliminary descriptions of such usages.

It has been shown that, for carrier synchronization, the frequency offset between the transmitter and receiver frequency references could be estimated according to (3.48) when multiple detected symbols are computed. The estimated frequency offset can be used to initialize the frequency register in the carrier synchronization block. If only one detected symbol is generated during initial acquisition, it is always possible to partition the reference symbol and signal sample sequences that generate the successful detection into two or more pairs of subsequences. Correlations are performed using the corresponding sample and symbol subsequences. The phase difference between these correlation outputs can be used to generate the estimate of the frequency offset.

As shown by (3.9), the detected symbol $d_{m,i}$ is an estimate of the channel coefficient $g_{n-m,i}$. The sampling time of each sample in the sample sequence that generates the $d_{m,i}$ with the largest magnitude is a timing estimate of the corresponding data symbol. When a few adjacent detected symbols all have large magnitudes, it is possible to obtain a better timing estimate by interpolating the timing of these detected symbols. These timing estimates can be used to initialize the timing synchronization block of the receiver. After the initialization, the timing error of the sample with respect to the data symbol is normally less than one symbol interval T. Optimal receiver timing selection and adjustment will be discussed further in Chapters 6 and 7.

The channel coefficient estimates generated from the initial acquisition process can also be used as the initial channel estimate for the demodulation of the received signal.

3.4.6   A High-Level View of System Acquisition

Above we have discussed various aspects of detecting the expected reference symbol sequence in the received signal as the first part of initial acquisition. Once the reference symbol sequence is detected and confirmed, the next step of initial acquisition is to acquire the necessary parameters of the acquired system so that a communication link can be established. We call this step *system acquisition*.

System acquisition is a procedure of mixed PHY layer and medium access control (MAC) layer operations. From the physical layer point of view, once the reference symbol sequence is detected, the received signal samples have aligned with these symbols in timing. Since the symbols, in turn, align with certain data frames in the transmitter signal, the receiver knows the frame boundaries in the received signal sample stream. Thus, the receiver should be able to extract the information carried by these data frames and to acquire the system parameters. These parameters can be used for establishing a communication link with the transmitter, either directly or in a bootstrap form according to the system design. This portion of the acquisition is beyond the PHY layer, but belongs to the MAC layer, operations.

While different systems within the PHY layer operations of initial acquisition as discussed above have many aspects in common, the procedures of system acquisition are different for different systems. Above, we have only provided a general concept of this stage of acquisition. More details can be found in the specifications of the systems under consideration. In the next two sections, examples of initial acquisition including the parts of system acquisition for widely used wireless communication systems are presented.

3.5 Initial Acquisition in DS-CDMA Systems

So far, we have presented the basics of initial acquisition and discussed some specific topics related to its performance, design, and implementation. In this and the next sections, we will describe initial acquisition in four wireless communication systems deployed commercially today. In this section, we consider the initial acquisition procedures of IS-95/cdma2000–1x and WCDMA wireless communication systems. Both of these systems employ DS-CDMA, or simply CDMA, technology, which was introduced in Section 1.5. Systems based on OFDM technology will be treated in the next section.

3.5.1 IS-95/cdma2000–1x

In this section, we describe the initial acquisition process of wireless communication systems based on IS-95 and cdma2000–1x standards. As cdma2000–1x has evolved from IS-95 and is backward compatible, the initial acquisition processes of the IS-95 and cdma2000–1x are essentially the same. To facilitate discussion, we first provide a simplified overview of their PHY layer, with an emphasis on the parts that are relevant to initial acquisition.

3.5.1.1 Overview of the IS-95/cdma2000–1x PHY Layer

A cdma2000–1x cellular network consists of many cells. Each of the cells usually consists of one to six sectors, which transmit forward-link signals to mobile devices and receive reverse-link signals from these devices.

Figure 3.1 depicts a simplified block diagram of a cdma2000–1x forward-link trans-mitter. Specifically, it shows the signal generation of the various forward-link channels, including the pilot, sync, traffic, paging, and other control channels, in a sector of a base

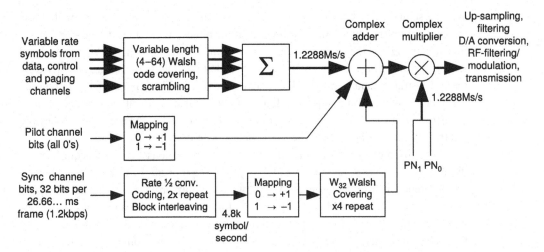

Figure 3.10 A simplified block diagram of a cdma2000–1x sector transmitter

station. To concentrate on the subject of initial acquisition that we discuss here, we consider only the relevant ones, i.e., the pilot and sync channels, in detail.

Cdma2000–1x is a *synchronous* CDMA system. In most cases, the transmissions from all of the base stations in the cells or sectors are synchronized using the information from the Global Positioning System (GPS) [14, 15]. The data symbols are spread before transmission. In cdma2000, the spreading process takes two steps. First, each data symbol is spread, or covered, by a Walsh code. The data symbols from different users or channels are covered by different Walsh codes. Walsh covered data symbols are added together before being further spread or scrambled by PN sequences.

Two types of PN sequences are used in the cdma2000–1x system: two short PN sequences and one long PN sequence. The two short sequences are the m-sequences generated by 15-bit shift registers. These two sequences, called PN_I and PN_Q, serve as the in-phase (I) and quadrature (Q) components of a quadrature spreading sequence after their elements are mapped from logical 0 to real value 1 and logical 1 to −1. The characteristic polynomials of the generators are

$$f_I(x) = 1 + x^2 + x^6 + x^7 + x^8 + x^{10} + x^{15} \tag{3.49}$$

and

$$f_Q(x) = 1 + x^3 + x^4 + x^5 + x^6 + x^9 + x^{10} + x^{11} + x^{12} + x^{15} \tag{3.50}$$

A period of the m-sequences generated by 15-bit shift registers is equal to $2^{15} - 1 = 32767$ bits. For the convenience of implementation, an additional logical zero is inserted after the consecutive 14 zeros, which occurs once every period for PN_I and PN_Q. The first logical one after the 15 consecutive zeros is defined as the *start* of either PN_I or PN_Q. The elements of the quadrature spreading sequence, which consists of PN_I and PN_Q, scramble the transmitter channel symbols. The scrambled symbols are called *chips*. The chip rate is equal to 1.2288 MHz as specified in the cdma2000–1x standard.

Thus, the duration of a chip, denoted by T_c, is approximately equal to 0.814 microseconds. Each period of the short sequences is equal to 80/3 or approximately 26.667 millisecond (ms).

All of the cdma2000–1x forward-link transmitters use the same short PN sequences. To distinguish the signals from different sectors and to reduce the interference between them, each sector's PN sequences are delayed by an integer multiple of 64 chips, called the *PN-offset* of the sector. The pair of PN sequences with no delay is called the *zero-shift PN sequences* and is defined as the time reference of the short PN sequences. The zero-shift PN sequences are aligned with the even-second time mark of real time.

To ensure the normal operations of a cdma2000–1x system, it is necessary to define an unambiguous system time to which all the transmitters align. The period of 26.667 ms of the short PN sequences is too short to be used uniquely as a global system time reference. Thus, the cdma2000–1x specification defines another PN sequence, which is $2^{42}-1$ chips long and corresponds to about 41.4 days in real time. With such a long sequence, no confusion could occur. This PN sequence is called the long PN sequence. It is a common reference of the system time for all of the users. The long PN sequence is also the basis for generating the user uplink spreading sequences and for scrambling the downlink data symbols transmitted from the sectors. The user devices must know the state of the long PN sequence to communicate with the network.

Every cdma2000–1x mobile device has three PN generators to create the local copies of these sequences. Two of them generate the zero-shift short PN sequences PN_I and PN_Q. A mask is applied to each of the generators to produce a shifted version of the short PN sequence [1, 9]. The third generator creates the long PN sequence. The states and masks of these generators are set at the sync-channel acquisition stage after the system information is acquired.

In a cdma2000–1x system, the data symbol rates of the data and control channels are variable. The system is designed such that for different channels' data rates, Walsh coverings with different lengths are used so that the samples of the Walsh covered data symbols are at the same rate as the chip rate, i.e., 1.2288 Mchips/second. Since the quadrature spreading sequence is also at the same rate, the spreading sequence does not increase the rate of the Walsh covered data symbols. Because the Walsh codes are not random sequences, their spectrums are not white. The main role of the PN spreading sequences is to make the transmitted sequence have a white noise-like spectrum, which is required by a DS-CDMA system for interference randomization.

Traffic, paging, and other control channels are organized into 20 ms frames. Four of these data frames form an 80 ms superframe, which aligns with three periods of the zero-shift short PN sequences.

The forward-link channels that are essential for initial acquisition are the pilot and sync channels. The pilot channel transmits all binary zero sequences covered by Walsh code 0, which has all binary zero values. After the covering and mapping, the pilot channel is all (real value) 1s at the chip level. As a result, the pilot channel signal from any sector is the same as the signal generated by the quadrature spreading sequence defined by PN_I and PN_Q, and thus, is known to the receiver.

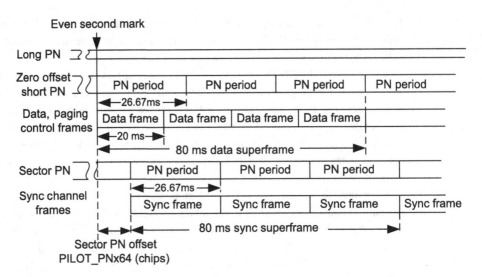

Figure 3.11 cdma2000–1x frame timing diagram

To ensure a robust initial acquisition performance, the sync channel has a low data rate of 1200 bits/s. The data is coded with a rate ½ convolutional code and with symbol repetition of a factor of 2. The output symbol rate is 4800 symbol/second. The sync channel is organized in frames and superframes. Each sync frame is 26.667 ms long and aligns with its sector's short PN sequences. Three sync frames form an 80 ms sync superframe. The sync superframe has a delay relative to the data superframe. This delay is equal to the delay of the PN sequences of the sector relative to the zero-offset short PN sequence. In a sync superframe, there are 128 sync data symbols, which are block interleaved within each frame. Each of the interleaved sync data symbols is covered by a Walsh 32 code word of length 64 Walsh codes and repeated four times, so that the chip rate is 1.2288 Mchips/s. The Walsh 32 code word has 32 consecutive 1s followed by 32 minus 1s.

The timing diagram of the cdma2000–1x frames of various channels is shown in Figure 3.11.

The initial acquisition process of cdma2000–1x includes two steps: pilot channel acquisition and sync channel acquisition.

3.5.1.2 Pilot Channel Acquisition

Pilot channel acquisition, also called *pilot search*, achieves the following objectives. First, the device needs to know if there are cdma2000–1x signals that it needs to acquire. Second, if such signals of interest exist, it determines the timing of the pilot sequence of the sector, from which the signal was transmitted. Third, once acquired, it generates initial channel estimates for the demodulation of the signals of other channels of this sector. Finally, it determines other characteristics of the signal, such as its carrier and timing frequencies and phases to be used in initializing the corresponding synchronization blocks.

In cdma2000–1x systems, the waveforms of the pilot channel signals transmitted from all of the sectors are the same, as they are generated from the same short PN sequences. The difference between these waveforms is the different time delays from each other. Thus, for finding the pilot channel signal of the nearby sector, the searcher does not need to know from which sector the signal is transmitted. The delay difference is resolved during the sync channel acquisition discussed below. Moreover, the pilot channel is usually transmitted at a power higher than other channels to facilitate the pilot search process and to provide better channel estimates for demodulation. For example, it typically consumes 15 percent of the total power transmitted from a sector.

The process of searching for the pilot channel in cdma2000–1x systems is just a special case of the general initial acquisition process described in Section 3.2. We will describe its operations briefly and discuss its special properties and characteristics in more detail.

When performing initial acquisition, the searcher block in a cdma2000–1x mobile device generates a copy of the quadrature short PN sequence. It also samples the received signal at a rate of $2/T_c$ or higher. One or more segments of the PN sequence are correlated with the segments of the sample sequence to generate a decision metric with or without noncoherent integration. The decision metric is compared to a predetermined threshold to decide if the search succeeds or fails as described in Section 3.2. If the search fails, the searcher continues to perform another set of correlations using different segments of the received signal sample sequence. Otherwise, if the searcher declares a possible match, it may perform additional verifications. It declares that the expected signal indeed exists if the verifications are successful. The information acquired is passed to the system acquisition function for further processing.

The parameters of the pilot search, i.e., the coherent integration length and number of noncoherent integrations needed, are determined by the required SNR and initial XO frequency offset. The selection principles have been discussed in Section 3.3.3.

Once the detection is successful and multiple signal sample segments are used, the phase differences between the outputs of the correlations that yield the successful detection are proportional to the frequency offset. Thus, they can be used for estimation and correction of the XO's frequency error.

To summarize, the special property of the cdma2000–1x system is that it has a continuous pilot channel. Whenever the pilot channel signal from any sector exists, a segment of the received signal contains a part of the short PN sequence. Thus, the searcher can perform the correlation for the acquisition of the pilot channel in two different but equivalent ways. It can use a single segment of the PN sequence to correlate with the segments of the sample sequence with different delays as described in Section 3.2. Alternatively, it can use multiple different segments of the short PN sequence to correlate with the same segment of the sample sequence.

Historically, the first approach is often used because waiting for new samples to be received gives the hardware enough time to process. Currently, however, the second approach is preferred due to the improved processing capability of the receiver hardware. In this case, the searcher just needs to buffer a set of the received signal samples and then to perform all of the processing internally on the same segments of samples using different segments of the short PN sequence. The correlations performed must cover all of the sampling phases within one chip interval. For example, the sampling rate is usually chosen to be equal to $2/T_c$. Each PN sequence segment could correlate with two sample segments with a time offset of $T_c/2$ to yield two detected symbols corresponding to the two sample timing phases.

The information obtained from the searcher is fed to the *RAKE receiver* as the initial values for performing data symbol despreading and demodulation. The structure, operations, and properties of a RAKE receiver have been described in Section 1.5.3. More information on the RAKE receiver can be found in other references including [1].

Once the successful pilot acquisition is achieved, the receiver has acquired the timing of the sector's 26.667 ms frames of the pilot and sync channels. However, additional information is needed to establish the communication link between the device and the sector. We will discuss how the device obtains the needed information in the next subsection.

3.5.1.3   Sync Channel Acquisition

As shown in Figure 3.11, the sync channel frames are aligned with the pilot channel, which aligns with the short PN sequences of the sector. Once the receiver acquires the sector's pilot channel, it knows the sync frame boundaries. Given that the sync messages are interleaved within a frame, deinterleaving can start once the data of a sync frame is received. Moreover, the state of the convolutional encoder does not reset between the sync channel frames [15]. Hence, the receiver can start decoding as soon as it receives the deinterleaved data.

The sync channel is transmitted continuously. In order to recover the sync message correctly, it is necessary to determine its starting point. To facilitate determining the boundaries of the sync messages, the first bit of each sync frame is defined as the *start of the message* (SOM) bit [15]. A sync message may consist of one or more superframes and thus multiple frames. Only the SOM bit in the first frame of the sync message has a value of one. All of the SOM bits in the following frames that belong to the same message are zeros. Thus, the SOM bit with a value of one in a sync frame marks the beginning of a sync message. This message is completed when another SOM bit with a value of one is found because it marks the beginning of the next sync message. A sync message with SOM bits' locations is depicted in Figure 3.12.

Each of the sync messages contains the information of a number of system parameters. The most important parameters related to the PHY layer are *sector pilot*

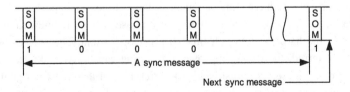

Figure 3.12 A sync message

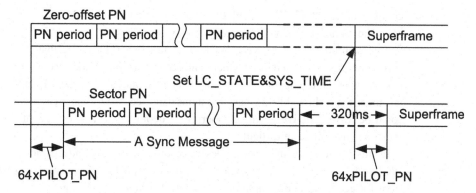

Figure 3.13 Setting long PN state and system time in a device

PN offset (PILOT_PN), Long-Code State (LC_STATE), and the *system time (SYS_TIME)*. PILOT_PN provides the sector PN offset in units of 64 chips. The receiver sets the states and masks of its short PN generators according to the value of $64 \times$ PILOT_PN. Once it is done, the direct and masked outputs of the PN generators are aligned with the zero-offset PN sequences and the PN sequences with sector-specific delay, respectively. In addition, the receiver sets the long PN generator state and system time according to LC_STATE and SYS_TIME given by the sync message. To allow enough time to set this state, these values given by the message become valid 320 ms in the future. Specifically, at 320 ms minus $64 \times$ PILOT_PN chips after the end of the sync superframe that contains the sync message, the long PN generator's state is set to the value of LC_STATE. The long PN state setting time is aligned to the start of a zero-offset PN superframe. The device's system time is also set to SYS_TIME at the same time. The timing relations for setting these values are shown in Figure 3.13.

At this time, the sync channel acquisition is complete. The wireless device is ready to receive paging and other control channel messages from the network and to send information and requests to the network over the access channels.

3.5.2 WCDMA (UMTS)

Following the success of the IS-95 CDMA mobile communication system standardized in North America, NTT-DoCoMo started to develop a wideband version of the DS-CDMA cellular system, called WCDMA, in Association of Radio Industries and Businesses in Japan. The intention was to take the advantages of the CDMA technology demonstrated by IS-95 while addressing some of its shortcomings to develop a better mobile communication system. This effort paralleled a similar effort by European wireless communication companies including Ericsson and Nokia. In the late 1990s ETSI adopted WCDMA as its third-generation mobile communication standard. This effort was furthered by the 3GPP partnership of standard organizations with ARIB and ETSI as members. WCDMA became a part of 3GPP's UMTS standards. Soon afterwards, UMTS was accepted by the international standard organization, the ITU, as a component of its IMT-2000 standard. As of today, WCDMA is the most widely adopted 3G standard and is often used synonymously with UMTS.

The main differences between WCDMA and cdma2000–1x are that WCDMA is an asynchronous system and has a wider bandwidth. Due to its wider bandwidth, WCDMA has the potential to support a higher data rate and can be more flexible in allocating its resources than cdma2000–1x-based systems can. Because it is asynchronous, WCDMA could be less restrictive when considering its deployment. However, it also makes the initial acquisition more difficult.

In this section, we first briefly describe some properties and characteristics of the WCDMA system that are relevant to its initial acquisition. Then we will describe the steps for acquiring and completing a WCDMA system acquisition.[2]

3.5.2.1   Basic Properties and Characteristics of WCDMA Forward Link

The nominal bandwidth of a WCDMA system is 3.84 MHz. The carrier spacing can be up to 5 MHz in order to leave some gap as the guardbands between the signals of adjacent carriers. The system's channel modulation symbol rate, i.e., its chip rate, is equal to 3.84 Mchips/second.

The data transmitted from a cell on a WCDMA downlink are organized into radio frames. They are 10 ms long with 38400 chips in each frame. A frame is evenly divided into 15 data slots. Each slot consists of 2560 chips with a time span of approximately 666.67 μs. According to the WCDMA standard, every transmission unit is called a cell, which is equivalent to a sector in the cdma2000–1x family of standards.

WCDMA is a DS-CDMA system, i.e., its data symbols are spread before transmission. Similar to cdma2000, the data symbols are first spread in time using *channelization codes*, which are the same as the Walsh-Hadamard codes in cdma2000, for eliminating or reducing the interference between channels. WCDMA uses variable data symbol rate spreading. The channelization codes have different lengths from 4 to 256 chips for spreading data with different rates. In addition, a larger spreading factor can be achieved by symbol repetition. After spreading, the output symbol rate is equal to the chip rate, i.e., 3.84 MS/s (million samples per second).

This spreading step is called *time spreading* because it only increases the data symbol rate in the time domain. In the frequency domain, the Walsh codes are tones with harmonics. Thus, spreading in the frequency domain is needed to make the spread signals white noise-like. It is achieved by scrambling the chips using scrambling codes, which have white spectrums.

When transmitting the primary common pilot channel (P-CPICH) and many other channels, including various data, paging, and control channels, a common *primary scrambling code* spreads channelized data after they are added together. Since different channelization codes are orthogonal to each other, the interference among these channels is eliminated or greatly reduced. There are also other scrambling codes used in WCDMA for other channels. The synchronization channels (SCHs) are scrambled by

---

[2] The WCDMA forward link can operate in transmit diversity or nontransmit diversity modes. To simplify description, only operations in the nontransmit diversity mode are discussed in this section.

their own codes. We will consider only the prime scrambling codes and the scrambling codes for the synchronization channels in this section.

All data and pilot channel scrambling codes are complex valued. They are constructed based on two real-valued m-sequences with the length of $2^{18}-1$. The construction procedures of the scrambling codes are described in [16]. The scrambling codes for the synchronization channels are constructed differently.

There are 8192 data and pilot channel scrambling codes defined in the WCDMA specification. Each of the scrambling codes is 38,400 chips long and used to spread the transmitted data of every radio frame. These scrambling codes are divided into 512 sets, each of which consists of a primary scrambling code and 15 secondary scrambling codes. Thus, there are 512 primary scrambling codes, which are further organized into 64 scrambling code groups. Each code group consists of eight primary scrambling codes. Each cell uses only one primary scrambling code in a code group.

Besides the P-CPICH, the channels pertinent to the WCDMA initial acquisition are the primary synchronization channel (P-SCH) and the secondary synchronization channel (S-SCH).

The P-SCH sends a primary synchronization code (PSC) every slot. It is 256 chips long and located at the beginning of the slot. The PSC is a 256-bit BPSK sequence with good aperiodic correlation properties. It is rotated by 45 degrees to become a complex sequence. Every cell in the network sends the same PSC.

The S-SCH in a cell sends an S-SCH sequence in every frame. Each of the S-SCH sequences consists of 15 code words selected from a set of 16 secondary synchronization codes (SSCs), denoted by $\{C_{ssc,i}\}$, $i = 1, \ldots, 15$. The SSCs are also 256 chips long, located at the beginning of every slot and sent sequentially. The sequences of 15 SSC codes sent by S-SCHs in different cells are distinct, i.e., they differ in order and/or contain different SSCs. Every S-SCH sequence is unique under the cyclic shifts of the SSCs.

The SSCs, similar to the PSC, are 256-bit BPSK sequences rotated by 45 degrees to become complex sequences. They are uncorrelated with each other and with the PSC.

There are 64 different SSC sequences, which correspond to 64 scrambling code groups. The locations of the PSC and SSCs in a WCDMA frame and slots are shown in Figure 3.14.

Details of PSC and SSCs and their constructions are described in [16]. The initial acquisition in WCDMA is carried out in three steps as described below.

3.5.2.2   Acquisition of Primary Sync Channel

The first step of initial acquisition is to acquire P-SCH. Because different cells transmit the same PSC, and it is in all of the slots, the receiver only needs to search for the unique PSC regardless of the cell where it is located. This simplifies the hypothesis testing process. At this stage, however, the receiver may not have any prior information about the locations of the slot boundaries. It searches all the possible offsets within a slot. Given that a slot consists of 2560 chips, the searcher needs to perform up to 5120 hypothesis tests to cover one slot for samples generated every $T_c/2$. On average, the number of tests needed to cover a slot is equal to 2560.

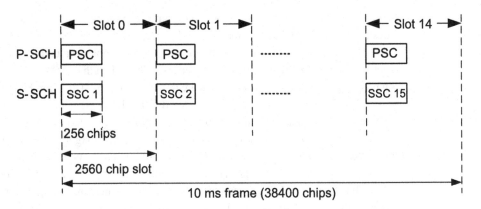

Figure 3.14 P-SCH, S-SCH locations in a WCDMA frame

The process of hypothesis testing is the same as that described in Section 3.3. The length of PSC used as the reference sequence in the correlation is equal to 256 chips. Thus, the maximum coherent integration length is 256 samples, and it is doable if the frequency offset is low. As analyzed in Section 3.4.4, performing such integration is possible if the accuracy of the XO is less than 2 ppm for 1.9 GHz frequency band. If the frequency offset is much higher than that, the coherent integration length should be reduced. In the latter case, multiple detected symbols can be generated from a single PSC. These detected symbols can be noncoherently combined to generate a composite decision metric. Moreover, additional detected symbols can be generated from the segments of the samples separated by one or multiple 2560 chips. Such detected symbols with correlation peaks can be combined noncoherently for additional H_1 hypothesis test verifications. The timing of the correlation peak can be used to compute the path location for the initialization of a RAKE finger.

Once a correlation peak is found and verified as indeed yielding a valid H_1 hypothesis, it is determined that a PSC exists at the time position of the sample sequence involved in the correlation. The timing of the first chip of the sequence that yields the detection peak marks the start of the slot. The timing of the first chip in the next slot is 2560 chips later, and so forth.

At this point, the boundaries of the slots are determined. Since the SSCs of S-SCH are also located at the beginnings of the slots, for S-SCH acquisition, SSCs can be detected by simply using the sample sequence that yields the valid H_1 hypothesis in P-SCH detection.

If a large frequency offset exists, multiple coherent integrations can be performed on the segments of that 256-chip sample sequence that was used for PSC detection. Let us assume that the 256 samples of the sequence are divided into four segments. These four segments of the PSC can generate four detected symbols d_i, $i = 0, 1, 2$, and 3. As discussed in Section 3.4.4, these detected symbols can be used to compute an estimate of the frequency offset. After the H_1 hypothesis is confirmed, an estimate of the frequency offset based on the four detected symbols can be computed by applying (3.48), such that

$$\hat{f}_{offset} = \frac{1}{3}\sum_{i=0}^{2} \frac{\text{Im}\left[d_i^* d_{i+1}\right]}{2\pi|d_i|^2 64T_c}$$ (3.51)

The receiver can use this estimate to correct the frequency offset due to the inaccurate XO and significantly reduce the residual frequency offset. As a result, the integration for SSC detection could fully utilize the 256-sample sequence and the performance of S-SCH detection can be improved.

3.5.2.3 Acquisition of Secondary Synchronization Channel

Once P-SCH is acquired, the receiver of the user equipment (UE) knows the slot boundaries of the forward-link channels.[3] However, it is still necessary to determine their frame boundaries. This is done by the detection of the S-SCH's code words in frames, the second step of initial acquisition.

As specified in [16], in every frame, each Node-B transmits a cell specific sequence that consists of 15 SSCs chosen from 64 SSC sequences.[4] Each of the 15 SSCs in an SSC sequence belongs to a set of 16 SSCs. Each SSC is transmitted in the same slot that contains a PSC. The objective of S-SCH acquisition is to determine which sequence of SSCs is transmitted.

Considering the ith slot, the receiver correlates all of the 16 SSCs with the samples sequence that generated the largest PSC correlation peak during P-SCH detection to generate 16 correlation outputs. The SSC produces the largest correlation magnitude among the 16 correlations and greater than a predefined threshold is declared as the SSC of the ith slot, denoted as $SSC^{(i)}$. The SSCs so estimated in 15 consecutive slots constitute an estimated SSC sequence, up to a cyclic shift.

To remove the shift ambiguity, the 15 possible cyclic shifts of the estimated SSC sequence are compared to all of the 64 valid SSC sequences. By design, there can be only one match because the 64 SSC sequences are uniquely defined under cyclic shifts. Once such a valid SSC sequence is found, and the first SSC in the valid sequence is equal to $SSC^{(k)}$, the UE receiver determines that the frame starts from the kth slot occurred during the detection. More sophisticated detection methods can also be employed to improve the S-SCH detection performance [17, 18].

As specified by the WCDMA standard, each SSC sequence corresponds to one of the 64 primary scrambling code groups, each of which consists of eight primary scrambling codes. Once the UE knows the cell's code group, it needs to determine which one of the eight primary scrambling sequences is used by the cell. The determination is achieved by the P-CPICH acquisition.

3.5.2.4 Acquisition of Primary Common Pilot Channel

Recall that the forward channels are organized as 10 ms frames, each of which consists of 38,400 chips. A cell's primary scrambling sequence is also 38,400 chips long. It scrambles the data in each frame and reused in every frame. Since P-CPICH is

---

[3] In 3GPP terminology, a UE refers to a device that is used by an end user to communicate.

[4] In 3GPP terminology, a Node-B is equivalent to a base station (BTS), in other cellular networks.

unmodulated, its waveform is the same as the primary scrambling sequence of the cell. Acquiring P-CPICH is equivalent to detecting the primary scrambling sequence. From the S-SCH search, the UE knows the scrambling sequence group of the cell. However, given that each group contains eight primary scrambling sequences, it is necessary to determine which one of these eight sequences is used by this cell.

Similar to the S-SCH determination, the primary scrambling sequence of the cell can be determined by correlating the eight candidates with the sample sequence and selecting the largest peak. However, as the scrambling sequences are quite long, it is neither necessary nor possible to correlate the entire sequence to a 38,400 long sample sequences. In practice, it is sufficient to correlate a shorter segment of the scrambling sequence with the corresponding sample sequence. Specifically, it is convenient to use the first 256 chips of the primary scrambling sequence to correlate with the segment of the samples of the first slot that yielded successful P-SCH and S-SCH detections.

The segment of a primary scrambling sequence that produces the largest correlation peak that is also greater than a predefined threshold is determined as the sequence used by the cell. Further verification may be performed to increase the confidence level of the decision.

If everything is verified successfully, the UE knows all the information it needs to communicate with the acquired Node-B. Specifically, it would first decode the *primary common control physical channel* (P-CCPCH) to obtain system information before establishing a dedicated connection with the Node-B.

3.6   Initial Acquisition in OFDM Systems

In this section, we look into the initial acquisition processes of the LTE WAN and the 802.11a/g Wi-Fi LAN systems.

3.6.1   LTE Initial Acquisition

Long-Term Evolution is the fourth generation (4G) of wireless wide area network standards. Hence, it is also called *4G LTE*. Whereas 3G systems mainly employ CDMA technology, the key technology of 4G systems is OFDM. LTE is a packet data system by design. It can support downlink data rates up to 300 Mbps, or even higher, if multiple wireless carriers are aggregated together. In contrast, 3G systems, such as 3G HSPA+, are limited to a downlink data rate of up to around 168 Mbps unless more complex multicarrier system designs are employed. Note that these data rates are their peak rates. The average throughput is significantly lower when averaged over different channel conditions in actual operations.

In this section, we will mainly describe the PHY layer aspects of LTE initial acquisition. Since both WCDMA and LTE were developed in 3GPP, it is not surprising that they have many aspects in common including their initial acquisition procedures.

To facilitate discussion, we will first describe the fundamental characteristics of the LTE physical layer with emphases on its channels and on its signaling related to initial acquisition. A detailed discussion of the initial acquisition process including implementation considerations will follow.

It should be noted that LTE has a *frequency division duplex* (FDD) mode and a *time division duplex* (TDD) mode. We will concentrate our discussion on the FDD mode, but most of the discussion is also applicable to the TDD mode.

3.6.1.1 A Brief Description of the FDD LTE Downlink Physical Layer

To be flexible in various deployment scenarios for supporting different data rates, as specified in the LTE standard [19], LTE supports six transmission signal bandwidths: 1.4, 3, 5, 10, 15, and 20 MHz. To simplify system and receiver designs, the OFDM signals for all of these bandwidths have the same subcarrier spacing, which is equal to 15 kHz. Thus, the OFDM symbol length without the cyclic prefix (CP) is always equal to 1/15 kHz $\approx$ 66.67 μs.[5] The modulation symbols on the subcarriers could be QPSK, 16QAM, and 64QAM.[6] The frequency-domain parameters of LTE are summarized in Table 3.1.

The FFT/DFT sizes listed in Table 3.1 are typically used in implementation to generate the signal at a certain bandwidth. It is possible, sometimes even desirable, to use a larger or smaller FFT/DFT size by using a different number of guard subcarriers with zero modulation symbols. For example, the LTE signal at a bandwidth of 1.4 MHz is usually generated using a 128 iFFT with 55 guard subcarriers, i.e., 28 and 27 guard carriers on the negative and positive frequency sides, respectively. Alternatively, it can also be generated by performing a 256 iFFT with 183 guard carriers, 92 and 91 on the two sides. In LTE terminology, the reciprocal of the OFDM channel sampling rate, i.e., $1/f_s$, is defined as the *standard time unit*, denoted by T_s, which is the time duration of an iFFT output sample, i.e., an *OFDM channel sample*.

Due to the guard carriers in OFDM signal, the actual signal bandwidth is narrower than f_s, the sampling frequency, or sampling rate. Therefore, the sampling rate satisfies the Nyquist sampling theorem, and no sub-T_s sampling is necessary for initial acquisition and other receiver functions.

In the time domain, LTE data are organized as 10 ms radio frames, or simply frames. A frame is further divided into 10 subframes, each of which is 1 ms long. A subframe consists of two 0.5 ms slots. As specified in [19], two types of CPs may be used: normal CP (5.21/4.61 μs) and extended CP (16.67 μs). Each slot contains seven OFDM symbols with normal CPs and six OFDM symbols with extended CPs. In the case of the normal CP, the number of standard time units T_s in a slot is not divisible by seven, the number of OFDM symbols in a slot. As a result, in a slot with seven OFDM symbols, the first OFDM symbol has a longer CP, which is equal to 10/128 of the main portion of an OFDM symbol, or 5.21 μs. The remaining six symbols have shorter CPs,

---

[5] These parameters are for the data transmission mode. In the broadcasting mode, i.e., MBSFN, the subcarrier spacing and symbol length with no CP are 7.5 kHz and 133.33 μs, respectively.

[6] Since LTE release 12, 256QAM is also supported.

Table 3.1 LTE Frequency-Domain Parameters

Bandwidth (MHz)	1.4	3	5	10	15	20
Active subcarriers	73	181	301	601	901	1201
Typical FFT/DFT size	128	256	512	1024	1356	2048
Sampling rate (f_s)	1.92	3.84	7.68	15.36	23.04	30.72

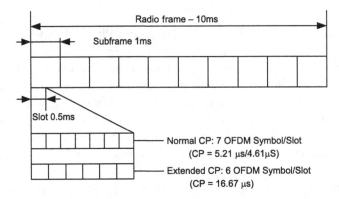

Figure 3.15 LTE time-domain frame structure

each of which is equal to 9/128 of the main portion, or 4.61 μs. The length of the extended CP is equal to ¼ of the main portion of the OFDM symbol. The structure of a time-domain LTE frame is shown in Figure 3.15.

To improve the granularity, the data for transmission are divided into resource blocks and resource elements. A resource block, which is the smallest unit that can be allocated to a user, consists of 12 OFDM subcarriers, which span 180 kHz in frequency and one slot (0.5 ms) in time. One subcarrier in an OFDM symbol is called a *resource element*.

3.6.1.2   Synchronization Signals

The LTE standard defines two synchronization signals: primary synchronization signal and secondary synchronization signal. They are used in initial acquisition and for acquiring the signals from neighboring cells. To facilitate initial acquisition by any UE receiver, the synchronization signals occupy the narrowest available channel bandwidth of LTE, i.e., 1.4 MHz with 72 subcarriers, among which 62 are nonzero data subcarriers. For wider bandwidth transmission, the synchronization signals occupy the center 73 subcarriers including a zero subcarrier at the zero frequency.

The synchronization signals are transmitted every 5 ms in the 0th and 10th slots of every frame. Each transmission of synchronization signals includes a PSS and an SSS, each of which occupies the 72 or 73 center subcarriers in an OFDM symbol. The PSS is in the last OFDM symbol of the slot that contains synchronization signals. The SSS is in the OFDM symbol prior to the one containing the PSS.

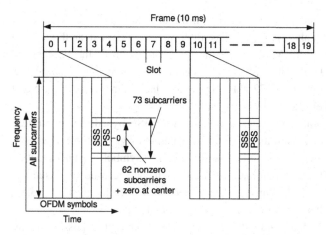

Figure 3.16 Synchronization signals in an FDD LTE frame

The locations of the PSS and SSS in an OFDM signal frame that has bandwidth wider than 1.4 MHz are shown in Figure 3.16.

Similar to WCDMA, the LTE initial detection is a multistep process. A UE receiver first performs a frequency scan to identify the right bandwidth and center frequency of the available LTE signals. It then detects the PSS to acquire the half-frame boundary and determines the *physical-layer cell-identity* $N_{ID}^{(2)}$ (0 to 2). In the third step, it detects the SSS and determines the *physical-layer cell-identity group number* $N_{ID}^{(1)}$ (0 to 167).

3.6.1.3   Primary Synchronization Signal and Its Detection

The PSS occupies 73 center subcarriers of the last OFDM symbol in the 0th and 10th slots of every frame. Among these subcarriers, the middle one at zero frequency has zero value. The 31 subcarriers on either side of the zero frequency are nonzero and consist of two 31 long segments of Zadoff–Chu sequences. As discussed in Section 3.4.1, after converting to the time domain, the Zadoff–Chu sequences have constant amplitudes and thus a low PAPR. In addition, their autocorrelations are impulses. The PSS is a slightly altered Zadoff–Chu sequence, but it still has low PAPR with an approximate impulse auto-correlation function.

Three PSSs are used to identify three $N_{ID}^{(2)}$ (cell id) values. They are given by [19]:

$$
d_u(n) = \begin{cases} e^{-j\frac{\pi u n(n+1)}{63}} & n = 0, 1, ..., 30 \\ e^{-j\frac{\pi u(n+1)(n+2)}{63}} & n = 31, 32, ..., 61 \end{cases} \tag{3.52}
$$

It can be seen from (3.52) that $d_u(n)$ is actually a 63 long Zadoff–Chu sequence with the center element set to zero. The values of u in (3.52) have a one-to-one correspondence to $N_{ID}^{(2)}$. Specifically, if $N_{ID}^{(2)}$ equals to 0, 1, or 2, u equals 25, 29, or 34, respectively.

Even though the PSS is defined in the frequency domain, when the CIR is unknown its detection is preferably done in time domain due to implementation complexity and performance considerations. To perform the detection of the PSS in the time domain, the received signals, after anti-alias filtering, are sampled at the 1.4-MHz channel sampling-rate of 1.92 MS/s. Since the data bandwidth of the PSS is within ± 472.5 kHz with 75 kHz guardbands on both sides, the anti-alias filter should have a passband wider than or equal to ± 472.5kHz and a stop-band narrower than or equal to ± 547.5 kHz. Alternatively, the 1.92 MS/s samples can be generated from digital samples at a higher sampling rate after down sampled to 1.92MS/s by a decimation filter with similar characteristics as the anti-alias filter. The generated signal sample sequences are correlated with the three possible time-domain reference symbol sequences that correspond to $N_{ID}^{(2)}$ (0 to 2). These time-domain reference sequences can be generated by performing 128-point iFFTs of the three PSSs in frequency-domain given by (3.52) with zero appended to either side to become 128 symbols long.

Because the PSSs' bandwidth is equal to 945 kHz, it is possible to use length-64 correlations for PSS detection with the received signal further down-sampled to 0.96 MS/s. The reference symbol sequences are generated by taking 64-point iFFTs of the PSSs given by (3.52) with one zero added to the negative frequency side. Using such a low sampling rate may result in some performance degradation due to noise aliasing, especially under low SNR conditions.

The detection hypothesis tests are performed for every one-sample-shift of the sample sequences until a success is found. However, if rough OFDM symbol timing is determined by using the CP correlation method, which will be discussed in Chapter 6, the number of searches can be significantly reduced.

The squared magnitudes of the correlation peaks, i.e., the detected symbols, are compared to a predetermined threshold to make a decision. The process is essentially the same as that described previously in Sections 3.2 and 3.3 and can be analyzed similarly. As the channel is likely to be multipath, it would be advantageous to make the decision based on the sum of the squared magnitudes of the large correlation peaks that are close to each other. The correlation peaks that yield the final successful decision are the estimates of the multipath channel taps as discussed in Section 3.2.3.

For the detection procedure discussed above, the signal's coherence time should be longer than 67 μs. Otherwise, there is an integration loss. For example, as discussed in Section 3.4.4, if there is a 2 ppm XO frequency error, the frequency offset is 5200 Hz at 2.6-GHz carrier frequency. There would be a loss of 1.8 dB for a full correlation of 128 samples, which have a time span of 67 μs.

Detection that is based on the correlation method using the entire OFDM symbol with a time span of 67 μs would not work well for a large frequency offset. For example, if the XO error is 4 ppm, there would be a loss of more than 6 dB as shown in Figure 3.4. If such a performance degradation is not acceptable, additional remedies would be necessary. For the time-domain correlation method, it is possible to perform multiple correlations, each of which uses a fraction of the reference symbol and the received signal sample sequences. The correlation results can be noncoherently combined as discussed in Section 3.3.2. Possible issues with this approach are noncoherent integration loss and nonideal aperiodic correlation properties of the time-domain symbol sequences of the PSS.

Another method applicable in the case of large frequency offset is to divide the entire range of the frequency offset into multiple frequency bins. The frequency offset in each bin is corrected by analog or digital means as discussed in Section 3.4.4.2. Hypothesis tests are performed on the frequency-corrected data sample sequences for each bin. This frequency-binning method works with no performance loss. Its main caveat is the increased computation complexity of detection, which is proportional to the number of bins.

Once the PSS detection is successful, the half-frame boundary and the ID $N_{ID}^{(2)}$ are known to the UE. The UE's receiver can continue to perform SSS detection. It is advisable to correct the frequency offset as much as possible before other receiver operations. The frequency offset estimation and compensation methods discussed in Section 3.4.4.2 are applicable to any of the PSS detection methods described above.

3.6.1.4   Secondary Synchronization Signal and Its Detection

The SSS is in the same slot and occupies the same subcarriers as the PSS in the OFDM symbol that is the one before the symbol containing the PSS. The 62 nonzero subcarrier symbols are constructed from two length-31 m-sequences, each of which has a different cyclic shift depending on the physical-layer cell ID group number $N_{ID}^{(1)}$. They are scrambled by sequences defined according to the physical-layer cell ID number $N_{ID}^{(2)}$.

There are 168 such SSS OFDM symbols ($N_{ID}^{(1)} = 0, 1, ..., 167$) with combinations of different shifts constructed based on two 31 long m-sequences, $\tilde{s}$ and $\tilde{z}$. The elements of the SSS symbols are scrambled according to $N_{ID}^{(2)}$ with the scramble sequence constructed based on another 31 long m-sequence $\tilde{c}$. The generators' characteristic polynomials of the m-sequences $\tilde{s}$, $\tilde{z}$, and $\tilde{c}$ are $x^5 + x^2 + 1$, $x^5 + x^4 + x^2 + x + 1$ and $x^5 + x^3 + 1$, respectively. Overall, a total of 504 cell IDs is given by $N_{ID}^{cell} = 3N_{ID}^{(1)} + N_{ID}^{(2)}$. The SSSs in the slot 0 and 10 in a cell are constructed based on the same $N_{ID}^{(1)}$ and $N_{ID}^{(2)}$ but interleaved differently. The details of constructing the SSSs are documented in [19].

After the PSS is detected, the UE knows the ID $N_{ID}^{(2)}$ of the cell and, thus, the scrambling codes. The detection of SSS needs to determine which one out of the 168 possible SSS sequences matches the received signal sample sequences at the time position derived from the location of the detected PSS. Since the SSS sequences in the 0th and 10th slots are different, it would be sufficient to detect one SSS sequence per cell ID per frame. The detection can be done either in the time domain or in the frequency domain. The procedure of detection by correlation in the time domain is similar to PSS detection. However, it could be more efficient computationally to perform detections in the frequency domain than the time domain and it is probably preferable. This approach is described below.

Even though the UE has learned the location of the PSS OFDM symbol, there are still two possibilities of the SSS OFDM symbol's locations. This is because there are two types of CPs, the normal CP and the extended CP, as shown in Figure 3.17. After the PSS is detected, the UE must determine the type of CPs in the LTE signal for SSS detection by using either the time-domain or the frequency-domain method.

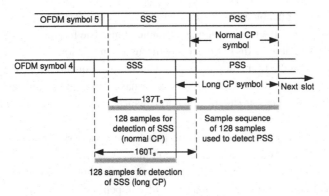

Figure 3.17 Possible SSS symbol positions

As shown in the figure, for normal and extended CPs, the sequences for detection are 137 and 160 samples, respectively, ahead of the sequence that was used in the successful detection of the PSS. The brute force method is to test the 128 sequences at both locations. However, that would increase the complexity of the detection by a factor of two. Another method is to perform two correlations of the samples at the two possible CP locations with the corresponding samples of the last part of the sequence used for PSS detection. The magnitudes of the correlations should be normalized by the correlation length. The one with greater value indicates that it is generated by the assumed type of CP in the LTE signal [20].

To perform SSS detection in the frequency domain, the samples in the FFT window at the expected SSS location, as shown in Figure 3.17, are converted to the frequency domain by a 128-point FFT. The elements of the FFT output are multiplied by the corresponding expected SSS modulation symbols and then coherently combined to yield a detected symbol for hypothesis testing. To perform coherent combining, the phases of the FFT output elements must be corrected, so that they are aligned with each other. This requirement can be addressed by using the results obtained from the PSS detection process.

For SSS detection, the FFT window of samples should be ahead of the sample sequence that generates the first large correlation output in PSS detection. They are separated by the number of samples equal to the CP length. The case for the normal CP is shown in Figure 3.18.

Using the FFT window placement described above and assuming that jth SSS is transmitted, we can express the kth element, $u_S(k)$, $0 < |k| \leq 31$, of the FFT output for SSS detection as

$$u_S(k) = H_S(k)S_j(k) + z(k) \tag{3.53}$$

where $H_S(k)$ is the kth subcarrier coefficient; $S_j(k)$ is the symbol of the transmitted SSS with index j modulated on the kth subcarrier; and $z(k)$ is the additive noise in $u_S(k)$. The detected symbol with coherent integration for the hypothesis testing of the ith SSS that is transmitted has the form

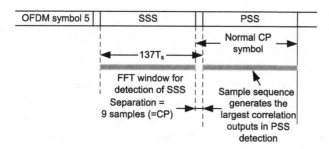

Figure 3.18 FFT window placement for SSS detection (normal CP)

$$d_{S,i} = \sum_{0<|k|\leq31} S_i^*(k)\widehat{H}_S^*(k)u_S(k) \tag{3.54}$$

where $\hat{H}_S(k)$ is the estimated kth subcarrier coefficient. The operation performed by (3.54) is repeated for all possible S_i's. If a detected symbol $d_{S,i}$ has the largest magnitude among all $d_{S,i}$'s generated in the test, S_i is determined as the SSS sent in this slot.

The estimates of the subcarrier coefficients are computed from the results of the PSS detection as follows. The sample sequences used in PSS and SSS processing are separated by the CP length in the FDD case. By taking the FFT of the sample sequence for the PSS processing, the kth output element can be expressed as

$$Y_P(k) = H_P(k)P_l(k) + z(k) \tag{3.55}$$

where $P_l(k)$ is the symbol of the transmitted PSS with index l modulated on the kth subcarrier

As PSS has been detected so that $P_l(k)$ is known, the estimate of $H_P(k)$ can be computed by

$$\hat{H}_P(k) = P_l^*(k)Y_P(k)/|P_l(k)|^2 = P_l^*(k)Y_P(k) \tag{3.56}$$

for $|P_j(k)| = 1$. If the PSS and SSS are adjacent to each other as in the FDD case, we can assume that $\hat{H}_S(k) = \hat{H}_P(k)$. This direct method to compute $H_P(k)$ is straightforward and computationally efficient. However, it is possible to reduce the estimation error by additional processing in the time domain and converting it back to the frequency domain after some processing as shown in Section 5.5.1.2.

There is yet another way to perform SSS detection. This method uses the differential correlation between pairs of adjacent FFT output bins and adjacent reference SSS elements to perform noncoherent detection that is described in [20]. This method should be quite robust, but it also introduces additional noise. As a result, the SNR of the detected symbols is reduced by 3 dB or more.

All of these methods should work in principle. Their performances may vary depending on the various operating environments. Evaluation by simulation under various conditions can help to select the proper approach.

Because the cross-correlation properties between SSS sequences are not ideal, additional verification steps may be needed to reduce the false alarm probability. A detailed description of such detections of SSS can be found in many references, e.g., [20, 21].

In the discussion above, our main concern is the initial frequency offset due to local oscillator frequency error. Doppler frequency may also introduce additional errors. However, except in special cases, the frequency error due to Doppler is relatively low. For example, the vehicle speed of 350 kilometers per hour, relative to light speed, is about 0.35 ppm. It would affect the demodulation performance but may not cause a serious impact on initial acquisition.

After both PSS and SSS are detected, the UE starts to detect the physical broadcast channel (PBCH). Similar to PSS and SSS, PBCH has a bandwidth of 1.4 MHz. Hence, the UE can decode the information without knowing the transmission bandwidth from the cell identified by PSS and SSS. PBCH broadcasts a few essential system parameters for the initial access of the cell. Finally, the UE performs detection of the System Information Block (SIB) that provides further information such as TDD DL/UL configuration. Once the UE has learned these parameters, all of its loops can run normally, and it can establish communication with the cell and the entire network. At this point, the LTE initial system acquisition process is completed.

3.6.2   802.11a/g Wi-Fi System: Frame and System Acquisitions

In a Wi-Fi system, an *access point*, the equivalent of a base station, Node-B or eNB in cellular systems, sends data in individual data packets, or frames. A Wi-Fi device, called a *station* (STA) in 802.11 terminology, needs to acquire the signal every time it receives a data frame. Thus, there is no significant difference between conducting initial acquisition and receiving a regular data frame from the PHY point of view. In this section, we describe the operation procedure of frame acquisition in 802.11 OFDM Wi-Fi systems.

There have been a few generations of 802.11 OFDM Wi-Fi standard evolutions since 802.11a/g were defined. However, the basic frame acquisition remains the same. There are some changes to support wider bandwidth and higher data rate. Nonetheless, the implementation of 802.11a/g signaling for frame acquisition, called legacy mode, still exists in all newer standards such as 802.11n and 802.11ac to support backward compatibility. We will use 802.11a/g as an example to offer an overview of the OFDM Wi-Fi systems and to describe their frame acquisition process.

3.6.2.1   Data Frame and Preamble

As specified in 802.11 standard [22], 802.11a/g can be implemented in three channel bandwidths of 20MHz, 10MHz, and 5MHz. Their transmitter channel sample rates are equal to the corresponding bandwidth when implemented by using 64-point FFTs. Each OFDM symbol has 52 active subcarriers. Among them, four are used as pilots and

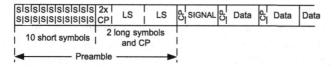

Figure 3.19 PPDU frame structure

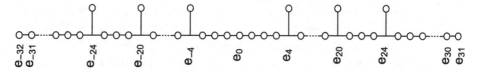

Figure 3.20 Data vector for generating short preamble sequences

48 are for data transmission. The length of its CP, called *guard interval* in 802.11 terminology, is a quarter of the length of the main portion of an OFDM symbol. Each OFDM symbol that includes a CP consists of 80 (64+16) channel samples. The time spans of OFDM symbols are 4 μs, 8 μs, and 16 μs for 20, 10, and 5 MHz channels, respectively.

In 802.11 systems, the upper-layer data frames are converted to PHY data frames by the *Physical Layer Convergence Procedure* (PLCP). Such data packets transmitted over the PHY layer are called *PLCP protocol data units* (PPDUs). A typical 802.11a/g legacy mode PPDU frame is shown in Figure 3.19.

As can be seen from the figure, an 802.11 PPDU contains a preamble portion and a data portion. The preamble portion consists of 10 *short preamble symbols* and 2 *long preamble symbols* with a CP, which is twice as long as a CP of a regular OFDM symbol. Following the preamble is a SIGNAL OFDM symbol that is followed by data OFDM symbols.

The short preamble symbol sequence transmitted at the beginning of a PPDU is generated by performing a 64-point iFFT of a preassembled data vector. This data vector has 64 elements, among which only 12 are not zeros. These nonzero elements are in the BPSK form with random +1 or −1 values and rotated by 45 degrees. There are three zeros between any two nonzero elements including a zero in the middle, i.e., at zero frequency. In other words, among the 64 elements $\{e_i\}$, $i = -32 \ldots 31$, only the elements with indices $4k$, $k = \pm1, \ldots, \pm6$ are not zero elements (Figure 3.20). The iFFT output generated from such an input data vector is periodic with four periods. A period of the iFFT output is used as a short preamble symbol that consists of 16 samples. The 160 samples long short preamble contains 10 of such short preamble symbols, i.e., it is periodic with 10 periods and spans 8 μs for signals with 20 MHz bandwidth.

Following the short preamble is a long preamble. The long preamble contains two long preamble symbols. The long preamble symbol is constructed by performing a 64-point iFFT of data vector with 64 elements, among which 52 are not zeros. These nonzero elements are the symbols modulated on the OFDM subcarriers with indices from -26 to -1 and 1 to 26. The elements at zero frequency and on both sides all have zero values. The 64 samples at the output of iFFT are repeated once to form the long preamble after prepending a CP of 32 samples, which is equal to the last 32 samples of the long preamble symbol. The total length of the long preamble is also 160 samples long.

As described above, the elements at both ends of the 64 long data vectors are zeros for both short and long preamble generation. These zero elements are also called guard symbols to form the guard carriers of the signals in OFDM transmission.

Next to the long preamble is an OFDM symbol named SIGNAL. It contains, in addition to other information, a Rate field and a Length field. The Rate field informs the STA about the data rate, which can be from 6 to 54 Mbps for the 20 MHz channel, and the Length field indicates the amount of data in octets in the PPDU received from the MAC layer.

The detailed construction of the preambles and SIGNAL symbol can be found in [22].

3.6.2.2   Short Preamble Processing for Frame Acquisition

To acquire a data frame, an STA receiver first searches for the short preamble of the PPDU. Because the short preamble is periodic with a period of 16 samples, without frequency offset and ignoring the additive noise, the samples of the received signal that contains the short preamble are also periodic. With a frequency offset introduced during transmission, two of such samples separated by 15 other samples have a phase difference between them due to the frequency offset. It is easiest to detect such a periodic signal with frequency offset by the *correlation-and-accumulation* method originally proposed in [23] and described in [24].

To implement this approach, the received signals are sampled at the OFDM sample rate. For the signal with 20-MHz channel bandwidth, the sampling rate is 20 MS/s. The samples, denoted by y_k, are stored in a data buffer. The current sample y_k is multiplied by the complex conjugate of the sample received 16 samples earlier, i.e., y^*_{k-16}. The generated products are stored in a delay line. The stored product values in a sliding window with length L, which is usually equal to the number of the symbols in a period of the short preamble, i.e., 16, are accumulated to yield an accumulated value $R(n)$ at the sampling time nT_s, where T_s is the OFDM channel sample interval. Namely,

$$R(n) = \sum_{k=n-15}^{n} y^*_{k-16}y_k \tag{3.57}$$

This operation is often called (delayed) correlation and accumulation.

It can be seen from (3.57) that many samples involved in generating $R(n_1)$ and $R(n_2)$ are the same if n_1 and n_2 are close to each other. Therefore, they are correlated. The $R(n)$s can be down-sampled, e.g., storing one out of every eight of the accumulated values, to save computation and memory, and there is no loss of information.

In addition, the squared magnitudes of the samples y_k are stored in another delay line. Their values are also accumulated in a sliding window with the same length, i.e., 16, to yield an estimate of the signal samples' energy, i.e.,

$$V(n) = \sum_{k=n-15}^{n} |y_k|^2 \approx 16E\left[|y_k|^2\right] \cong 16E_y(n) \tag{3.58}$$

Using the same terminology as in the previous sections, we name the quotient of $R(n)$ divided by $V(n)$, denoted by $d(n)$, the *detected symbol* for short preamble acquisition at nT, i.e.,

$$d(n) = R(n)/V(n) \tag{3.59}$$

Its real part $\text{Re}[d(n)]$ or its squared magnitude $|d(n)|^2$ can be used as the decision metric, denoted by $D(n)$, to compare with a threshold for detecting the existence of the short preamble similar to the general initial acquisition process described earlier. The decision metrics $D(n)$ are stored in a buffer for determining timing, i.e., the position of FFT windows. Similar to $R(n)$, the values of $V(n)$, $d(n)$, and $D(n)$ can all be down-sampled if so desired. Since the short preamble has multiple periods and each $d(n)$ is computed using the samples in one of the periods, multiple detected symbols can be combined into a composite decision metric to result in a more reliable decision.

The performance of short preamble detection can be analyzed in a way similar to how the general initial acquisition procedures were analyzed earlier in this chapter. A brief discussion is given as follows.

The detected symbol, $d(n)$, can be approximately expressed as

$$d(n) = \frac{R(n)}{V(n)} \approx \sum_{k=n-15}^{n} y_{k-16}^* y_k \Big/ 16E_y \cong \frac{1}{16} \sum_{k=n-15}^{n} \bar{y}_{k-16}^* \bar{y}_k \tag{3.60}$$

where $\bar{y}_k = y_k/\sqrt{E_y}$ is the kth normalized received signal sample and $E\left[|\bar{y}_k|^2\right] = 1$.

If no short preamble exists, all of the terms inside summation in (3.60) are zero-mean and independent. Applying the central limit theorem, $d(n)$ can be approximated as a zero mean Gaussian variable with a variance of 1. Under these approximations, $\text{Re}[d(n)]$ has a zero mean normal distribution and $|d(n)|^2$ has a central chi-square distribution with two degrees of freedom.

When the short preamble exists and in the case of a single-path channel with a channel coefficient g_0, the mean of $\bar{y}_k$ can be expressed as

$$\bar{y}_k = g_0 + z_k \tag{3.61}$$

Its SNR is equal to $|g_0|^2/\sigma_z^2$. In addition, we have:

$$\bar{y}_{k-16}^* \bar{y}_k = |g_0|^2 e^{j2\pi 16 T_s f_{offset}} + g_0 z_{k-16}^* + g_0^* z_k + z_{k-16}^* z_k \tag{3.62}$$

In (3.62), z_{k-16} and z_k are the noise components in $\bar{y}_{k-16}$ and $\bar{y}_k$, respectively. If the SNR is relatively high, the variance of $g_0 z_{k-16}^*$ and $g_0^* z_k$ are much larger than the variance of $z_{k-16}^* z_k$. By ignoring the last term in (3.62), the SNR of $\bar{y}_{k-16}^* \bar{y}_k$ in (3.57) is approximately equal to

$$\frac{|g_0|^4}{|g_0|^2 \left(\sigma_z^2 + \sigma_z^2\right)} = |g_0|^2/2\sigma_z^2 \tag{3.63}$$

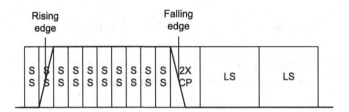

Figure 3.21 The ensemble average of the decision metrics

where σ_z^2 is the variance of $z(k)$. Comparing (3.63) to the SNR of $\bar{y}_k$, we observe that there is a loss of at least 3 dB in SNR due to the delayed correlation.

The decision metrics defined by $D(n) = \text{Re}[d(n)]$ and $D(n) = |d(n)|^2$ are similar to the coherent and noncoherent decision metrics, respectively, for the binary hypothesis tests discussed and analyzed in Sections 2.5.4 and 3.2.3. It can be shown that the decision metric generated by the delayed correlation in (3.57) has an SNR of at least 3 dB lower than the corresponding metric generated by the correlations between the reference symbols and the samples. However, the procedure and analysis presented in Sections 3.2 and 3.3 can be applied to short preamble acquisition with minor modifications.

In addition, during the detection of the 802.11a/g signal, the short preamble sequence is used to determine the coarse timing of the data frame as shown below.

The ensemble average of the decision metrics, $D(n) = |d(n)|^2$ after removing the noise effects, is depicted in Figure 3.21.

As can be seen from the figure, the average of the decision metric starts to rise after the first short symbol of the preamble is received. It starts to fall after the last short symbol has passed. The rising and falling edges can be used as the indicators of the data frame's boundaries. However, the detected positions of the rising and falling edges could have errors from their actual positions of up to about half of the length of a short symbol, i.e., eight samples, due to the noise in the short preamble. Such accuracy is not sufficient for determining the position of the FFT window for demodulation. More accurate FFT window positions can be determined by using the long preamble symbols, as to be shown in the next subsection.

Short preamble symbols can also be used for frequency offset estimation. For the sampling rate of 20 MS/s, the coherent integration time of 16 samples is equal to 0.8 μs. It can be shown that the frequency offset that the estimation based on the short symbols can determine is about 300 kHz, i.e., 125 ppm for 2.4 GHz band or 60 ppm for 5 GHz band. From our analysis in Section 3.4.4, such a frequency offset, if it exists, is too high for effective demodulation of the 802.11a/g signals. From (3.60) and (3.62), a coarse frequency offset estimate can be computed from each decision metric $d(n)$ by $f_{offset} = \text{Arg}[d(n)]/32\pi T_s$. When the angle is small, it can be approximated by $\text{Im}[d(n)]/(|d(n)| \times 32\pi T_s)$. The estimates of frequency offset from multiple $d(n)$'s are averaged to generate a more accurate estimate, which can be used to correct the frequency offset by analog or digital means. The performance of channel estimation and fine timing acquisition with the long preamble, as well as the performance of data demodulation, can be improved if the frequency offset is corrected based on its estimate from the short preamble detection.

Finally, the sample energy estimate $V(n)$ given by (3.58) is proportional to the received signal power. It can be used to set up the AGC gain for ensuring that the sampler is operating at the desired range of the input signal.

In practical implementations, it may be desirable to train the AGC before starting short preamble detection in some cases. It would take two to three short preamble symbols for the AGC to converge to its steady-state value. As a result, only seven to eight periods of the short preamble may be observed and only the falling edge of the decision metric shown in Figure 3.21 can be used for determining the position of the short preamble. At the same time, since the received signal after the AGC has a fixed average magnitude, $V(n)$ is roughly constant. Its value can be determined from the AGC design rather than computed according to (3.58). It is possible to reduce the computation and storage requirements when this AGC-based method is used.

The correlation-and-accumulation method is simple to implement and can provide reasonable performance in an AWGN environment. However, it is quite susceptible to periodic interference that has a period equal or close to the period of the short preamble. One way to mitigate this problem is to perform correlations of the short preamble symbols as the reference symbol sequence with the received signal samples similar to that described in Section 3.2. However, this method would significantly increase the complexity of the detector.

3.6.2.3   Long Preamble Processing for Channel and Timing Estimation

As shown above, an STA uses the short preamble to detect the existence of the 802.11a/g OFDM signal and to obtain the rough timing information about its symbol boundaries. However, to demodulate and decode the data packets, a good channel estimate and more accurate timing information are needed. These tasks can be accomplished by using the long preamble provided in PPDU.

A long preamble-based channel estimate may be generated in the following steps:

1.   Determine an approximate position of an FFT window of 64 received signal samples in the long preamble region determined from the short preamble detection process;

2.   Perform a 64-point FFT of the samples inside the window;

3.   Multiply the complex conjugates of the long preamble subcarriers' modulation symbols to the corresponding FFT output elements to generate a *frequency-domain channel estimate* (FCE);

4.   Perform an iFFT of the FCE generated in step 3 to yield a time-domain CIR estimate and determine the location of the main channel taps; and

5.   Determine the desired FFT window position for data demodulation based on the CIR estimate from step 4.

Let us first look into how to use a 64-point FFT of the long preamble samples to obtain the channel estimate. The approximate position of the long preamble can be determined from the rising edge, the falling edge, or their average, of the decision metrics as shown

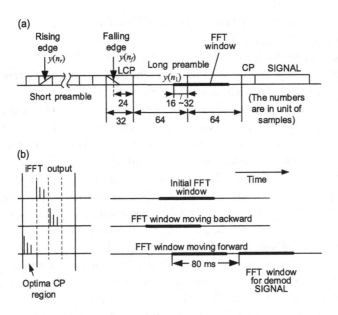

Figure 3.22 Channel estimation timing diagram

in Figure 3.21 and Figure 3.22(a). For a robust operation, the FFT window in step 1 should start from about 16 to 32 samples prior to the expected beginning of the second long preamble symbol. An accurate FFT window position for data demodulation is determined based on the channel estimate by performing steps 2, 3, and 4.

As will be shown in Chapter 6, the iFFT output from steps 2, 3, and 4 is indeed an estimate of the time-domain CIR of the channel. Moreover, in order to perform the proper FFT operations for data demodulation, the coefficients of the channel estimate should be located in the first 16 samples of the iFFT outputs. If the channel estimate from step 4 does not satisfy this condition, its position should be adjusted by changing the FFT window position for channel estimation in step 5. Moving the FFT window's position in one direction moves the estimated channel coefficients in the opposite direction by the same amount, as shown in Figure 3.22(b). For data demodulation, the FFT window adjustment is performed by computing the required amount of the FFT window's movement from the window's initial position and by setting the FFT window's position.

The FFT window for demodulation of the SIGNAL symbol should be placed at 80 samples, i.e., the FFT window length plus the CP, after the FFT window is determined according to the above steps as shown in Figure 3.22(b). Moreover, the estimate of the CIR is converted to the frequency domain to be used in the demodulation and decoding of the SIGNAL and of the data symbols that follow. Once the SIGNAL symbol is decoded and the necessary transmission parameters are known, the receiver is ready to decode the other OFDM data symbols.

In the above discussion, we have considered using only one long preamble symbol for timing and channel estimation. A more accurate channel estimate could be generated by using both of the long preamble symbols when possible.

3.6.2.4 Initial System Acquisition

The required PHY layer operations for initial system acquisition are no different from the regular data frame acquisition as stated above. However, initial system acquisition involves additional MAC and above-layer operations and can be performed in passive or active mode [22].

In the passive mode, an STA searches all available frequency channels for beacon frames. The beacon frames are sent from APs in the infrastructure mode at 10 frames per second and provide data rates, time stamps, possibly also the AP's ID (SSID/BSSID) and other system information. The STA acquires the beacon frame the same way it acquires any other type of frame. Once a beacon frame is acquired, the STA starts the connection procedure with the AP.

In the active mode, an STA sends a probe to the AP requesting beacon frames and waits for the AP to respond.

In both cases, after the STA is able to communicate with the AP, *Authentication and Association procedures* begin. The system acquisition is completed once these two procedures are successfully concluded.

3.7 Summary and Remarks

Initial acquisition is the first step of establishing a communications link between a transmitter and a receiver. Its performance directly affects the reliability and the general performance of the communication system. Initial acquisition starts with a PHY layer procedure for the receiver to detect the existence of the expected signals from the network. The procedure involves the transmitter-sending reference symbols known to the receivers and the receivers searching for such expected signals. Once such signals are found, the receiver proceeds to perform other system acquisition tasks.

While initial acquisition is necessary for any digital communication systems, it is particularly crucial for wireless communications. This is so because wireless communication systems are operated more likely under dynamic channel conditions and in a random access mode. In such environments, the performance of initial acquisition greatly affects the overall system's efficiency and robustness.

In this chapter, we discussed various aspects of initial acquisition in digital communications. In particular, we introduced the theoretical foundation of the hypothesis testing, which is the core component for performing the reference signal detection. The optimality and performance of the hypothesis testing were analyzed, and practical implementation issues were considered with an emphasis on wireless communication systems. Examples of the design and implementation of initial acquisition in practical CDMA and OFDM wireless communication systems were described in detail in the last part of this chapter.

The main conclusions and remarks on what is discussed in this chapter are listed below.

- To facilitate initial acquisition by devices, transmitters periodically send reference symbols known to the devices.
- A device receiver searches for the reference signals by correlating the expected reference symbols with the received signal samples generated from its front end.

- Hypothesis testing is performed by comparing the magnitude of the correlator output to a predetermined threshold. If the magnitude exceeds the threshold, the hypothesis test is declared successful. Further verification may be performed to confirm that the expected signal indeed exists. Otherwise, another test should be performed and sample sequences at a different time offset and/or different reference symbols should be used.

- The acquisition process continues until the expected signal is found or until it is determined that the expected signal does not exist due to too many failures.

- If the receiver front end is implemented as a matched filter of the received signal, the correlation and hypothesis testing-based method described above is optimal for signal detection as a maximum likelihood test. It satisfies the Cramer–Rao bound if the overall channel is nonfrequency selective. Under relaxed conditions, it would be suboptimal.

- The initial acquisition performance is characterized by false alarm probability (P_F) and detection probability (P_D). If the pdfs of the correlator output with and without the existence of the reference symbols are known, these probabilities can be computed analytically.

- The decision based on the thresholding of the correlation output is optimal in the sense that if the threshold is selected based on the expected P_F, the resulting P_D is the highest for the given P_F according to the Neyman–Pearson lemma.

- If the channel coherence time is longer than the time span of the samples involved in the correlation, coherent integration, also called predetection integration, is achieved.

- If the channel coherence time is not long enough relative to the time span of the samples involved in the correlation due to the frequency offset caused by local oscillator frequency error and/or Doppler effect, the correlation length must be reduced. In such cases, noncoherent integration, also called post-detection integration, can be used to improve the acquisition performance.

- Noncoherent integration introduces a combining loss relative to coherent integration. For example, when doubling the number of samples involved, coherent integration achieves a gain of 3 dB, while the gain from doubling the number of noncoherent integration is between 1.8 to 2.5 dB. In other words, there is a loss of 0.5 to 1.2 dB relative to coherent integration. As a first order approximation, 2 dB can be used as a nominal gain of noncoherent integration when the number of involved data samples is doubled, i.e., the loss is equal to 1 dB relative to coherent integration.

- The performance loss of coherent integration due to frequency offset is analyzed. The analytical results are useful for system design and for determining the trade-offs of using coherent and noncoherent integration when frequency offset and/or Doppler effect exist.

- The ideal data symbol sequence for initial acquisition should have an impulse autocorrelation function. While such sequences exist, it is difficult to meet other system design goals. Two practically used sequences for this purpose, the maximum linear shift register sequences, or m-sequences, and the Zadoff–

Chu sequences, a type of chirp sequences, are described. They have many desirable properties, such as impulse-like autocorrelation and low peak-to-average ratios.

- During initial acquisition, the overall system performance is highly affected by the false alarm and detection probabilities that the detector yields. These probabilities, in turn, highly depend on the accuracy of the estimation of the variance of the observed noise and interference. To improve the estimation accuracy, it is most effective to use as many samples as possible that contain noise and interference in the estimation.

- False alarms occur when there is no expected reference signal. Hence, the total signal variance, or power, can be used as the estimate of the variance of noise and interference. Such an estimate of noise and interference may not be accurate when the expected signal exists. As a result, the detection probability may not be as good as that in the ideal case. However, this is the only practically feasible approach for noise variance estimation, unless the true value of the variance can be estimated. The performance degradation due to this inaccuracy was analyzed.

- Another important factor that affects initial acquisition is the accuracy of the local oscillator that generates the demodulation reference frequency. The impact due to the oscillator inaccuracy was analyzed, and the remedies to address this issue were discussed.

- Four examples of the initial acquisition in practical systems deployed in the real world today – two CDMA based systems, cdma2000–1x and WCDMA, and two OFDM based systems, LTE and 802.11a/g – were presented in the last two sections of this chapter. Their system design and considerations for the realization of their initial acquisition functions were described. The implementations of initial acquisition for each of the systems were discussed in detail.

References

[1] A. J. Viterbi, *CDMA Principle of Spread Spectrum Communications*, Reading, MA: Addison-Wesley, 1995.

[2] G. N. Watson, *A Treatise on the Theory of Bessel Functions*, Cambridge University Press, Cambridge, 1944.

[3] M. Skolnik, *Introduction to Radar Systems*, Boston: McGraw Hill, 1980.

[4] L. L. Scharf, *Statistical Signal Processing*, Reading, MA: Addison Wesley, 1991.

[5] H. V. Poor, *An Introduction to Signal Detection and Estimation*, 2nd edn, New York: Springer, 1994.

[6] J. G. Proakis, *Digital Communications*, 4th edn, Boston: McGraw Hill, 2001.

[7] J. Neyman and E. S. Pearson, "On the Problem of the Most Efficient Tests of Statistical Hypotheses," Philosophical Transactions of the Royal Society A: Mathematical, *Physical and Engineering Sciences*, vol. 231, pp. 289–337, February 1933.

[8] S. W. Golomb, *Shift Register Sequences*, Holden-Day, 1967.

[9] M. K. Simon, J. M. Omura, R. A. Scholtz and B. K. Levitt, *Spread Spectrum Communications Handbook*, rev. edn, Boston: McGraw Hill, 1985, 1994.

[10] D. C. Chu, "Polyphase Codes with Good Periodic Correlation Properties," *IEEE Transactions on Information Theory*, vols. IT-18, no. 4, pp. 531–2, July 1972.

[11] R. L. Frank and S. A. Zadoff, "Phase Shift Pulse Codes with Good Periodic Correlation Properties," *Information Theory, IRE Transactions*, vols. IT-8, no. 5, pp. 381–2, October 1962.

[12] G. R. Lennen, "Receiver Having a Memory Based Search for Fast Acquisition of a Spread Spectrum Signal," US Patent 6,091,785, July 18, 2000.

[13] J. G. Proakis and M. Salehi, *Digital Communications*, 5th edn, New York: McGraw-Hill, 2008.

[14] J. S. Lee and L. E. Miller, *CDMA System Engineering Handbook*, Norwood, MA: Artech House, 1998.

[15] 3rd Generation Partnership Project 2, "Physical Layer Standard for cdma2000 Spread *Spectrum Systems*," 2004.

[16] 3GPP TS 25.213, "Technical Specification Group Radio Access Network; Spreading and Modulation (FDD)," 2010.

[17] M. S. Korde and A. S. Gandhi, "An Improved Method for Secondary Code Synchronization in WCDMA," *International Journal of Scientific Research Engineering & Technology*, vol 1, no. 3, pp. 1–6, June 2012.

[18] D. Liao, D. Qiu, and A. K. Elhakeem, "New Code Synchronization Algorithm for the Secondary Cell-Search Stage in WCDMA," in *International Conference on Communication Software and Networks (ICCSN)*, Macau, China, 2009.

[19] 3GPP TS 36.211, "Technical Specification Group Radio Access Network; Evolved Universal Terrestrial Radio Access (E-UTRA); Physical Channels and Modulation," Sophia-Antipolis, France, 2008–2014.

[20] J. I. Kim, J. S. Han, H. J. Roh, and H. J. Choi, "SSS Detection Method for Initial Cell Search in 3GPP LTE FDD/TDD Dual Mode Receiver," in *Ninth International Symposium on Communications and Information Technology*, Icheon, Korea, 2009.

[21] B. Li and X. Wang, "Efficient SSS Detection for Neighbor Cell Search in 3GPP LTE TDD Systems," in *IEEE International Conference on Communications (ICC)*, Tokyo, Japan, 2011.

[22] IEEE LAN/MAN Standards Committee, "Part 11: Wireless LAN Medium Access Control (MAC) and Physical Layer (PHY) Specifications," New York, 2012.

[23] T. M. Schmidl and D. C. Cox, "Robust Frequency and Timing Synchronization for OFDM," *IEEE Transactions on Communication*, vol. 45, no. 12, pp. 1613–21, December 1997.

[24] J. Heiskala and J. Terry, *OFDM Wireless LANs: A Theoretical and Practical Guide*, Carmel, IN: Sams Publishing, 2002.

[25] T. Star, M. Sorbara, J. M. Cioffi, and P. J. Silverman, *DSL Advances*, Upper Saddle River, NJ: Prentice Hall, 2003.

4   Basics of Phase-Locked Loop Techniques

4.1   Introduction

Phase-locked loops (PLLs) are an essential component for performing carrier and timing synchronization in digital communication systems. Fundamentally, PLLs are close-loop feedback systems that can perform accurate tracking and estimation of the quantities of interest but require only low precision in their implementation. Most PLLs used for synchronization for digital communications are first- and second-order feedback systems. Their analysis is relatively simple and within the reach of engineers with a basic understanding of linear systems. At the same time, such analysis, while being simple, can provide insights into communication systems' functional blocks that have PLLs as their components. Thus, it is a valuable complement to simulation techniques, which are widely used today for understanding and evaluating PLLs.

The purpose of this chapter is to present the basic analytical and practical aspects of PLLs that are related to synchronization functions, not to cover all of the PLL-related topics. Due to their widespread use, PLLs have been studied extensively in many technical fields of electrical engineering and beyond. There exists a rich body of literature regarding PLLs, including a number of textbooks and reference books [1, 2, 3, 4]. Interested readers are referred to those references for more information.

It should be emphasized that the various general aspects of PLLs given in the literature and discussed in this chapter are also applicable to other types of similar feedback control loops, which may or may not involve phases of sinusoidal waveforms. The term *phase* is just a representative of the quantities involved in such loops, as can be seen from some of the examples that will be given later.

Technologies for PLL implementation have evolved greatly since the PLL technique was introduced in the early 1930s [5]. In the early days, PLLs were implemented exclusively using analog components. Digital technology has been used since the 1960s in mixed-signal PLLs. In recent years, all digital and software-based PLL implementations became dominant, especially in the field of digital communications. Our treatment of the analysis and implementation of PLLs will focus primarily on digital PLLs with analog PLLs serving as references.

This chapter is organized as follows.

After this introductory section, Section 4.2 provides an overview of PLL techniques and applications. The history and evolution of PLL technology will be discussed. The classification of PLLs is presented according to the implementation technologies and the system characteristics. Key parameters and characteristics of PLLs will be briefly discussed. Classic analog PLLs (APLLs) and their implementations are presented in Section 4.3.

Sections 4.4 and 4.5 are the foci of this chapter. Section 4.4 provides a fundamental view of digital PLLs (DPLLs), which are most widely used in today's digital communications for performing carrier and time synchronization that is discussed in the later chapters of this book. The fundamental aspects of DPLLs are presented. The properties of DPLLs with practical importance and the expressions of these properties useful in practice are presented. Section 4.5 focuses on the implementation considerations and details of DPLLs, in particular, the DPLLs implemented by DSPs and/or application-specific integrated circuits (ASICs).

The analysis of properties and characteristics of analog and digital PLLs are presented separately. The basic analysis of APLLs is provided for completeness and as reference. Considering that readers are most likely to deal with DPLLs in their practice, we analyze DPLLs directly from the basics, rather than applying the analysis of APLLs to the corresponding DPLLs. The materials regarding DPLLs are self-contained. Readers can study DPLLs without prior knowledge about APLLs. The information provided about APLL would be useful in the cases when comparisons of properties between APLLs and DPLLs are of interest.

In Section 4.6, we consider a few applications of PLLs. The examples directly related to carrier and timing synchronization that will be discussed in Chapters 6 and 7 are briefly mentioned. An introduction to the application of PLLs to frequency synthesis will be given as it is widely used in digital communication systems.

Finally, Section 4.7 summarizes what is discussed in this chapter and provides remarks on the topics of significance.

4.2   An Overview of Phase-Locked Loop Techniques

4.2.1   History and Evolution of PLLs

PLL technology is closely related to modern digital communications. At its beginning, the PLL was invented for the reception of radio communication signals [5]. Since then, while it has found applications in other technical areas such as servo control in mechanical applications, telecommunication has been its focus. As of today, PLLs and related techniques are indispensable components of digital communication transmitters and receivers. In particular, they play important roles in carrier and timing synchronization. PLLs are also essential components in frequency synthesis to generate desired carrier frequencies for performing modulation and demodulation in transmitters and receivers.

In the early days, PLLs were exclusively implemented by using analog circuitries and components. Such PLLs are called APLLs. In the late 1960s, digital components started being used in PLL implementations. PLLs using both analog and digital components are called *mixed-signal PLLs*. Later, *all-digital PLLs* (ADPLLs), which are implemented

without any analog components, appeared. However, in digital communication applications, *voltage-controlled oscillators* (VCOs), which operate primarily in the analog
domain, continue to be used as phase controllers in the PLL. VCOs began to be phased
out in the late 1990s though the technology is still used in some implementations today.
Thus, many PLLs used in digital communications are mixed-signal PLLs.

In their early days, DPLLs were built using digital logic circuits. With the advances in
DSP technology, DPLLs that are based exclusively on software implementation gained
in popularity. These are also called *software PLLs* (SPLLs) or *numeric PLLs* (NPLLs).
In addition to the use of SPLL, another trend in digital receiver designs is to replace
accurate yet expensive *voltage-controlled temperature-compensated crystal oscillators*
with basic crystal oscillators without temperature compensation. This trend reduces the
cost of receiver implementations but also poses challenges to communication system
and modem engineers, especially in the synchronization area. We will consider these
challenges and present their solutions in this book when we discuss the designs and
implementations of PLLs for such applications.

4.2.2   System Description and Major Components

The high-level block diagram of a PLL is depicted in Figure 4.1.

As shown in the figure, the structure of a PLL is relatively simple. It consists of three
components: a *phase error detector*, or simply phase detector (PD), a *loop filter*, and an
output phase controller, or simply phase controller. Their basic functions are
described below.

Phase Error Detector
A PD has two input ports. The first port receives the input signal, from which a
reference phase θ_{in} of a sinusoidal waveform is derived. The second port receives the
PLL's output signal with phase θ_{out}. The function of the PD is to generate the phase
difference between input and output signals, i.e., $\theta_{in}-\theta_{out}$. There are different forms of
PDs, and they depend on the applications and the technologies used in the
implementation.

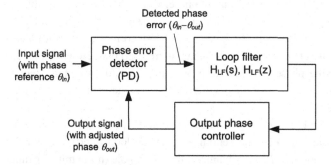

Figure 4.1 A high-level block diagram of a PLL

Loop Filter

The loop filter can be analog or digital with a transfer function, denoted by $H_{LF}(s)$ or $H_{LF}(z)$. A loop filter usually has a low-pass characteristic and likely contains *perfect*, or *ideal*, integrators, which have poles at zero frequency, i.e., DC. It usually also contains zeros in order to improve the PLL's stability [1].

Output Phase Controller

The phase controller generates the output of PLL and adjusts its phase. It can have either the analog or the digital form depending on the application and implementation. The most popular form of phase controllers is the VCO for APLLs and mixed-signal PLLs. In software PLLs, the phase controller can be a piece of software that generates the PLLs' output phase in numerical form. We will describe the details of this phase controller when we consider its implementation.

The function of a PLL that consists of these components is to adjust the phase of its output toward, and to lock to, the phase of its input. Since the reference phase derived from the input signal likely contains noise and/or interference, the loop filter is commonly a low-pass filter, whose function is to reduce the variance, or jitter, caused by these disturbances. As a result, the overall transfer function of the PLL also has a low-pass characteristic, usually with a narrow passband, to reject most of the jitter in the reference phase of the input.

4.2.3   Order and Characteristics of PLLs

As a type of feedback control system the PLLs can be classified according to the order of the system transfer functions. The order of a PLL is determined by its loop filter. In practice, first- and second-order PLLs are most widely used. Third- and higher-order PLLs are less common and used for specific application environments when necessary and appropriate.

In general, there should be at least one integrator in a PLL. Because the phase controller contains an integrator, the loop filter of a *first-order PLL* is simply a constant gain. The denominator of the resulting PLL's transfer function is a first-degree polynomial of s or z. The first-order PLL is commonly used to track a constant initial input phase error or a step change in its input phase. Once it passes the transition period, i.e., after its learning stage, the PLL enters the steady state, or becomes "locked."

In steady state, a first-order PLL can completely correct a constant phase error. The mean of the phase error is equal to zero, i.e., there is no bias. However, it cannot track a linearly increasing phase error, i.e., a phase ramp, due to a constant frequency offset, without a steady-state phase bias. To track such a phase ramp, a second-order PLL should be used.

A second-order PLL can track a linearly changing phase with the zero-mean output phase error in steady state. The denominator of the loop filter's transfer function is a first-degree polynomial of s or z. Namely, there is a pole in the transfer function of the loop filter. The denominator of the resulting PLL's transfer function is a second-degree polynomial of s or z. The second-order PLLs are usually used to track sinusoidals with constant frequency offsets, such as for achieving carrier and timing synchronization in digital communication systems.

If the quantity to be tracked changes with time according to the square law, using a third-order PLL can achieve a zero-mean steady-state phase error. One possible usage is for tracking the phase of a sinusoid with a linearly changing frequency. However, for synchronization applications, third-order and higher-order PLLs are seldom used because such situations rarely occur. Even when it does happen, it will not last for a long time. A second-order loop with good tracking characteristics should be sufficient to handle such situations. Possible applications of third- or higher-order APLLs will be discussed in Section 4.3.4.

Various aspects of analog and digital PLLs will be presented in Sections 4.3 and 4.4. In this book, when dealing with the characteristics of different kinds of PLLs, we will mainly rely on linear system analysis. It is relatively simple but usually sufficient for evaluating PLLs' behaviors and impacts on system performance related to synchronization. Linearized models will be first established as the foundation for PLL analysis. The insight gained from the analysis and their implications in practical applications will then be discussed and summarized.

4.2.4   PLL Applications in Digital Communication Systems and the Trends

The PLL is an essential component of carrier and timing synchronization in digital communication systems. The carrier synchronization block in a wireline modem receiver is essentially a second-order PLL, and so is the timing synchronization block in most wireline and wireless receivers. We will discuss the details of the design and implementation of these synchronization blocks in Chapters 5 and 6 of this book.

By its original definition, a PLL deals with the phase of sinusoidal waveforms. However, in digital communications, there are other applications, which are based on the same general principles as PLLs but do not deal with "phase" in the conventional sense. In these applications, a feedback loop is used to adjust one or more parameters of a transmitter or receiver function to approach a reference value and to "lock" to it. Therefore, all of the theories developed for the PLL can be applied to these applications.

For example, the *frequency-locked loop* (FLL) and the AGC are closely related to the PLLs discussed in this section. Their operations follow the same principle as PLLs even though there are no "phases" directly involved. The analysis and characteristics of the PLL are directly applicable to these applications.

As mentioned above, PLLs also play an important role in *frequency synthesis*, which is an essential component in digital transmitters and receivers. Briefly, the function of frequency synthesis is to generate a frequency reference for a transmitter exciter or a local receiver frequency down-converter based on another frequency source. An introduction to frequency synthesizers is presented in Section 4.6.2.

Due to the cost, size, and power reductions in integrated circuits (ICs) as well as the advances in DSP technology over the past few decades, most PLLs in communication systems have transitioned from analog to digital. More recently, because of competitive pressures on cost reduction of mobile devices, there is also a trend toward using nonadjustable XOs with no or only simple temperature compensation instead of VC-TCXO. The trend further increases the DSP/ASICs' roles in PLL implementation. The majority of this chapter is devoted to DPLL rather than to APLL.

4.3   Analog PLLs

APLLs were developed for and used in various applications before DPLLs. They have been extensively studied and analyzed. Although DPLLs are more prevalent in today's applications of digital communications, we begin our discussion with the basics of APLLs to establish a general understanding of PLLs and to present some fundamental results. Then our emphasis in the remainder of this chapter will be on DPLLs.

4.3.1   Basic Structures and Transfer Functions of APLLs

An APLL is an analog linear feedback system. A conceptual block diagram of a basic APLL is shown in Figure 4.2.

As shown in the figure, a typical APLL has an input port and an output port, and it consists of three components:

- A phase error detector
- A loop filter with transfer function $H_{LF}(s)$
- A voltage-controlled oscillator

The input to an APLL is a sinusoidal waveform with the phase θ_{in}, and its output x_{out} is a sinusoid generated by the VCO with the phase θ_{out}. Both of the phases are, in general, functions of time. Their Laplace transforms are denoted by $\Theta_{in}(s)$ and $\Theta_{out}(s)$, respectively. Ideally, once the APLL has converged, the phase difference between the input and output is zero on average, i.e., they are *locked* to each other.

The PD in an APLL can be implemented in different forms. Below we give brief descriptions of the two types of PDs that are commonly used in APLLs. More information can be found in many references including [1, 2].

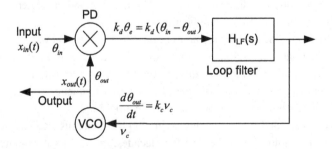

Figure 4.2 A conceptual block diagram of an APLL

The simplest way to implement a PD is to use an analog multiplier. Let us define the input signal and output signals by

$$x_{in}(t) = A \sin(\omega t + \theta_{in})$$
$$x_{out}(t) = B \cos(\omega t + \theta_{out}) \cong B \sin(\omega t + \theta_{out} + \pi/2) \qquad (4.1)$$

The output of the multiplier is

$$x_{in}(t) \times x_{out}(t) = \frac{AB}{2}[\sin(\theta_{in} - \theta_{out}) + \sin(2\omega t + \theta_{in} + \theta_{out})] \qquad (4.2)$$

The first term on the right side of (4.2) has the same sign as $\theta_e = \theta_{in} - \theta_{out}$, as long as $|\theta_e| < \pi$. The second term is a sinusoid with frequency $2\omega t$, which can be removed through low-pass filtering by the PLL's loop filter. Hence, the output of such a PD has the same sign as that of the phase difference between the input and output sinusoidals. Moreover, the output is proportional to the phase difference if it is close to zero. After the PLL has converged, θ_e is zero on average. From (4.1) and (4.2), it can be seen that there is a $\pi/2$ phase offset between the input and output sinusoidals in steady state.

In a mixed-signal implementation, the PD can be implemented by using exclusive OR (EXOR) or flip-flop logic circuits. The two sine waves $x_{in}(t)$ and $x_{out}(t)$ are first converted to rectangular waveforms by amplifying and limiting their magnitudes. The converted rectangular waveforms are fed to the inputs of the EXOR or flip-flop. The output is low-pass filtered and is further processed to generate the phase error θ_e, which is proportional to the phase difference between the two inputs.

The phase error θ_e is filtered by the loop filter, which is an analog filter with a transfer function $H_{LF}(s)$. The output of the loop filter is sent to the VCO, which serves as the phase controller of the APLL.

In analog PLLs and in many mixed-signal PLL implementations, the output signals, which are sinusoidal waveforms, are commonly generated by *voltage-controlled oscillators*. Using VCOs as the phase controllers, we achieve phase adjustment by changing the frequency of the VCO, which is proportional to the control voltage v_c at the input control port of the VCO. The phase change is equal to the frequency change integrated over time and, thus, is proportional to v_c integrated over time with a scaling factor k_c. Hence, the VCO behaves as a perfect integrator in the PLL's loop transfer function with a phase controller gain k_c.

When the APLL is turned on, the phase of the VCO output, θ_{out}, may be different from the reference phase θ_{in} of the APLL's input. The phase difference is detected by the PD and sent to the loop filter. The PD has a gain of k_d in a unit of volts/radian. Thus, the input to the loop filter is equal to $k_d(\theta_{in} - \theta_{out})$ volts.

The output of the loop filter is its input convolved with the time impulse response of the loop filter, $h_{LF}(t)$, i.e., $k_d(\theta_{in} - \theta_{out}) * h_{LF}(t)$, where $*$ denotes the convolution operation. It is sent to the VCO as control voltage v_c with the Laplace transform of

$$V_c(s) = k_d[\Theta_{in}(s) - \Theta_{out}(s)]H_{LF}(s) \qquad (4.3)$$

The frequency of the sinusoidal waveform at the output of the VCO can be expressed as $f_{out} = k_c v_c$. Since frequency is the derivative of phase with respect to time, we have

$$\frac{d\theta_{out}}{dt} = f_{out} = k_c v_c \qquad (4.4)$$

By taking Laplace transform of (4.4) and combining it with (4.3), we have

$$s\Theta_{out}(s) = k_c V_c(s) = k_c k_d [\Theta_{in}(s) - \Theta_{out}(s)] H_{LP}(s) \qquad (4.5)$$

From (4.5) and with the loop closed as shown in Figure 4.2, the overall PLL loop transfer function, $H_{APLL}(s)$, is equal to

$$H_{APLL}(s) = \frac{\Theta_{out}(s)}{\Theta_{in}(s)} = \frac{k_c k_d H_{LF}(s)}{s + k_c k_d H_{LF}(s)} \qquad (4.6)$$

It is clear from (4.6) that the PLL's order and transfer function $H_{APLL}(s)$ are determined by the loop filter's transfer function $H_{LF}(s)$. Below we discuss the most commonly used APLLs: the first- and second-order APLLs.

4.3.2   First-Order APLL

If the loop filter is a constant gain $H_{LF}(s) = k_1$, we have

$$H_{APLL,1}(s) = \frac{k_c k_d k_1}{s + k_c k_d k_1} \cong \frac{K}{s + K} \qquad (4.7)$$

where K is called the *loop gain*, or DC loop gain, of a first-order APLL.

If the initial phase error is not equal to zero, or if there is a step change in the input phase during operation, a first-order APLL exponentially converges to the input phase with a time constant of $1/K$. However, if the input phase changes linearly with time, i.e., as a phase ramp, due to a nonzero frequency offset between the input and the VCO output, there would be an irreducible phase error in steady state as has been documented in the literature, e.g., [1]. In this case, a second-order APLL can be used to track the linearly changing phase with a zero-mean output phase error in steady state as discussed below.

4.3.3   Second-Order APLL

There are two main shortcomings of first-order APLLs. First, a first-order APLL cannot track the phase of a sinusoid with a constant frequency offset without a steady-state output phase bias. Secondly, the roll-off, i.e. attenuation versus frequency, of the first-order APLL is only 20 dB for every tenfold increase in frequency, i.e., 20 dB per decade. These two issues can be addressed by using a second-order APLL.

A second-order APLL has the same general form as the first-order APLL shown in Figure 4.2. Its difference from the first-order APLL is that the transfer function of the loop filter has a pole instead of simply being a constant. For the second-order PLL to have a low-pass characteristic, its loop filter must also have a low-pass frequency response. The simplest first-order low-pass filter can be implemented as a passive RC

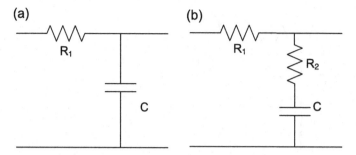

Figure 4.3 Examples of passive loop filters in second-order PLLs

filter, as shown in Figure 4.3(a). It has a transfer function of $H(s) = \frac{1}{1+s\tau_1}$ with a single real pole at $-1/\tau_1$, where $\tau_1 = R_1C$. Despite the simplicity, such an RC filter is rarely used in APLLs because of the difficulty in maintaining stability when the loop gain is large [1].

A slightly modified version of the passive RC loop filter is shown in Figure 4.3(b). The transfer function of this type of filter is $H(s) = (1 + s\tau_2)/(1 + s\tau_1)$, where $\tau_1 = (R_1+R_2)C$ and $\tau_2 = R_2C$. Due to the zero at $-1/\tau_2$, it is more flexible in selecting the parameters of the loop filter in this form to attain better stability of the second-order loop [1]. The close-loop transfer function of the second-order APLL with the pole-zero passive filter[1, 6] is

$$H_{APLL,2p}(s) = \frac{k_d k_c (s\tau_2 + 1)/\tau_1}{s^2 + s(1 + k_d k_c \tau_2)/\tau_1 + k_d k_c/\tau_1} \tag{4.8}$$

In the literature, APLLs' transfer functions are often expressed in terms of their *natural frequency* ω_n and *damping factor* ζ. The loop transfer function $H_{APLL,2p}(s)$ in (4.8) can be rewritten in terms of ζ and ω_n as

$$H_{APLL,2p}(s) = \frac{s\omega_n(2\zeta - \omega_n/K) + \omega_n^2}{s^2 + 2\zeta\omega_n s + \omega_n^2} \tag{4.9}$$

where the loop gain is $K = k_d k_c$, $\omega_n = \sqrt{K/\tau_1}$ is known as the loop's *natural frequency* and $\zeta = (1/K + \tau_2)\sqrt{K/\tau_1}/2$ is known as the loop's *damping factor*. Both ζ and ω_n are commonly defined in textbooks on linear systems for the transfer functions of second-order systems. For applications to synchronization in communication systems, the damping factor ζ is an important parameter of the APLL.

The loop filter constructed with passive elements shown in Figure 4.3b contains a pole at $-1/\tau_1$. As τ_1 is usually large but finite, the loop filter has a pole near but not at the origin. Such an APLL is called a *Type 1* second-order APLL.[1] It can track the phase

---

[1] The term *type* of a PLL is defined as the number of perfect integrators in the PLL. Because a VCO has one perfect integrator, the loop filter must contain $n - 1$ perfect integrators in a type n PLL. In contrast, the order

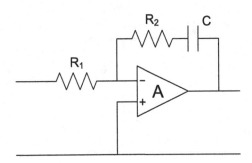

Figure 4.4 Second-order PLL loop filter with active components

ramp better than a first-order APLL but still has a nonzero phase bias in steady state. To achieve a zero-mean output phase error for an input signal with a phase ramp, we need a loop filter that has a pole at the origin. An APLL with a loop filter that has a pole at the origin, i.e., contains a perfect integrator, is called a *Type 2* second-order APLL, because its loop transfer function has two perfect integrators.

The loop filter of a type 2 second-order APLL can be implemented by using active components. Figure 4.4 shows one such implementation, which uses an operational amplifier (op-amp) with a gain $-A$. For $A \gg 1$, its transfer function is given by

$$H(s) = \frac{1 + s\tau_2}{s\tau_1 + 1/A} \approx \frac{1 + s\tau_2}{s\tau_1} \tag{4.10}$$

where $\tau_1 = R_1 C + (R_1 + R_2)C/A \approx R_1 C$ and $\tau_2 = R_2 C$.

The close-loop transfer function of a second-order APLL with the transfer function of the loop filter given by (4.10) is

$$H_{APLL,2a}(s) = \frac{K(s\tau_2 + 1)/\tau_1}{s^2 + sK\tau_2/\tau_1 + K/\tau_1} \tag{4.11}$$

where $K = k_d k_c$. Expressing (4.11) in terms of the natural frequency and damping factor, we have

$$H_{APLL,2a}(s) = \frac{2\zeta\omega_n s + \omega_n^2}{s^2 + 2\zeta\omega_n s + \omega_n^2} \tag{4.12}$$

where $\omega_n = \sqrt{K/\tau_1}$ and $\zeta = \tau_2\sqrt{K/\tau_1}/2$.

As can be seen from (4.10), the loop filter has a pole at the origin. With such a loop filter, the second-order APLL is a *Type 2* APLL because it contains two perfect integrators. A Type 2 APLL can track a constant phase ramp with a zero-mean-output phase error in steady state.

of a PLL is equal to the poles in its loop transfer function. The order of a PLL may be different from its type if the loop filter does not contain poles at the origin.

4.3.4   Third- and Higher-Order APLLs

As mentioned, the most widely used APLLs are the first- and second-order ones. Third- or higher-order APLLs are rarely needed for synchronization in digital communications. One advantage of these APLLs, however, is that they can provide higher attenuations at higher frequencies. As we will show in the next subsection, in general, the attenuation of an nth order APLL increases at $20n$ dB per decade. This feature would be useful in practical applications if the noise in the input signal has strong high-frequency components. More information regarding higher-order APLLs and their designs and applications can be found in [2, 6].

4.3.5   Parameters and Characteristics of APLLs

APLLs are often characterized by their loop bandwidths, transition responses including time constants and damping factors, and pull-in ranges. In this section, we simply state these parameters and characteristics of first- and second-order APLLs without derivations as they have been thoroughly studied and well documented in the literature.

4.3.5.1   Loop Bandwidth

In the applications of APLLs, it is important to determine the APLLs' response to signals and noise. Such responses are characterized by their loop bandwidths.

There are different types of APLLs' loop bandwidth. Two of the most popular are *3 dB bandwidth* and *noise bandwidth*. These two bandwidths are closely related.

The 3-dB bandwidth characterizes the distortion introduced to the input signal by an APLL. Because APLLs have low-pass characteristics in the frequency domain, the 3-dB bandwidth, B_{3dB}, of an APLL is equal to the frequency, at which the squared magnitude of the frequency response of the APLL is equal to 0.5. In contrast, the noise bandwidth is defined as the total noise power at the output of the APLL, when the noise at the input is white with unit power density. Note that the APLL's frequency response at zero frequency is equal to one.

By definition, the frequency response of a linear circuit with the transfer function of $H(s)$ is $H(j\omega)$. The single-sided noise bandwidth is defined by

$$B_n = \frac{1}{2\pi} \int_0^\infty |H_{APLL}(j\omega)|^2 d\omega \tag{4.13}$$

From (4.7), the frequency response of a first-order PLL can be shown to be

$$H_{APLL,1}(j\omega) = \frac{K}{j\omega + K} \tag{4.14}$$

Its 3-dB bandwidth satisfies

$$|H_{APLL,1}(j\omega)|^2 = \frac{K^2}{K^2 + \omega^2} = 0.5 \tag{4.15}$$

It can be shown that

$$\omega_{3dB} = K \ (\text{rad.}) \ \text{or} \ f_{3dB} = K/2\pi \ (\text{Hz}) \tag{4.16}$$

The noise bandwidth of the first-order APLL is equal to

$$B_n = \frac{1}{2\pi} \int_0^\infty \frac{K^2}{\omega^2 + K^2} d\omega \overset{\omega = Ku}{=} \frac{1}{2\pi} K \int_0^\infty \frac{1}{u^2 + 1} du = \frac{1}{2\pi} K \arctan(u)|_0^\infty = \frac{K}{4} (\text{Hz}) \tag{4.17}$$

The frequency response of the second-order Type 2 APLL with $s = j\omega$ can be expressed as

$$H_{APLL,2}(j\omega) = \frac{2j\zeta\omega_n\omega + \omega_n^2}{-\omega^2 + 2j\zeta\omega_n\omega + \omega_n^2} \tag{4.18}$$

By letting $|H_{APLL,2}(j\omega)|^2$ be equal to 0.5 and solving the equation, we obtain

$$\omega_{-3dB} = \omega_n \left[(1 + 2\zeta^2) + \sqrt{(2 + 4\zeta^2 + 4\zeta^4)} \right]^{1/2} \tag{4.19}$$

The noise bandwidth of the second-order Type 2 APLL can be computed by using (4.13). As summarized in [1], its (single-sided) noise bandwidth can be expressed as

$$B_n = 0.25K(1 + \tau_2/K) \cong 0.5\omega_n(\zeta + 1/4\zeta) \tag{4.20}$$

From (4.19) and (4.20), it can be observed that, for the Type 2 second-order APLL, while both the 3-dB loop bandwidth and the noise bandwidth are proportional to ω_n, they are also roughly proportional to the damping factor ζ, when ζ is large. Thus, to fairly compare the convergence behaviors of second-order APLLs with different values of ζ, we should let their noise bandwidths be the same. It can be seen from (4.20) that the noise bandwidth is approximately proportional to the loop gain K when K is large. Thus, the loop gain K would be a good measure of a second-order Type 2 APLL's noise bandwidths under this condition.

4.3.5.2   Roll-Off Characteristics of APLLs

Roll-off is defined as the steepness of the frequency-response attenuation's slope of a filter, or any linear circuit, beyond the passband of the circuit. The unit of the roll-off is usually *dB/decade*, i.e., the number of dBs of additional attenuations when the frequency of the signal or noise increases by 10 times. Since APLLs are often used to reduce the phase jitter with high-frequency components, roll-off is an important parameter of APLLs.

As can be observed from (4.14), the magnitude of a first-order APLL's transfer function is inversely proportional to the value of frequency. When the frequency of noise components increases 10 times, the attenuation increases by 20 dB. Thus, the roll-off of a first-order APLL is equal to *20 dB/decade*. This is not very steep if the input phase jitter has strong high-frequency components.

There are two poles in a second-order APLL. It can be shown from (4.18) that the attenuation of a second-order APLL is proportional to the square of the frequency of the signal or noise components. Consequently, the roll-off of a second-order APLL is equal to 40 dB/decade.

In general, an nth order APLL has n poles and its roll-off is $20n$ dB/decade. High-order APLLs are useful for rejection of noise with strong high-frequency components.

4.3.5.3   Transition Behavior

Other important properties of an APLL are its capability and behavior in tracking input phase changes. Two types of phase changes that are of most practical and theoretical interest are a *step-phase change* and a *linear-phase change*, i.e., a *phase ramp* due to the frequency offset. The transition behavior of an APLL is characterized by the response of a phase error $\theta_e = \theta_{in} - \theta_{out}$ after an input phase change occurs.

From (4.6), we have $\Theta_{out}(s) = H_{APLL}(s)\Theta_{in}(s)$. The Laplace transform of θ_e is given by

$$\Theta_e(s) \cong \mathcal{L}(\theta_e) = \Theta_{in}(s) - \Theta_{out}(s) = [1 - H_{APLL}(s)]\Theta_{in}(s) \qquad (4.21)$$

When there is a step-phase change of $\Delta\theta$, the Laplace transform of θ_{in} is $\Theta_{in}(s) = \Delta\theta/s$. The Laplace transform of θ_{in} with a phase ramp corresponding to an angular frequency offset of $\Delta\omega$ is $\Delta\omega/s^2$.

First-Order APLL

Let us first consider the case that the input reference phase has a step change of $\Delta\theta$. From (4.21) and (4.7) we have

$$\Theta_{1e,\text{step}}(s) = [1 - H_{APLL,1}(s)]\frac{\Delta\theta}{s} = \left[1 - \frac{K}{s+K}\right]\frac{\Delta\theta}{s} = \frac{\Delta\theta}{s+K} \qquad (4.22)$$

By taking the inverse Laplace transform of (4.22), the phase response is given by

$$\theta_{1e,\text{step}}(t) = \Delta\theta e^{-Kt}U(t) \qquad (4.23)$$

where $U(t)$ is the unit step function, i.e.,

$$U(t) = \begin{cases} 0 & t < 0 \\ 1 & t \ge 0 \end{cases} \qquad (4.24)$$

In response to a phase ramp change, we have

$$\Theta_{1e,\text{ramp}}(s) = [1 - H_1(s)]\frac{\Delta\omega}{s^2} = \left[1 - \frac{K}{s+K}\right]\frac{\Delta\omega}{s^2} = \frac{\Delta\omega}{s(s+K)} = \left(\frac{1}{s} - \frac{1}{s+K}\right)\frac{\Delta\omega}{K} \qquad (4.25)$$

By taking the inverse Laplace transform of (4.25), the phase error's response to the phase ramp can be shown to be

$$\theta_{1e,\text{ramp}}(t) = \left(-\frac{\Delta\omega}{K}e^{-Kt} + \frac{\Delta\omega}{K}\right)U(t) = \frac{\Delta\omega}{K}(1 - e^{-Kt})U(t) \qquad (4.26)$$

The phase error $\theta_{1e,\text{ramp}}(t)$ in (4.26) contains a constant term and a term that exponentially decays with time. Thus, in steady state, there is an irreducible phase error, which is equal to $\Delta\omega/K = 2\pi\Delta f/K$.

Second-Order APLL

We use the second-order APLL with the active loop filter shown in Figure 4.1 as an example. Its close-loop transfer function is given by (4.11), or equivalently (4.12). From (4.21) and (4.12), it can be shown that, in response to a step-phase change, the phase error of the second-order PLL, in the Laplace transform domain (s-domain), is

$$\Theta_{2e,\text{step}}(s) = [1 - H_{APLL,2a}(s)]\frac{\Delta\theta}{s} = \frac{s}{s^2 + 2\zeta\omega_n s + \omega_n^2}\Delta\theta = \frac{s}{(s - p_1)(s - p_2)}\Delta\theta$$

(4.27)

where p_1 and p_2 are the two roots of the polynomial in the denominator. They are equal to

$$p_1, p_2 = -\omega_n\zeta \pm \omega_n\sqrt{\zeta^2 - 1}$$

(4.28)

By taking the inverse Laplace transform of (4.27), the phase error responses of Type 2 second-order PLL can be shown to be

$$\theta_{2e,\text{step}}(t) = \begin{cases} \Delta\theta e^{-\zeta\omega_n t}\left(\cos\left(\sqrt{1-\zeta^2}\omega_n t\right) - \frac{\zeta}{\sqrt{1-\zeta^2}}\sin\left(\sqrt{1-\zeta^2}\omega_n t\right)\right) & \zeta < 1 \\ \Delta\theta(1 - \omega_n t)e^{-\omega_n t} & \zeta = 1 \\ \Delta\theta e^{-\zeta\omega_n t}\left(\cosh\left(\sqrt{\zeta^2-1}\omega_n t\right) - \frac{\zeta}{\sqrt{\zeta^2-1}}\sinh\left(\sqrt{\zeta^2-1}\omega_n t\right)\right) & \zeta > 1 \end{cases}$$

(4.29)

If ζ is less than 1, a second-order PLL is called underdamped. It is called critically damped if ζ is equal to 1 and overdamped if ζ is greater than 1.

In response to a phase ramp due to a frequency offset of $\Delta\omega$, the phase error of the second-order PLL in the s-domain can be expressed as

$$\Theta_{2e,\text{ramp}}(s) = [1 - H_{2a}(s)]\frac{\Delta\omega}{s^2} = \frac{1}{s^2 + 2\zeta\omega_n s + \omega_n^2}\Delta\omega = \frac{1}{(s - p_1)(s - p_2)}\Delta\omega \quad (4.30)$$

where p_1 and p_2 are defined by (4.28).

By taking the inverse Laplace transform of (4.27), we obtain the second-order PLL's time-domain error response to the phase ramp as

$$\theta_{2e,\text{ramp}}(t) = \begin{cases} \dfrac{\Delta\omega}{\omega_n\sqrt{1-\zeta^2}}e^{-\zeta\omega_n t}\sin\left(\sqrt{1-\zeta^2}\;\omega_n t\right) & \zeta < 1 \\ \dfrac{\Delta\omega}{\omega_n}e^{-\omega_n t}\omega_n t & \zeta = 1 \\ \dfrac{\Delta\omega}{\omega_n\sqrt{\zeta^2-1}}e^{-\zeta\omega_n t}\sinh\left(\sqrt{\zeta^2-1}\;\omega_n t\right) & \zeta > 1 \end{cases} \quad (4.31)$$

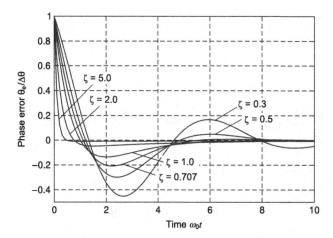

Figure 4.5 Response to phase step change of second-order PLLs with $\omega_n = \omega_0$

The results given by (4.29) and (4.31) have been previously reported in [7] and [1].

Figure 4.5 plots the Type 2 second-order APLL's response to a step input phase change. All of the PLLs have the same natural frequency $\omega_n = \omega_0$ but with different damping factors. Plots based on equations (4.29) given in [7] and [1] are under the same assumptions. In Figure 4.5, it appears that the APLLs with larger damping factor ζ converge faster than the ones with smaller ζ's. However, this conclusion is not based on a fair comparison and thus misleading as explained below.

As shown by (4.20), a large damping factor results in a large noise bandwidth. Hence, there would be a large phase jitter at the APLL output. For applications to digital communications, the phase jitter at the APLL output is an important parameter that affects receiver performance. Thus, to fairly compare different APLLs, it is necessary to make sure that their noise bandwidths are the same. Below, we compare the convergence characteristics of APLLs with different damping factors but with the same noise bandwidth, i.e., the same output phase jitter variance.

Figure 4.6 shows the responses of second-order APLLs to a step-phase change with ζ's equal to 0.3, 0.5, 0.707, 1.0, 2.0, and 5.0 with corresponding ω_n's equal to ω_0 times 1.1, 1.25, 1.179, 1.0, 0.588, and 0.247, respectively. According to (4.20), the noise bandwidths with different ζ's are approximately equal to each other.

In Figure 4.6, we observe that except for the highly underdamped case of $\zeta = 0.3$, the initial convergence rates of the other damping factors are quite close but with different degrees of overshoots. However, it is not the case for the responses to the phase ramp, i.e., frequency offset, in the input signal as shown in Figure 4.7.

In Figure 4.7, it can be seen that when there is a frequency offset, phase convergence is very slow for the overdamped APLLs with $\zeta = 2.0$ and $\zeta = 5.0$. This is because, in order to keep the loop noise bandwidth in line with other cases, we have made the value

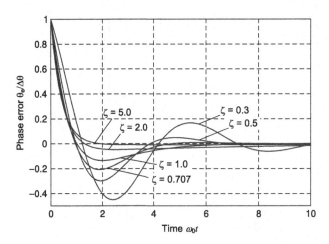

Figure 4.6 Response to phase step change of second-order PLLs (with the same noise bandwidth)

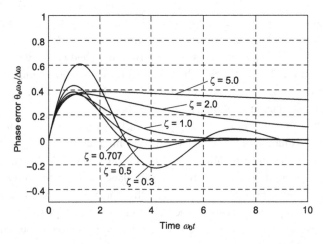

Figure 4.7 Response to phase ramp of second-order PLLs (with the same noise bandwidth)

of the natural frequency ω_n small to compensate for the large ζ. It can be shown that the time-constant for tracking the frequency offset is inversely proportional to ω_n of the loop. With a small ω_n, it would take a long time for the second-order APLL to converge to the offset frequency.

Given all of these factors, it is desirable to use a damping factor near 1, i.e., around the critically damped, or in the slightly underdamped, regions.

The output phase error in response to a phase ramp function in the input converges to zero for the Type 2 second-order PLL when time goes to infinity. Namely, the average phase error is zero in the steady state. However, in the steady state, there is always a nonzero-mean output phase error for a Type 1 second-order PLL in response to a frequency offset in the input signal [1].

4.4   Digital PLLs

Along with the general trend in the implementation of transmitters and receivers in telecommunications, the technologies employed in the implementation of PLLs have been gradually migrating from analog to digital since the 1970s. At the beginning, only PDs were implemented using digital circuits, such as exclusive-OR gates or flip-flops. The most important change in PLL implementation occurred when analog continuous time loop filters were replaced by filters based on discrete-time DSP. Phase controllers employed in digital PLLs remained being implemented by using variations of VCOs, including digitally controlled VCOs, for a long time. More recently, phase controller functions are commonly implemented based on discrete-time digital processing in receiver designs. PLLs employing both analog and digital components are called *mixed-signal PLLs*. PLLs fully based on digital technologies are called *all-digital PLLs*, or *ADPLLs*. They are usually implemented by DSPs or ASICs.

In this and the following chapters, the emphasis is on PLLs with loop filters based on discrete-time DSP. Such PLLs are generally referred to as DPLLs. DPLLs are used in the majority of recent designs and implementations of receivers and transmitters.

While there are many similarities between APLLs and DPLLs, there are also differences between them. The realization and analysis of DPLL are relatively simple and straightforward. Due to the popularity of DPLLs in the implementations of modern digital communication systems, students, engineers, and researchers in this technical field mostly need to deal with DPLLs. Thus, it would be beneficial to analyze DPLLs based on the fundamentals without referring to the analysis of APLLs.

In this section, we analyze the characteristics of first- and second-order DPLLs, including their time response to a step-phase change and a phase ramp, i.e., frequency offset, in the input signals. Similar analyses of APLLs are well known as given in [7] and [1]. For DPLL, corresponding results have been derived but are less accessible. In many references, DPLLs were analyzed by mapping them to APLLs and applying the corresponding APLL analysis. To facilitate analyzing and investigating DPLLs' behavior, we derive the corresponding expressions for DPLLs directly. These derivations should be easy to follow by readers who are familiar with DSP.

The information about APLL provided in Section 4.3 would be useful in the case where a comparison of properties between APLLs and DPLLs is of interest. At the end of this section, we establish the relationships between APLLs and DPLLs through an approximation of derivatives by difference equations. The information is mainly for facilitating the understanding of DPLL for people who are already familiar with APLLs. However, the relationships given are only accurate if the poles of DPLLs are close to the unit circle or, equivalently, the time constants of the DPLLs are much longer than the sampling time interval, e.g., 10 to 50 times or more. At the same time, the expressions obtained from the discrete-time analysis would have no such constraints and, thus, are more general and accurate.

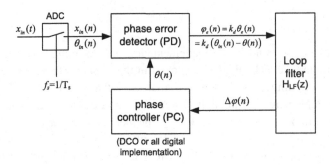

Figure 4.8 A conceptual block diagram of DPLL

4.4.1   Structure and Components of DPLLs

A high-level block diagram of a basic DPLL is shown in Figure 4.8.

DPLLs are operated in the discrete-time domain with a sampling frequency of f_s, namely, the sample interval is equal to $T_s = 1/f_s$.[2] Similar to an APLL, a DPLL consists of three components: a PD, a digital loop filter with transfer function of $H_{LF}(z)$, and a phase controller.

The function of the PD is to detect the difference between the reference phase $\theta_{in}(n)$ of the input signal $x_{in}(n)$ and the phase $\theta(n)$ of the phase controller's output. The detected phase difference $\theta_e(n)$ is converted to its digital representation $\varphi_e(n)$. The PD's gain k_d is equal to $\varphi_e(n)/\theta_e(n)$. The implementations of the PD can take different forms depending on the application, in which DPLL is used.

The loop filter is a discrete-time digital processing unit, which can be implemented by hardware or software in ASICs or DSPs. Based on the input of phase error $\varphi_e(n)$, the loop filter generates the required phase increment $\Delta\varphi(n)$ for the DPLL's next output.

The phase controller contains an integrator, which accumulates the phase increment value $\Delta\varphi(n)$ to generate the desired phase $\theta(n+1)$ at the next sample time. The phase controller is usually implemented in one of two forms: as a digitally controlled oscillator (DCO), which is a VCO with a digital interface, or as a digital integrator and a digital phase generator implemented in DSP/ASIC software. When used as a phase controller, a DCO behaves as an integrator, so that no explicit integrator is implemented.

4.4.2   First-Order DPLL

To begin with, we study the first-order DPLL, which is the simplest PLL with a discrete-time loop filter.

4.4.2.1   System Model and Details of Operation

A block diagram of the system model of a first-order DPLL is depicted in Figure 4.9.

---

[2] Note that the meaning of T_s and f_s in this chapter are different from the meaning of the same notations when considering the OFDM technology.

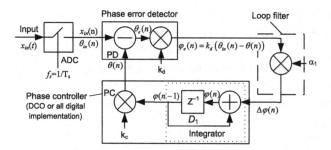

Figure 4.9 System block diagram of a first-order DPLL

The first-order DPLL consists of a PD with a gain equal to k_d, a loop filter, whose transfer function is a constant coefficient α_1, and a phase controller, which contains an integrator and has a gain k_c. The operation of the first-order DPLL is described below.

At time nT, the input signal $x_{in}(t)$ is converted to digital samples $x(n) \cong x(nT)$, at a rate of f_s by an ADC. The difference between the phase $\theta_{in}(n)$ of $x_{in}(n)$ and the phase of the phase controller output $\theta(n)$ is denoted by $\theta_e(n)$. The PD detects the phase difference and converts it to the digital representation $\varphi_e(n)$ with a detector gain k_d.

Because $\varphi_e(n)$ is the digital representation of the phase error, k_d has a unit of the *number unit* of $\varphi_e(n)$ per radian, or simply 1/radian. The number unit is defined based on the digital representation of $\varphi_e(n)$. For example, it can be the least significant bit (LSB) for integer representations or 1.0 for fractional representations of fixed-point fractional numbers. The phase error $\varphi_e(n)$ is scaled by a constant α_1, which is called the DPLL's *first-order loop coefficient*.

The scaled phase error $\Delta\varphi(n) = \alpha_1\varphi_e(n)$ is integrated by the (perfect) integrator in the phase controller to generate an output $\varphi(n)$, such that

$$\varphi(n) = \varphi(n-1) + \Delta\varphi(n) \tag{4.32}$$

which is the digital representation of the DPLL's output phase at $(n+1)T_s$, i.e., $\theta(n+1)$.

The integrator output, $\varphi(n)$, is converted by the phase controller to the output phase $\theta(n+1) = k_c\varphi(n)$ in the form acceptable by the PD. The digital phase error $\varphi_e(n+1)$ is computed by the PD for the next cycle of the DPLL's operation. The phase controller's gain k_c has a unit of radian per number unit, which depends on the digital representation of $\varphi(n+1)$, or simply radian.

Phase controllers could be implemented by using a DCO or implemented in all digital forms using an ASIC or DSP. The details of their implementation will be described in Section 4.5.1.3.

When the operation of DPLL starts, say at sampling time 0 (in the unit of T_s), the delay elements of the first-order DPLL contain an initial value $\varphi(-1)$, which may be zero or a preset value. It is converted to an initial output phase $\theta(0) = k_c\varphi(-1)$ at the output of the phase controller. The difference between the phase of the input signal at time 0 and $\theta(0)$ is computed by the PD to generate the initial phase error $\theta_e(0) = \theta_{in}(0) - \theta(0)$ and converted to the digital form $\varphi_e(0) = k_d\theta_e(0)$. The digital phase error $\varphi_e(0)$ is scaled by the first-order loop coefficient a_1 and added to register D_1 in the integrator.

To understand the details about how the first-order DPLL operates, let us consider the following example.

Assume that the initial input phase $\theta_{in}(0) > 0$, $\varphi(-1) = 0$, and there is no additive noise. At time 0, $\varphi_e(0) > 0$. The accumulated value in D_1 of the integrator for the next cycle of DPLL operation would be greater than D_1's initial value, i.e. $\varphi(0) > \varphi(-1)$. Consequently, $\theta(1) > \theta(0)$. If the input phase $\theta_{in}(n)$ is constant, at sampling time $1 \times T_s$ we have $\theta_e(1) = \theta_{in}(1) - \theta(1) < \theta_e(0)$ and $\varphi_e(1) < \varphi_e(0)$. In other words, the phase error is reduced. This is also the case at the next sampling time and so on. In each step, the phase error $\theta_e(n)$ is smaller than its previous value $\theta_e(n-1)$ and keeps moving towards zero. The situation is similar if the initial phase error is negative. In either case, the phase error gradually converges to zero after a sufficiently long time.

Once the phase error $\theta_e(n)$ reaches zero, the value D_1 no longer changes. Thus, $\theta(\infty)$ is equal to the input phase $\theta_{in}(n)$.

If there is noise in the input signal, the analysis given above is still correct statistically.

When the input phase is a ramp function caused by a frequency offset, it increases by a fixed amount, e.g., $\Delta\theta$, every sampling instant. To compensate for the phase increments, the value $\varphi(n)$ in D_1 in the steady state must increase by $\Delta\theta/k_c$ every sample time when it updates. In other words, we must have $\theta_e(n)k_d a_1 = \varphi_e(n)a_1 = \Delta\theta/k_c$, i.e.,

$$\theta_e(n) = \varphi_e(n)a_1 = \Delta\theta/(k_c k_d a_1) = \Delta\theta/k_1 \tag{4.33}$$

From (4.33), we conclude that, if the input signal has a phase ramp of $\Delta\theta/T_s$ the first-order DPLL has a constant steady-state error of $\Delta\theta/k_1$, where $k_1 = a_1 k_c k_d$ is defined to be the *loop gain* of the first-order DPLL. The same conclusion can be shown from the analytical treatment given below.

4.4.2.2   Transfer Function

Based on the description of the first-order DPLL given above, the linearized transfer function in z-domain is derived as follows.

The transfer function of a perfect integrator is $1/(1 - z^{-1}) = z/(z - 1)$. Thus, we have

$$\Phi(z) = \frac{a_1}{1 - z^{-1}}\Phi_e(z) \tag{4.34}$$

where $\Phi(z)$ and $\Phi_e(z)$ are the z-transforms of $\varphi(n)$ and $\varphi_e(n)$, respectively. We also have

$$\Phi_e(z) = k_d(\Theta_{in}(z) - \Theta(z)) = k_d\left(\Theta_{in}(z) - k_c z^{-1}\Phi(z)\right) \tag{4.35}$$

By combining (4.34) and (4.35), and after simplification, we obtain *the loop transfer function* of the first-order DPLL as

$$H_{\text{DPLL},1}(z) = \frac{\Theta(z)}{\Theta_{in}(z)} = \frac{k_c k_d a_1 z^{-1}/(1 - z^{-1})}{1 + k_c k_d a_1 z^{-1}/(1 - z^{-1})} = \frac{k_c k_d a_1 z^{-1}}{1 - (1 - k_c k_d a_1)z^{-1}} \cong \frac{k_1 z^{-1}}{1 - (1 - k_1)z^{-1}}$$

(4.36)

4.4.2.3   Frequency and Time Domain Characteristics

As shown by (4.36), the first-order DPLL is a first-order linear system. It is stable as long as the loop gain $0 < k_1 < 2$. Since k_1 is usually chosen to be much smaller than 1, such a first-order DPLL is stable. Below we analyze the properties of the first-order DPLL.

Loop Bandwidths

The loop bandwidths characterize the DPLLs' frequency-domain response to signal and noise. As with the APLL, we consider two popular ones: the *3-dB loop bandwidth* and the *noise bandwidth*. For first-order DPLLs, it is possible to obtain closed-form expressions of these two types of loop bandwidth. However, it is not always the case for higher-order DPLLs.

The 3-dB bandwidth, $B_{3\text{dB}}$, characterizes the distortion introduced to the input signal by a DPLL. It is defined as the frequency at which the DPLL introduces a 3-dB attenuation of the input signal relative to the gain at zero frequency, which is equal to 1. Since a first-order DPLL has a low-pass frequency characteristic, the single-sided 3-dB bandwidth must satisfy $|H_{\text{DPLL},1}(e^{j\omega})|^2\big|_{\omega=\omega_{3dB}} = 0.5$, where $\omega = \omega_a T_s$ is the angular frequency of the discrete-time signal, $-\pi \leq \omega < \pi$, and ω_a is the angular frequency of the analog, i.e., continuous-time, signal.

Based on (4.36), the frequency response of the first-order DPLL can be expressed as

$$H_{\text{DPLL},1}(e^{j\omega}) = \frac{k_1}{e^{j\omega} - (1 - k_1)}$$

(4.37)

Its 3-dB bandwidth point satisfies

$$|H_{\text{DPLL},1}(e^{j\omega})|^2 = \left|\frac{k_1}{e^{j\omega} - (1 - k_1)}\right|^2 = \frac{k_1^2}{(1 - k_1)^2 + 1 - 2(1 - k_1)\cos\omega} = 0.5 \quad (4.38)$$

and $\cos\omega_{3dB}$ satisfies

$$\cos\omega_{3dB} = \frac{(1 - k_1)}{2} + \frac{1 - 2k_1^2}{2(1 - k_1)} = 1 - \frac{k_1^2}{2(1 - k_1)}$$

(4.39)

For example, for $k_1 = 0.1$, from (4.39) it can be shown that $\omega_{3dB} = 0.10546$. Moreover, if $k_1 \ll 1$, by using a Taylor series expansion of $\cos\omega$ and taking only the first two terms, we obtain $\omega_{3dB} \approx k_1/\sqrt{1 - k_1} \approx k_1$. $B_{3\text{dB}}$ is equal to ω_{3dB} divided by 2π, i.e., $B_{3\text{dB}} \approx k_1/2\pi$.

The second type of bandwidth of the DPLL is the *equivalent loop noise bandwidth*, or simply *noise bandwidth*, which is a useful parameter for applications of DPLLs in digital communications. The *two-sided* noise bandwidth B_n is defined by

$$B_n = \frac{1}{2\pi} \int_{-\pi}^{\pi} \frac{|H_{DPLL}(e^{j\omega})|^2}{|H_{DPLL}(e^{j\omega})_{\omega=0}|^2} d\omega = \frac{1}{2\pi} \int_{-\pi}^{\pi} |H_{DPLL}(e^{j\omega})|^2 d\omega \qquad (4.40)$$

The unit of noise bandwidth given by (4.40) is expressed in terms of the frequency of the discrete-time signal f, $-1/2 \le f < 1/2$. It is equal to the ratio of the phase noise power at the output of the DPLL divided by the total power of the white input phase noise. Because the total bandwidth of the discrete-time signal is equal to 1, the total power of the white noise is equal to the noise power density in the DPLL's input signal. Thus, this definition of B_n of DPLLs is consistent with the B_n of APLLs defined by (4.13).

For the first-order DPLL, by substituting (4.36) into (4.40) and noting that the denominator is equal to one, we have

$$B_n = \frac{1}{2\pi} \int_{-\pi}^{\pi} |H_{DPLL,1}(e^{j\omega})|^2 d\omega = \frac{1}{2\pi} \int_{-\pi}^{\pi} \frac{k_1^2}{(1-k_1)^2 + 1 - 2(1-k_1)\cos\omega} d\omega \qquad (4.41)$$

Using the known result in integration of trigonometry functions, i.e.,

$$\int \frac{dx}{a + b\cos x} = \frac{2}{\sqrt{(a+b)(a-b)}} \tan^{-1}\left(\sqrt{\frac{a-b}{a+b}} \tan\left(\frac{x}{2}\right)\right) \qquad (4.42)$$

we obtain

$$B_n = \frac{k_1}{2 - k_1} \qquad (4.43)$$

It can be approximated by $B_n \approx k_1/2$, if $k_1 << 1$. The noise bandwidth as defined is called the two-sided noise bandwidth. The single-sided noise bandwidth is equal to $0.5k_1/(2 - k_1)$ or approximately equal to $k_1/4$ for $k_1 << 1$. Compared to the 3-dB bandwidth derived above, it is slightly larger than B_{3dB}.

For describing DPL's noise behavior, we could define a hypothetical equivalent rectangular filter with a magnitude equal to 1 and a passband width of B_n. The variance of the DPLL's output noise is equal to the noise variance at the output of the equivalent filter for white input noise. The relationship between the frequency responses of the DPLL and the noise equivalent filter is depicted in Figure 4.10.

To summarize, if the phase jitter in the input is white, the jitter variance of the first-order DPLL's output would be reduced by $k_1/(2 - k_1)$, or approximately $k_1/2$, of the variance of the input jitter.

Tracking of Step-Phase Change

The tracking characteristics reflect the behavior of the output of PLLs in response to the phase change in the input. Let us first consider the case when the phase of the input signal has a step change with a magnitude of $\Delta\theta$ at time 0.

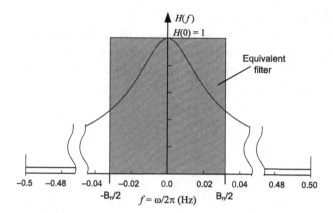

Figure 4.10 Noise equivalent filter of DPLL

From (4.36) it can be shown that the z-transform of $\theta_e(n)$, the phase error between the input phase $\theta_{in}(n)$ and the first-order DPLL's output phase $\theta(n)$, can be expressed as

$$\Theta_e(z) = \Theta_{in}(z) - \Theta(z) = (1 - H_{\text{DPLL},1}(z))\Theta_{in}(z) = \left(\frac{z-1}{z-(1-k_1)}\right)\Theta_{in}(z) \quad (4.44)$$

If $\theta_{in}(n)$ undergoes a step change with a magnitude of $\Delta\theta$, i.e., $\Delta\theta U(n)$, where $U(n)$ is the discrete unit step function defined by

$$U(n) = \begin{cases} 0 & n < 0 \\ 1 & n \geq 0 \end{cases} \quad (4.45)$$

we have $\Theta_{in}(z) = \Delta\theta\, z/(z-1)$. From (4.44) the z-transform of the phase error is

$$\Theta_e(z) = \Delta\theta\left(\frac{z-1}{z-(1-k_1)}\right)\frac{z}{z-1} = \frac{\Delta\theta z}{z-(1-k_1)} \quad (4.46)$$

By taking the inverse z-transform of $\Theta_e(z)$, we obtain the expression of the time-domain phase error $\theta_e(n)$, which is given by

$$\theta_e(n) = (1-k_1)^n \Delta\theta \times U(n) \quad (4.47)$$

For $k_1 \ll 1$, we can approximate $\theta_e(n)$ by

$$\theta_e(n) \approx e^{-k_1 n}\Delta\theta \times U(n) \quad (4.48)$$

Its (analog) time constant is approximately equal to T_s/k_1 (seconds).

From (4.48), it is clear that, in response to a step change in the input phase, the error between the input and output phases of the first-order DPLL exponentially decays to zero. Thus, the average steady-state phase error is equal to zero. In other words, there is no bias in the estimated phase for the first-order DPLL in steady state.

Tracking of Phase Ramp/Frequency Offset

We now consider the response of the first-order DPLL to a phase ramp due to a frequency offset. In this case, assuming that the phase of the input signal increases by $\Delta\theta_{ramp}$ every sample, i.e., the input signal has an analog frequency offset $\Delta f_a = \Delta\theta_{ramp}/2\pi T_s$, the z-transform of the input phase ramp is

$$\Theta_{in} = \frac{\Delta\theta_{ramp}z}{(z-1)^2} \tag{4.49}$$

The phase error in the z-domain can be expressed as

$$\Theta_e(z) = \Delta\theta_{ramp}\left(\frac{z-1}{z-(1-k_1)}\right)\frac{z}{(z-1)^2} = \frac{z\Delta\theta_{ramp}}{[z-(1-k_1)](z-1)} \tag{4.50}$$

By performing the inverse z-transform of (4.50), we obtain

$$\theta_{e,ramp}(n) = \frac{\Delta\theta_{ramp}}{k_1}[1-(1-k_1)^n]U(n) = \frac{2\pi\Delta f_a}{k_1 f_s}[1-(1-k_1)^n]U(n) \tag{4.51}$$

The expression (4.51) shows that the error at the output of the first-order DPLL is zero at the beginning, and it gradually increases and saturates at $\Delta\theta_{ramp}/k_1 = 2\pi T_s\Delta f_a/k_1$ radian. Thus, the first-order DPLL has a steady-state phase error, which is proportional to the frequency offset and is inversely proportional to the loop gain k_1. This analytical result agrees with what we obtained heuristically in Section 4.4.2.1.

Lock-In Range

Another parameter that characterizes the ability of a DPLL to track a sinusoidal input signal is the lock-in range. It is defined as the maximum frequency offset of the input signal that the DPLL can track and lock in, even if there is a nonzero steady-state phase error. The DPLL's lock-in range depends on the characteristic of its PD.

In general, the detector's gain k_d is not a constant, i.e., the gain could be a function of the phase error in the input signal. In addition, the maximum phase error that it can detect may be limited. Beyond the limit, the PD would not work properly such that its output may no longer have the same sign as the phase error. Below, we will consider a simple case of the PD when deriving the DPLLs' lock-in range.

First, we assume that the PD's gain is always equal to a constant k_d for the entire range of the phase errors that the PD can detect. Second, we assume that the maximum phase error that can be detected with the detector gain k_d is equal to $\Delta\theta_{max}$. With these two assumptions, the steady-state phase error given by (4.51) due to the frequency offset in the input signal must be less than $\Delta\theta_{max}$, i.e.,

$$\Delta\theta_{ramp}/k_1 = 2\pi T_s f_a/k_1 = 2\pi f_a/k_1 f_s < \Delta\theta_{max} \tag{4.52}$$

or

$$f_{a,max} < \Delta\theta_{max}k_1 f_s/2\pi, \tag{4.53}$$

If the PD has $\Delta\theta_{max} = \pi$, then,

$$f_{a,\,max} < k_1 f_s/2, \tag{4.54}$$

The maximum phase error that can be detected by PDs used in APLLs is usually limited to $\pm\pi$. However, it is possible that PDs based on digital implementation could have a detection range larger than $\pm\pi$. This becomes important when we consider the tracking range of second-order DPLLs.

Practical PDs usually have characteristics that are more complex. For example, their detection gain may depend on the input phase error and other factors. The lock-in and tracking range can be determined through simulation and/or analysis that takes into account the characteristics of the PDs used in the DPLLs.

4.4.3 Second-Order DPLL

The first-order DPLL described in Section 4.4.2 is capable of correcting a constant input phase error with no steady-state bias. However, it is not able to correct a frequency offset without a nonzero average phase error in the steady state. To compensate for a phase ramp in the input signal, a second-order DPLL should be used. Due to its ability to correct both phase and frequency offsets and its simplicity of implementation, second-order DPLLs are most widely used in digital communication systems. In this section, we will look into the various aspects of second-order DPLLs.

4.4.3.1 System Model and Operation

A block diagram showing the system model of a second-order DPLL in the discrete-time domain is shown in Figure 4.11. By comparing Figure 4.11 and Figure 4.9, we see that the second-order DPLL shares many components with the first-order DPLL.

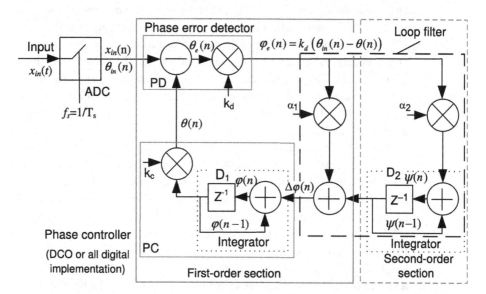

Figure 4.11 Block diagram of a second-order DPLL.

Specifically, the second-order DPLL can be viewed as being constructed by adding a second-order section to a first-order DPLL. The PD and the phase controller in the second-order DPLL are the same as their counterparts in the first-order DPLL. The loop filter of the second-order DPLL consists of a first-order loop coefficient α_1 as in the first-order DPLL, a second-order loop coefficient α_2 and a perfect integrator. The operations of the second-order DPLL are described below.

At time nT_s, $\theta_e(n)$, the phase difference between $\theta_{in}(n)$ and θ (n) is detected, converted to its digital representation $\varphi_e(n)$, and scaled by k_d by the PD as in the first-order DPLL. The phase error $\varphi_e(n)$ is added to the inputs of the integrators inside the phase controller and in the second-order section after being scaled by constants α_1 and α_2, respectively. We call α_1 the *first-order loop coefficient* and α_2 the *second-order loop coefficient* of the second-order DPLL.

The first-order section in the second-order DPLL shown in Figure 4.11 is almost identical to the first-order DPLL. The only difference is that the output of the loop filter includes an additional term from the second-order section. Since the operations of the first-order DPLL have been discussed in detail in Section 4.4.2.1, here we will only examine the additional functions of the second-order section and refer the rest to Section 4.4.2 as needed.

In the second-order section of the DPLL, the digital phase error $\varphi_e(n)$ is scaled by the second-order loop coefficient α_2 before being added to the integrator in the section. The integrated value from register D_2 becomes a part of the loop filter's output $\Delta\varphi(n)$, which is added to the input of the integrator in the phase controller.

In a second-order DPLL, α_2 is usually much smaller than the first-order loop coefficient α_1. When the input phase has a step change, the second-order DPLL behaves similarly to a first-order DPLL. It can be observed from the figure that, if α_2 is equal to zero, the second-order DPLL degenerates to a first-order DPLL. Since the first-order DPLL can already correct the constant phase error with no bias in steady state, the need for the second-order section is for tracking the frequency offset in the input signal.

Let us consider the case when the input phase increases by $\Delta\theta_{ramp} = 2\pi\Delta f_a T_s$ every DPLL update cycle due to a frequency offset Δf_a. To correct this phase increment without introducing a steady-state phase error, it is necessary and sufficient if an amount of $n \times \Delta\theta_{ramp}$ is subtracted from the input phase at the nth DPLL update. This can be achieved by the second-order DPLL if its register D_2 contains a value $\psi(n) = \Delta\theta_{ramp}/k_c$.

Assume that the initial values in both registers D_1 and D_2 of the second-order DPLL are zeros and $\Delta\theta_{ramp}$ is positive. As time progresses, the output of the PD becomes positive as the phase error increases by $\Delta\theta_{ramp}$ every T_s. Both registers D_1 and D_2 contain positive values due to the accumulation of the phase errors. This process continues until the output of the integrator of the phase controller fully compensates for the phase increments. Once the compensation becomes perfect and stabilized, the PD output becomes zero on average and the value of D_2, which is equal to $\Delta\theta_{ramp}/k_c$, no longer changes. As can be seen from Figure 4.11, every T_s, this amount of $\Delta\theta_{ramp}/k_c$ is added to D_1 and $\Delta\theta_{ramp}$ is subtracted from the input phase by the PD. Thus, the DPLL enters the steady state, and the average phase error is equal to zero.

The above discussion only qualitatively describes the characteristics of the second-order DPLL. A quantitative analysis based on the DPLL's discrete-time transfer function is given below.

4.4.3.2 Transfer Function

As stated above, a second-order DPLL contains a first-order section, which is essentially a first-order DPLL, and an additional second-order section containing a perfect integrator. The input to the second-order section is the phase error $\varphi_e(n)$ scaled by the second-order loop coefficient α_2. The scaled phase error $\alpha_2\varphi_e(n)$ is accumulated in the integrator of the second-order section. Using the same derivation that led to (4.34), the z-transform of the integrator's output $\psi(n)$ in the second-order section is equal to

$$\Psi(z) = \frac{\alpha_2}{1 - z^{-1}} \Phi_e(z) \tag{4.55}$$

With one sample delay, $\psi(n)$ is added to the integrator in the phase controller together with $\alpha_1\varphi_e(n)$. The z-transform of the output from the phase controller can be expressed by

$$\Theta(z) = \frac{k_c z^{-1}}{1 - z^{-1}} \left[\alpha_1 \Phi_e(z) + z^{-1}\Psi(z)\right] = \frac{k_c z^{-1}}{1 - z^{-1}} \left[\alpha_1 + \alpha_2 \frac{z^{-1}}{1 - z^{-1}}\right] \Phi_e(z) \tag{4.56}$$

where $\Phi_e(z) = k_d(\Theta_{in}(z) - \Theta(z))$. Expressing (4.56) as a function of z, we have

$$\Theta(z) = \frac{k_c k_d}{z - 1} \left[\alpha_1 + \alpha_2 \frac{1}{z - 1}\right] (\Theta_{in}(z) - \Theta(z)) \tag{4.57}$$

By defining the *first-order and second-order loop gains*, $k_1 = k_c k_d \alpha_1$ and $k_2 = k_c k_d \alpha_2$, and by rearranging the terms, we can express (4.57) as

$$\{(z - 1)(z - 1 + k_1) + k_2\}\Theta(z) = [(z - 1)k_1 + k_2]\Theta_{in}(z) \tag{4.58}$$

Thus, the transfer function of the second-order DPLL is

$$H_{\text{DPLL},2}(z) = \frac{\Theta(z)}{\Theta_{in}(z)} = \frac{k_1 z + (k_2 - k_1)}{z^2 - (2 - k_1)z + (1 - k_1 + k_2)} \tag{4.59}$$

4.4.3.3 Frequency and Time Domain Characteristics

Based on the loop transfer function given by (4.59), we examine some of the second-order DPLL's characteristics that are pertinent to its applications to synchronization.

Stability and the Damping Behavior

In order for a discrete-time linear system to be stable, all of its poles must be inside the unit circle. Setting the denominator of the second-order DPLL equal to zero and solving the equation, we obtain the two roots of its transfer function, i.e., the position of the poles, as

$$z^2 - (2 - k_1)z + (1 - k_1 + k_2) = 0 \quad \Rightarrow \quad p_{1,2} = 1 - \frac{k_1}{2} \pm \frac{\sqrt{k_1^2 - 4k_2}}{2} \quad (4.60)$$

In practical DPLL designs, the loop gains are usually chosen to satisfy $0 < k_2 < k_1 < 1$. It can be shown that the second-order DPLL is stable with such k_1 and k_2.

The tracking and convergence behaviors of a second-order DPLL can be classified into three regions based on the sign of the discriminant, $\Delta = k_1^2 - 4k_2$, of the denominator of the DPLL's transfer function. When Δ is negative, i.e., $k_1^2 < 4k_2$, the two poles are complex conjugate of each other and the DPLL is called *underdamped*. In this region, the output of the second-order DPLL may exhibit oscillation behavior when its input phase or frequency changes. In the second region, Δ is equal to zero, i.e., $k_1^2 = 4k_2$, and the two poles in (4.60) are real and equal to each other. The DPLL is called *critically damped*. Finaly, when Δ is greater than zero, i.e., $k_1^2 > 4k_2 > 0$, the two poles are both real but with different values. The DPLL is in the *overdamped* region. If a DPLL is critically or overdamped, its output does not oscillate but may have an overshoot in some cases.

Similar to the treatment of an APLL, to quantitatively characterize the degree of damping, we define the *damping factor* ζ_d for second-order DPLLs to be equal to the square root of k_1^2 divided by $4k_2$. Namely,

$$\zeta_d = \sqrt{k_1^2/4k_2} = \frac{k_1}{2\sqrt{k_2}} \quad (4.61)$$

A second-order DPLL is underdamped, critically damped, or overdamped if its damping factor ζ_d is less than, equal to, or greater than one, respectively.

The behavior of the DPLL with $\zeta_d = 1$ defined by (4.61), i.e., it is critically damped, is consistent with the behavior of the APLL with $\zeta = 1$ shown in Section 4.4.4. However, the behavior of a DPLL with the damping factor ζ_d is not the same as the damping behavior of an APLL with $\zeta = \zeta_d \neq 1$. Nevertheless, similar to ζ, the value of ζ_d also characterizes the degree of under- or overdamping of a second-order DPLL.

Loop Bandwidths

Let $z = e^{j\omega}$, and the frequency response of a second-order DPLL derived from (4.59) is

$$H_{\text{DPLL},2}\left(e^{j\omega}\right) = \frac{k_1 e^{j\omega} + (k_2 - k_1)}{e^{j2\omega} - (2 - k_1)e^{j\omega} + (1 - k_1 + k_2)} \quad (4.62)$$

In the literature, there are not many analytical results available on the frequency domain characteristics for second-order DPLLs. In addition, unlike the first-order DPLL, obtaining the closed-form expressions is difficult, if not impossible. We resort to numerical tools to study these characteristics.

Figure 4.12 shows the frequency responses of first- and second-order DPLL over the entire positive frequency range from 0 to π. The leftmost curve is the first-order DPLL's frequency response given by (4.37). The three curves on the right of the figure are the frequency responses of second-order DPLLs with ζ_d's equal to 3, 1, and 0.7,

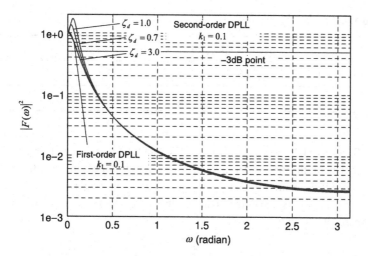

Figure 4.12 DPLL frequency responses (complete view)

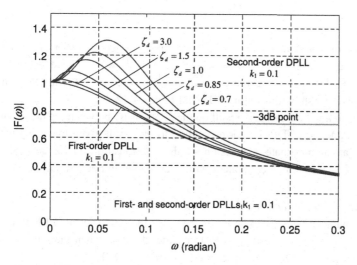

Figure 4.13 Frequency responses of DPLLs in passband region

respectively, plotted according to (4.62). All of the DPLLs have the same first-order loop gain k_1 equal to 0.1.

From Figure 4.12, we observe that with the same first-order loop gain, a DPLL that has a smaller damping factor ζ_d, i.e., a larger second-order loop gain k_2, has a larger DPLL loop bandwidth and a higher overshoot than the first-order DPLL. When the damping factor becomes large, i.e., k_2 is close to zero, the frequency response of the second DPLL converges to that of the first-order DPLL with the same first-order loop gain. The attenuations of first- and second-order DPLLs in the high-frequency region are very close.

The frequency responses of the DPLLs in the low-frequency area are plotted in Figure 4.13 to provide a more detailed view of the passband region. It can be seen from the figure that the 3-dB bandwidth of the first-order DPLL is slightly larger than k_1 as

Table 4.1 Noise Bandwidths of First- and Second-Order PLLs (k_1 = 0.1)

	Second-order DPLLs, k_1 = 0.1				
First-order DPLL, k_1 = 0.1	ζ_d = 3.0	ζ_d = 1.5	ζ_d = 1.0	ζ_d = 0.85	ζ_d = 0.7
B_n 0.05263	0.0541	0.0583	0.0655	0.0706	0.0795

shown in Section 4.4.2.3 from the analysis. The bandwidths of second-order DPLLs are larger than that of the first-order DPLL. Moreover, for the same k_1, the bandwidth becomes wider if the damping factor gets smaller due to a larger k_2.

The equivalent (two-sided) noise bandwidth of the second-order PLL can be computed numerically from (4.59) by

$$B_n = \frac{1}{2\pi} \int_{-\pi}^{\pi} \left| H_{DPLL,2}\left(e^{j\omega}\right) \right|^2 d\omega = \frac{1}{2\pi} \int_{-\pi}^{\pi} \left| \frac{k_1 e^{j\omega} + (k_2 - k_1)}{e^{2j\omega} - (2 - k_1)e^{j\omega} + (1 - k_1 + k_2)} \right|^2 d\omega$$

$$(4.63)$$

The results computed through numerical integration for second-order PLL with $k_1 = 0.1$, and various ζ_d's are given in Table 4.1. These results show that the noise bandwidth of a second-order DPLL gets larger as ζ_d decreases. This is consistent with the results of the 3-dB bandwidths shown in Figure 4.12 and Figure 4.13.

There is no known analytical result on the relationship between the noise bandwidths and the parameters of DPLLs corresponding to the result given by (4.20) for APLLs. However, an approximate expression of the relationship can be derived from the mapping of a DPLL to an equivalent APLL, as shown in Section 4.4.40.

Step-Phase Change Tracking Characteristics
Tracking characteristics describe the behavior of the output of a DPLL in response to phase changes of the input. Let us first consider the case when the phase of the input signal has a step change with a magnitude of $\Delta\theta$ at time 0.

From (4.59) it can be shown that the z-transform of the phase error samples $\theta_e(n)$ from the input $\theta_{in}(n)$ can be expressed by

$$\Theta_e(z) = \Theta_{in}(z) - \Theta(z) = (1 - H_{DPLL,2}(z))\Theta_{in}(z)$$

$$= \left(1 - \frac{k_1 z + (k_2 - k_1)}{z^2 - (2 - k_1)z + (1 - k_1 + k_2)}\right)\Theta_{in}(z)$$

$$= \frac{(z - 1)^2}{z^2 - (2 - k_1)z + (1 - k_1 + k_2)}\Theta_{in}(z) = \frac{(z - 1)^2}{(z - p_1)(z - p_2)}\Theta_{in}(z) \qquad (4.64)$$

where p_1 and p_2 are the two poles of the DPLL's transfer function as defined by (4.60).

When $\theta_{in}(n)$ undergoes a step change, $\Theta_{in}(z) = \Delta\theta \, z /(z-1)$. The z-transform of the phase error is

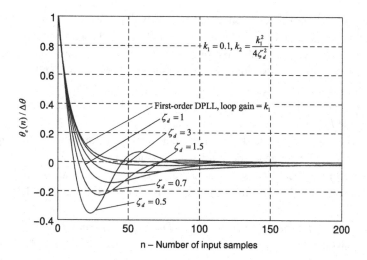

Figure 4.14 Output of second-order DPLLs in response to step change in input phase

$$\Theta_e(z) = \Delta\theta \frac{(z-1)^2}{(z-p_1)(z-p_2)} \frac{z}{z-1} = \frac{z(z-1)\Delta\theta}{(z-p_1)(z-p_2)} \qquad (4.65)$$

By performing a partial fraction expansion and taking the inverse z-transform of $\Theta_e(z)$, it can be shown that the expressions for the time-domain phase error $\theta_e(n)$ of over-damped, critically damped, and underdamped second-order DPLLs are

$$\phi_{e,step}(n) = \begin{cases} \Delta\theta(1 - k_1 + k_2)^{\frac{n}{2}}\left(\cos n\omega_n - \dfrac{k_1}{\sqrt{4k_2 - k_1^2}}\sin n\omega_n\right)U(n) & \zeta_d < 1 \\[4mm] \Delta\theta\left(1 - \tfrac{k_1}{2}\right)^n\left(1 - \dfrac{k_1}{2 - k_1}n\right)U(n) & \zeta_d = 1 \\[4mm] \Delta\theta\dfrac{1}{\sqrt{k_1^2 - 4k_2}}\Bigg[\left(-\dfrac{k_1}{2} + \dfrac{\sqrt{k_1^2 - 4k_2}}{2}\right)\left(1 - \dfrac{k_1}{2} + \dfrac{\sqrt{k_1^2 - 4k_2}}{2}\right)^n & \\[4mm] \qquad + \left(\dfrac{k_1}{2} + \dfrac{\sqrt{k_1^2 - 4k_2}}{2}\right)\left(1 - \tfrac{k_1}{2} - \dfrac{\sqrt{k_1^2-4k_2}}{2}\right)^n\Bigg]U(n) & \zeta_d > 1 \end{cases}$$

$$(4.66)$$

where $\omega_n = \arctan\left(\sqrt{4k_2 - k_1^2}/(2 - k_1)\right)$ following the notations in [8].

The output phase errors of a second-order DPLL in response to the input phase change $\Delta\theta$ are plotted in Figure 4.14 according to (4.66) for damping factors equal to 0.5, 0.7, 1.0, 1.5, and 3.0. As can be seen from the figure, it took about 100 iterations, i.e., with about 100 input samples, for the DPLL to completely converge when k_1 is equal to 0.1. Comparing the different convergence curves, we observe that the best convergence is achieved when ζ_d is equal to 0.7, i.e., when the DPLL is slightly underdamped.

The output phase error of a first-order DPLL with loop gain equal to k_1 is also plotted for comparison. At the beginning, the first-order loop converges together with all of the others. It takes a shorter time to converge completely than the second-order DPLLs with different damping factors. Thus, the first-order DPLL should work well for tracking a step-phase change. However, the first-order DPLL cannot compensate for a frequency offset with a zero-average steady-state error, but the second-order DPLL can, as we have discussed above. The convergence of the second-order DPLL with a large damping factor ζ_d, i.e., a small k_2, approaches the behavior of the first-order DPLL with the same k_1.

From (4.66) and Figure 4.14, it can be seen that the error between the input and output phases of the second-order DPLL in response to a step change of the input phase converges to zero. Thus, the average phase error is equal to zero in steady state, i.e., there is no bias in the estimated phase for the second-order DPLL.

Characteristics of Tracking Phase Ramp Due to Frequency Offset

We now consider the convergence behavior of the second-order DPLL in response to a phase ramp. In this case, assume that the phase of the input signal increases by $\Delta\theta_{ramp}$ every input sample due to a frequency offset of $\Delta f = \Delta\theta_{ramp}/2\pi$, which corresponds to an analog frequency offset $\Delta f_a = \Delta\theta_{ramp}/2\pi T_s = \Delta\theta_{ramp} f_s/2\pi$. For the initial value $\theta(0) = 0$, the z-transform of the input phase can be shown to be

$$\Theta_{in}(z) = \Delta\theta_{ramp} \frac{z}{(z-1)^2} \tag{4.67}$$

The phase error in the z-domain can be expressed as

$$\Theta_e(z) = \Theta_{in}(z) - \Theta(z) = \frac{\Delta\theta_{ramp}(z-1)^2}{z^2 - (2-k_1)z + (1-k_1+k_2)} \frac{z}{(z-1)^2} = \Delta\theta_{ramp} \frac{z}{(z-p_1)(z-p_2)} \tag{4.68}$$

where p_1 and p_2 are the two poles in $\Theta_e(z)$. By performing a partial fraction expansion and taking the inverse z-transform of $\Theta_e(z)$, the expressions of the time-domain phase errors $\theta_e(n)$ for the underdamped, critically damped, and overdamped second-order DPLLs in response to the phase ramp are

$$\phi_{e,ramp}(t) = \begin{cases} \Delta\theta_{ramp}(1-k_1+k_2)^{\frac{n}{2}} \dfrac{2}{\sqrt{4k_2-k_1^2}} \sin(n\omega_n)U(n) & \zeta_d < 1 \\[3mm] \Delta\theta_{ramp} n \left(1-\frac{k_1}{2}\right)^{n-1} U(n) & \zeta_d = 1 \\[3mm] \Delta\theta_{ramp} \dfrac{1}{\sqrt{k_1^2-4k_2}} \left[\left(1-\dfrac{k_1}{2}+\dfrac{\sqrt{k_1^2-4k_2}}{2}\right)^n - \left(1-\dfrac{k_1}{2}-\dfrac{\sqrt{k_1^2-4k_2}}{2}\right)^n \right] U(n) & \zeta_d > 1 \end{cases} \tag{4.69}$$

where ω_n has the same definition as that given in (4.66). The results given by (4.66) and (4.69) are equivalent to those obtained from a slightly differently configured DPLL shown in [9].

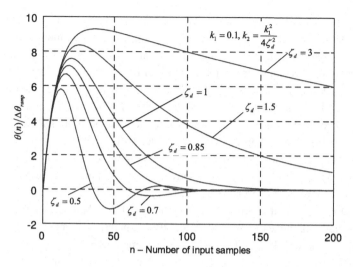

Figure 4.15 Output of second-order DPLLs in response to input phase ramp (frequency offset)

The output phase errors of a second-order DPLL in response to the input phase ramp (frequency offset) are plotted in Figure 4.15 according to (4.69).

The damping factors of the DPLLs plotted in the figure are equal to 0.5, 0.7, 0.85, 1.0, 1.5, and 3.0. The input phase increases $\Delta\theta_{ramp}$ radian per sample, which is equivalent to introducing a discrete-time frequency offset of $\Delta f = \Delta\theta_{ramp}/2\pi$. The y-axis shows the DPLL's output phase divided by $\Delta\theta_{ramp}$. Similar to the case of a step change in the input phase, the best convergence is achieved when ζ_d is equal to 0.5, 0.7, or 0.85, i.e. when the DPLL is slightly underdamped. With these damping factors, it takes about 100 iterations for the DPLL to converge completely when k_1 is equal to 0.1. Once the DPLL has converged, the average steady-state phase error is equal to zero.

It can also be seen from the figure that with large damping factors, the DPLL converges slowly in response to the input phase ramp. Therefore, it is not desirable to use DPLLs with large damping factors if there is a frequency offset, even though it converges quickly in response to step-phase change as shown in Figure 4.14.

Lock-In Range

Similar to a first-order DPLL, a second-order DPLL's ability to track the sinusoid with a frequency offset depends on the characteristic of its PD, in particular, the maximum phase error that the PD can detect. Since the phase error due to a frequency offset eventually converges to zero for a second-order DPLL, its lock-in range can be unlimited, at least in theory, as long as the PD can handle the maximum phase error that has occurred during the initial convergence process.

For analog PDs or PDs that are memoryless, the maximum phase error that they are able to handle cannot exceed $\pm\pi$. Because k_2 is usually much smaller than k_1, when tracking starts the phase increment value is mainly determined by k_1; k_2 helps only marginally. Thus, a second-order DPLL's lock-in range is similar to, or slightly better than, that of a first-order DPLL with a loop gain equal to the first-order loop gain of the second-order DPLL.

For example, from Figure 4.15, we observe that the maximum phase error $\Delta\theta(n)$ of the DPLL with $\zeta_d = 3$ and $k_1 = 0.1$ is about $9.2\Delta\theta_{ramp}$. If the maximum phase error that the PD can handle is equal to π, the DPLL can handle a phase ramp of $\pi/9.2$, or approximately 0.34 radian, per sample. It is equal to an analog frequency offset that is equal to $f_a = 0.34f_s/2\pi = 0.054f_s$. In comparison, with the same maximum phase error, the lock-in range for a first-order DPLL with $k_1 = 0.1$ is $\Delta\theta_{ramp} < k_1\pi = 0.314$, i.e., $0.05f_s$, according to (4.52). Thus, an overdamped second-order DPLL has a lock-in range only slightly larger than that of a first-order DPLL. The lock-in range improves when the damping factor gets smaller. For example, for $\zeta_d = 0.5$, the maximum phase error is about $5.8\Delta\theta_{ramp}$ as shown in the figure. Its lock-in range is $\Delta\theta_{ramp} \leq \pi/5.8 \approx 0.54$ radian per sample, i.e., $f_a < 0.086f_s$.

If the PD is implemented digitally, it may be possible to handle a phase error larger than π. The tracking range of a second-order DPLL can also be larger if "cycle slipping" is allowed. This is well known for APLLs as documented in [1].

When PDs with more complex detection behavior are used, the lock-in range of second-order DPLLs can be determined through simulation or analysis if we take into account the characteristics of the PDs used.

4.4.4   Mapping between DPLL and APLL

Above we have analyzed DPLL characteristics based on the transfer functions in the discrete-time domain. However, it may be of interest to map a DPLL into an equivalent APLL to take advantage of the abundant analytical and experimental results of APLLs existing in the literature. A few different methods can be used to map a discrete-time system to a continuous time system. These methods were originally used to transform analog filters into their corresponding digital filter forms. The most popular are the *bilinear transformation, approximation of derivatives by finite differences*, and *impulse invariant* methods [10]. In this section, we use the second option as an example to establish the relationship between DPLLs and their corresponding APLLs. As shown below, the conditions for achieving critical damping are the same before and after the mapping. The impulse invariant method described in [9] that involves nonlinear transformations is more accurate in some cases but is also more difficult to use.

4.4.4.1   Mapping a DPLL to APLL by Approximate Derivative Method

To map a z-domain system transfer function to an s-domain transfer function using the approximate derivative method [10], we express z as a function of s, such that

$$z = \frac{1}{1 - sT_s} \tag{4.70}$$

To simplify the notation, we let $T_s = 1$ in the derivation below with no loss of generality. The scaling can be brought back into the final results if necessary. Such a mapping between the z-domain to the s-domain is depicted in Figure 4.16.

By substituting z given by (4.70) into the transfer function of the second-order DPLL shown in (4.59), we obtain the analog form of the DPLL's transfer function as

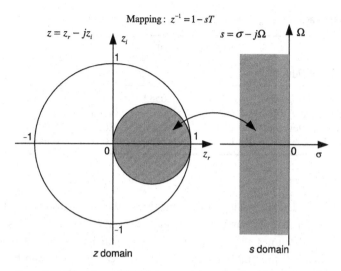

Mapping: $z^{-1} = 1 - sT$

$z = z_r - jz_i$

$s = \sigma - j\Omega$

z domain

s domain

Figure 4.16 Mapping of DPLL between z- and s-domains (approximate derivative method)

$$H_{DPLL,2}(s) = \frac{\dfrac{(k_2 - k_1)}{(1 - k_1 + k_2)}s^2 + \dfrac{(k_1 - 2k_2)}{(1 - k_1 + k_2)}s + \dfrac{k_2}{(1 - k_1 + k_2)}}{s^2 + \dfrac{(k_1 - 2k_2)}{(1 - k_1 + k_2)}s + \dfrac{k_2}{(1 - k_1 + k_2)}} \tag{4.71}$$

The denominator in the standard second-order linear system form is usually expressed as

$$s^2 + c_1 s + c_2 \cong s^2 + 2s\zeta\omega_{n,a} + \omega_{n,a}^2 \tag{4.72}$$

From (4.72), we obtain the natural frequency of the equivalent APLL as

$$\omega_{n,a} = \sqrt{\frac{k_2}{(1 - k_1 + k_2)}} \tag{4.73}$$

and the damping factor ζ is equal to

$$\zeta = \frac{(k_1 - 2k_2)}{2(1 - k_1 + k_2)} \bigg/ \sqrt{\frac{k_2}{(1 - k_1 + k_2)}} = \frac{k_1 - 2k_2}{2\sqrt{k_2(1 - k_1 + k_2)}} \tag{4.74}$$

From (4.72), (4.73), and (4.74), we can express (4.71) as

$$H_{DPLL,2}(s) = \frac{\omega_{n,a}^2 + 2s\zeta\omega_{n,a} - (2\zeta\omega_{n,a} + \omega_{n,a}^2)s^2}{s^2 + 2s\zeta\omega_{n,a} + \omega_{n,a}^2} \tag{4.75}$$

The system is critically damped if $\zeta = 1$. From (4.74), it can be shown that $\zeta = 1$ if $k_1 = 2\sqrt{k_2}$. This is the same condition for the second-order DPLL that achieves critical damping, i.e., $\zeta_d = 1$, as shown by (4.61). Namely, the approximate derivative method maps a DPLL with $\zeta_d = 1$ to an APLL also with $\zeta = 1$. However, if we map a DPLL with $\zeta_d' \neq 1$ to an APLL by using the same method, the damping factor ζ' of the resultant APLL will not be equal to ζ_d'.

Table 4.2 Numerical and Mapping Based Computations of Noise Bandwidths

k_1	Noise bandwidth (B_n)	Second-order DPLLs				
		$\zeta_d = 3.0$	$\zeta_d = 1.5$	$\zeta_d = 1.0$	$\zeta_d = 0.85$	$\zeta_d = 0.7$
0.1	Numerical integration	0.0541	0.0583	0.0655	0.0706	0.0795
	Approximate by (4.76)	0.0566	0.06	0.0658	0.0701	0.0780
0.01	Numerical integration	0.00521	0.00563	0.0063	0.0068	0.00763
	Approximate by (4.76)	0.00519	0.00560	0.00628	0.00676	0.00758

For the analog PLL approximation to be accurate, the time constant of the DPLL must be much longer than the sampling time interval, i.e., $k_1 \ll 1$. The time constant should be at least 10 times the sample interval or even larger. In other words, k_1 should be less than 0.1, preferably 0.02 or smaller. As can be seen from Figure 4.16, only part of the stable region in the z-domain is mapped into the entire stable region in the s-domain. The mapping is relatively linear only if the poles of DPLL are close to the point of $1+j0$, i.e., it has a long time constant. This requirement is not needed if the expressions of the DPLLs' transitional behavior derived in the last section are used.

4.4.4.2   Application: Noise Bandwidth Evaluation Based on DPLL to APLL Mapping

In the last section, we calculated the noise bandwidths of second-order DPLLs through numerical integration, because there is no counterpart of (4.20) available for DPLLs in a close form solution. However, it is possible to use the mapping method described above to obtain the approximate noise bandwidth of DPLLs without using numerical integration as discussed below.

By substituting (4.73) and (4.74) into (4.20), we obtain

$$B_n \approx \omega_{n,a}\left(\zeta + \frac{1}{4\zeta}\right) = 0.5\left(\frac{k_1 - 2k_2}{1 - k_1 + k_2} + \frac{k_2}{k_1 - 2k_2}\right) \tag{4.76}$$

The DPLLs' noise bandwidths computed by numerical integration and the approximate formula (4.76) for $k_1 = 0.1$ and 0.01 and different ζ_d's are compared in Table 4.2.

From Table 4.2 we observe that when $k_1 = 0.01$, the noise bandwidths computed according to (4.76) are quite accurate. The differences between the computed results and those from numerical integration are less than 0.7 percent. When $k_1 = 0.1$, the accuracy of the results obtained from (4.76) are slightly worse. The errors increase to about 5 percent. In both cases, the results obtained from (4.76) exhibit good agreement with the results from numerical integration.

It should be pointed out that since there are approximations involved in the derivation, the formula given by (4.76) is not analytically rigorous and is only approximately valid when $k_1 \ll 1$. Nevertheless, it is useful as a heuristic formula for approximate quantitative assessments of the noise bandwidth of second-order DPLLs.

4.5   Design and Implementation of Digital PLLs

In Section 4.4, we described the structures and transfer functions of first-order and second-order DPLLs. Their frequency domain and time domain characteristics were also analyzed. In particular, we derived their time response to step-phase changes and to phase ramp due to the frequency offset. These responses are derived directly from the discrete-time transfer functions of the DPLLs and are presented in closed form. Such formulations can precisely describe the behaviors of DPLLs and are useful for the investigation of their characteristics. Because these formulas are derived directly in the discrete-time domain, they are more accurate than the methods that examine the characteristics of the APLLs that are mapped from the DPLLs.

While analytical tools can provide insight into the various aspects of DPLLs, these tools are only accurate if the underlying model can precisely describe the DPLL being investigated. However, in real-world applications, many factors cannot be predicted at the design time. Each component of a DPLL may be realized differently based on the implementation platforms. Moreover, different requirements need to be met for different applications in their operation environments. Below, we address these practical issues of DPLL design and implementation.

In this section, we first describe a few possible realizations of the three major components of the DPLL: the PD, the loop filter, and the phase controller. Their approximate forms, trade-offs, and implementation specifics will also be considered. Secondly, we consider the DPLL system design and parameter selection regarding practical applications. Note that the design and implementation of these components are highly application dependent. We will discuss only some common features and requirements. Specific details are left to Chapters 5 and 6 where the applications of DPLLs are discussed. Finally, we provide an exemplary software DPLL design to illustrate some of these practical aspects.

For synchronization applications in modern communications systems, the DPLL is mainly implemented by discrete-time processing using software or hardware designed for performing such processing. We will devote most of our discussions in this chapter to such implementations. Information on pure hardware-based DPLL implementations, which are not discussed here, is available in the literature, including [1, 2, 6].

4.5.1   Practical DPLL Components and Their Characteristics

In this section, we consider the realizations of the three major components that constitute a DPLL: the phase error detector, the loop filter, and the phase controller.

4.5.1.1   Phase Error Detectors in DPLLs

DPLLs considered in this chapter are implemented by using discrete-time DSP techniques. The output of a PD must be in numerical form with the value representing the difference between the phases of the input signal and the output of the DPLL's phase controller.

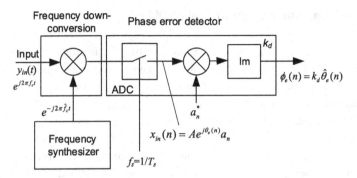

Figure 4.17 A digital phase error detector

If necessary, the digital representation of the phase difference can be generated by using the same PDs as in APLLs based on digital logic circuits by digitizing their output. For example, for PDs implemented by using XOR or flip-flops, the phase differences can be computed by counting the number of high-frequency clock cycles contained inside their rectangular output waveform. PDs implemented by using analog multipliers can be realized by digital multipliers of digital sample representations of analog waveforms. These approaches are useful for the DPLLs that are the direct replacements of APLLs because such DPLLs are still operated on analog sinusoids. These methods have been described in detail in the literature such as [2].

The inputs of PDs used in DPLLs for synchronization are most likely already in digital form. Below we will consider mainly the PDs with digital inputs.

An Example of Digital Phase Error Detector for Carrier Synchronization

For synchronization applications in digital communications, including carrier and timing synchronization, PDs are an integrated part of the synchronization blocks. We will discuss the details of the phase error detection techniques when considering synchronization techniques in the next two chapters. In this section, we provide an example to illustrate the basic operations of the PD. The block diagram of the PD used in carrier synchronization is shown in Figure 4.17.

As shown in the figure, the transmitted baseband signal modulated to carrier frequency f_c is received by the receiver after it has passed through a communication channel. This signal is shown as the input signal $y_{in}(t)$ in the figure. To simplify the discussion, we assume that the channel is a single path channel. The input signal is down converted in frequency to baseband in the frequency down-conversion block with the down-conversion frequency $\hat{f}_c$ generated from a local frequency synthesizer based on an XO. For single-carrier communication, the baseband signal is sampled by an ADC at a sampling frequency $f_s = 1/T$, where T is the data symbol duration. For the baseband signal generated from transmitted data symbols a_k's, the sampled output from the ADC corresponding to a_n can be expressed as $y_{in}(n) = Ae^{j\theta_e(n)}a_n + z(n)$, where A is the channel gain, $z(n)$ is the sampled additive noise and $\theta_e(n)$ is the phase error. The phase error can be expressed by

$$\theta_e(n) = \left(2\pi\left(f_c - \hat{f}_c\right)nT + \theta_0\right)_{\text{mod}2\pi} \qquad (4.77)$$

where $2\pi\left(f_c - \hat{f}_c\right)nT$ is the phase error caused by the frequency offset introduced during frequency down-conversion and θ_0 is the phase of the overall channel.

Assume that the data symbol a_n is known, being either a known transmitted (pilot) symbol or a decision made by the receiver. The phase error estimate at time nT is $\hat{\theta}_e(n) = \text{Arg}\left[a_n^* y_{in}(n)\right]$. In order to simplify implementation, the digital representation of the phase error is often approximately computed as

$$\varphi_e(n) = \text{Im}\left[a_n^* y_{in}(n)\right] \cong A\left|a_n^*\right|^2 \sin\left(\theta_e(n)\right) \qquad (4.78)$$

When the angle of $\theta_e(n)$ is small, $\varphi_e(n)$ is approximately equal to $A|a_n|^2\theta_e(n)$. Thus, the PD has the gain k_d equal to $A|a_n|^2$ at $\theta_e(n) \approx 0$ with a unit of 1/radian.

The objective of a receiver is to recover unknown data symbols a_n from received signal samples $y_{in}(n)$. Ideally, the phase error $\theta_e(n)$ should be equal to zero. A DPLL is employed to estimate and correct this phase error introduced during transmission and reception. From (4.77) it can be seen that the phase error $\theta_e(n)$ is caused by the frequency offset $f_c - \hat{f}_c$ and the constant phase error θ_0. As shown in the description and analysis of DPLLs in Section 4.4.3, such phase errors can be corrected by using a second-order DPLL.

Impact of the Variation of k_d on DPLL's Characteristics

One common issue in the design and implementation of DPLLs is how to determine and handle the PD's gain k_d. As shown in Section 4.4, k_d is a part of the DPLL's loop gains, which determine the DPLLs' characteristics. Thus, the designer of a DPLL must know the value of k_d to analyze the synchronization algorithm to be implemented. Moreover, loop gains k_1 and k_2 are the product of k_d, the phase controller's gain k_c and the loop coefficients α_1 and α_2, respectively. While the loop coefficients are determined at design time and do not change during DPLL operations, in many implementations of PDs, k_d is a function of the magnitudes and/or the SNRs of the input signals. The gain k_c may also change but usually not as significantly as k_d.

As in the example shown by (4.78), to simplify implementation, $A|a(n)|^2\sin[\theta_e(n)]$ is often used as the detected phase error instead of its true estimate $\text{Arg}[a^*(n)y_{in}(n)]$ of $\theta_e(n)$. As a result, k_d changes when the signal magnitude changes. In addition, k_d depends on the phase error $\theta_e(n)$.

For a first-order DPLL, the loop gain determines the noise bandwidth and convergence time. If the gain k_d changes due to the variations of the signal magnitude and phase, the convergence rates and the mean square error (MSE) of the first-order DPLL's output phase jitter will not be exactly as designed.

The situation is more complex for a second-order DPLL than for a first-order DPLL. As can be seen from the expressions of the first-order and second-order loop gains k_1 and k_2 in (4.58), both of these two loop gains are proportional to k_d. However, the damping factor ζ_d is proportional to the ratio of k_1 to the square root of k_2, as shown by (4.61). Thus, if the signal amplitude changes during real-time operations, the damping factor of the second-order DPLL also changes. Especially if the value of k_d decreases significantly, the DPLL could enter the heavily underdamped region and even exhibit undesirable oscillations.

To avoid this problem, it is desirable to decouple the PD gain and the signal amplitude. This can be done in a few different ways. First, the input signal magnitude can be normalized by the AGC or by numerical normalization before entering the PD. It is desirable that the time constant of the AGC or the normalization process is comparable to the time constant of the DPLL. This is because samples that have larger magnitudes usually have higher SNRs than samples with smaller magnitudes. In addition, a DPLL averages the input samples' phases within its time constant. Since samples with larger magnitudes weigh more during averaging, such phase combining is nearly optimal to yield low-output phase jitter. Normalizing the input on a sample-by-sample basis keeps the detector gain constant. However, it may increase the phase jitter level due to nonoptimal combining. Using exact expressions of phase differences, Arg $[a^*(n)y_{in}(n)]$, has a similar effect as using the sample-by-sample normalization.

Sign-Error PD

Another possible implementation of the PDs in DPLLs is to use the signs of the phase error estimates rather than the exact value of the phase error. Specifically, for the example discussed above, the estimated phase error is expressed as

$$\hat{\theta}_e \approx \text{sign}\left\{ \text{Im}\left[\left(A_{in}e^{j\theta_{in}} \right)\left(e^{j\theta_{out}} \right)^* \right] \right\} = \text{sign}[(\theta_{in} - \theta_{out})] \qquad (4.79)$$

This method is similar to the so-called *sign-error LMS algorithm* in adaptive filtering [11]. It is simple to implement, and the behavior of the DPLL that uses it is not affected by the magnitude of the input signal. However, the method introduces performance degradations due to the additional jitter in the detected phase error because of the one-bit quantization.

Figure 4.18 shows the simulation results of the response to a step-phase change with a conventional second-order DPLL and a DPLL that uses the sign-error method. Both of the DPLLs have $k_1 = 0.1$ and $k_2 = 0.005$. Noise with a variance of 0.0316 is added to the input signal. The simulation results show that both of the PLLs converge, and there is no phase error bias in steady state. With the loop gains used in the simulation, the DPLL with sign-error PD converges following a straight line until an overshoot occurs. However, it has a larger variance of steady-state errors due to the quantization errors than the normal second-order DPLL.

The sign-error algorithm works for both first-order and second-order DPLLs. It works well for tracking step-phase error as shown in Figure 4.18. Even though it also works for an input that has a frequency offset, there could be a bias in the steady-state phase error. Since the high nonlinearity of the detector makes the analysis difficult, simulations are usually used to evaluate the performance and behavior of the algorithm. The sign-error PD could be useful when the complexity of implementation is a critical factor and when a larger detector phase jitter can be tolerated, for example, when DPLLs have a long time constant, i.e., a narrow noise bandwidth.

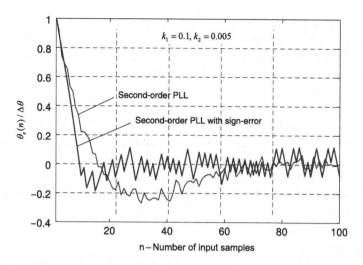

Figure 4.18 Simulated responses to step-phase error of second-order DPLLs (normal vs. sign-error method)

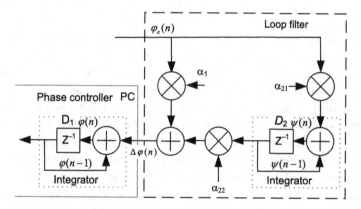

Figure 4.19 Diagram of second-order DPLL loop filter

4.5.1.2 Discrete-Time Loop Filter Implementation

Loop filters in DPLLs can be implemented by using basic digital logic circuits, software running on DSPs, or specifically designed hardware, i.e., ASICs, to perform the discrete-time signal processing functions. In modern communication systems, implementations of DPLLs for synchronization applications are mostly DSP or ASIC based. In this section, we focus on this type of DPLL loop filter realization. Information regarding DPLL loop filter implementations using basic digital logic circuits, such as counters, gates, flip-flops, etc., can be found in [2]. Since the first-order DPLL is a degenerated form of the second-order DPLL, we will consider only the latter here.

The block diagram of a second-order DPLL was shown in Figure 4.11. In Figure 4.19, we extract the loop filter portion of the DPLL with modifications that facilitate its implementation.

The implementation of the loop filter for a second-order DPLL is straightforward. It only contains a few adders, multipliers, and a delay element, i.e., the register D_2.

However, in order to implement the filter efficiently, especially on ASICs, the design should be optimized. The objective is to use the least amount of resources that introduce minimal or no performance degradation. This goal is achieved by slightly modifying the baseline second-order DPLL shown in Figure 4.11 and by making efficient use of each component.

Note that the contents of the registers in the integrators of DPLLs have clear physical meanings. In the loop filter, the value of register D_2 corresponds to the estimated frequency offset in the input signal. In ASIC or DSP implementations, the contents of the registers commonly represent fixed-point fractional numbers. The word lengths of the registers must be long enough to achieve the required accuracies but short enough to be implemented efficiently. Below, we look into two aspects of such an integrator implementation.

Frequency Offset Range of DPLL

In Figure 4.19 the second-order loop coefficient α_2 in Figure 4.11 is split into two parts, α_{21} and α_{22}, for controlling the range of frequency offsets that the DPLL can handle. The value $\psi(n)$ in register D_2 corresponds to the frequency offset. The maximum value of $\psi(n)$ should correspond to the maximum frequency offset. In steady state, the phase error $\varphi_e(n)$ is zero on average. As a result, the mean of the output from the multiplier, $\alpha_1\varphi_e(n)$, is equal to zero. Therefore, the mean value of $\Delta\varphi(n)$ sent to the phase controller is equal to $\alpha_{22}\psi(n-1)$. In the case of an all-digital phase controller implementation, $\Delta\varphi(n) = \alpha_{22}\psi(n-1)$ represents the phase increment to compensate for the phase change from nT_s to $(n+1)T_s$ due to the frequency offset. In a DCO-based phase controller implementation, $\Delta\varphi(n)$ determines the frequency offset of the DCO at time nT_s. It negates the phase change in the input signal from nT_s to $(n+1)T_s$, which is equal to the frequency offset of the DCO multiplied by T_s. In either case, for a given maximum value of $\psi(n)$, α_{22} determines the range of the frequency offset for which the DPLL can compensate.

For DCO-based phase controllers, the coefficient α_{22} maps the value of $\psi(n)$ to the frequency offset that the DCO generates. The maximum offset generated by the DCO that corresponds to the maximum value of $\psi(n)$ is the limit of the frequency offset that the DPLL can correct. If the characteristic of the DCO is known, the proper value of α_{22} can be determined by design according to the range of frequency offset that should be corrected by the DPLL.

Now let us consider a second-order DPLL with an all-digital phase controller as an example. When the analog domain offset frequency is equal to Δf_a and the sampling period of the DPLL is T_s, the phase increment per sample period is equal to $2\pi\Delta f_a T_s$. Such a frequency offset can be fully compensated for if $\Delta\varphi(n)k_c = 2\pi\Delta f_a T_s$. Moreover, we denote the maximum magnitude of $\psi(n)$ by A. To fully utilize the register's word length, the value of α_{22} is chosen such that A corresponds to the maximum frequency offset $\Delta f_{a,max}$ that the DPLL should handle. Namely, $A\alpha_{22}k_c$ should be greater than or equal to $2\pi\Delta f_{a,\,max} T_s$, i.e.,

$$\alpha_{22} \geq \frac{2\pi\Delta f_{a,\,max} T_s}{Ak_c} \qquad (4.80)$$

Once α_{22} is determined, α_{21} is set to be equal to α_2/α_{22}. More details on the two types of DPLLs' phase controllers will be given in Section 4.5.1.3. In addition, an example of how to select the value of α_{22} will be presented in Section 4.5.2.

Integrator Implementation

Above, we have shown that by properly selecting the value of α_{22}, $\psi(n)$ in D_2 can handle the maximum expected frequency offset with magnitude $|\Delta f_{a,\,max}|$. Below we consider how to implement the integrator and how to determine the word length of D_2.

The word length of $\psi(n)$, which generates $\Delta\varphi(n) = \alpha_{22}\psi(n-1)$ that is sent to the phase controller, determines the accuracy of the estimated offset frequency. However, it also affects the complexity of the phase controller implementation.

Assume that the value of $\psi(n)$ sent to the phase integrator is represented by an M_2 bits long fixed-point number. The accuracy of the frequency offset estimate is equal to $2^{-M_2}|\Delta f_{a,\,max}|$. Hence, M_2 should be chosen such that the offset frequency represented by half of the LSB of $\psi(n)$ meets the accuracy requirement. For example, if the maximum offset frequency is equal to ±3000 Hz and the residual offset frequency should be less than ±5 Hz, M_2 should be at least equal to 10 ($2^{10} > 3000/5$).

At the same time, the above requirement may not be sufficient to achieve the required accuracy of the frequency offset estimate. The product $\alpha_{21}\phi_e(n)$ may be below the LSB of $\psi(n)$ in D_2, given that both α_{21} and $\phi_e(n)$ are small numbers in steady state, and the residual offset frequency after compensation is small. Thus, the feedback loop may not close until the residual offset frequency becomes large enough to go beyond the accuracy of $\psi(n)$. The solution to this problem is to increase the word length of D_2 from M_2 to N_2. The N_2 bits of D_2 are partitioned into two parts with M_2 and N_2-M_2 bits, respectively, as shown in Figure 4.20.

The M_2 most significant bits (MSBs) of D_2 represent the value of $\psi(n-1)$ sent to the phase controller, such that the accuracy of phase increment is satisfied. Further, the entire N_2 bits are used for accumulation of the input phase error $\alpha_{21}\varphi_e(n)$. Namely, the lower N_2-M_2 bits are only for the error accumulation. The value of N_2 should be chosen such that the accumulator does not underflow when $\alpha_{21}\varphi_e(n)$ is added to D_2, for the

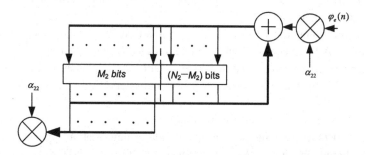

Figure 4.20 Integrator implementation and register bit partition

smallest possible residual frequency offset. To meet this condition, $\alpha_{21}\varphi_e(n)$ should be greater than the half LSB of the register, when the residual frequency offset is at the required accuracy level.

Because both α_{21} and $\varphi_e(n)$ are small numbers, the size of the multiplier is also small. Moreover, the value added to the accumulator does not need to be very accurate because of the feedback nature of the DPLL. It is common practice that α_{21} is selected to be in the form of a negative power of two. In this case, the multiplier can be replaced by a barrel shifter to reduce the complexity of implementation. This method of using different word lengths for the estimate of the frequency offset and for the error accumulation is a special case of the so-called "mixed-precision arithmetic" implementation of adaptive algorithms [12]. This approach can improve the estimation accuracy while keeping the implementation complexity low. The selection of word length N_2 of D_2 will be shown in the design example in Section 4.5.2.

Preventing Bias

To reduce the output phase jitter, it is desirable to reduce the DPLL's loop bandwidth, i.e., to increase the DPLL's DC gain. Hence, any DC bias in the input to the DPLL is amplified to generate a large bias at the output. As a result, the residual offset frequency or phase error could increase significantly.

To prevent such biases, attention must be paid at every point where the word length of the data needs to be reduced after accumulation. For example, it would likely happen when $\alpha_1\varphi_e(n)$ or $\alpha_{21}\varphi_e(n)$ are added to D_1 or D_2. Therefore, biases must be prevented from these operations.

As well known in numerical analysis, there are three basic ways to reduce the word length of a number, i.e., truncation, rounding and unbiased-rounding, each of which has its own bias characteristic. Rounding is preferred to truncation and is usually sufficient from the bias point of view. For DPLLs with the very high DC gain, unbiased rounding may be needed. A good review of different rounding/truncating methods can be found in [13].

4.5.1.3   Phase Controllers

The third component of the DPLL is the phase controller. Two types of phase controllers are commonly used in the DPLL. The first type of phase controller is a *digitally controlled oscillator*. DCOs are similar to the VCOs used in APLLs. However, since the output from the loop filter of a DPLL is digital, the phase controller must be able to take digital values as its input.

A VCO can be converted to a DCO by adding a DAC between the loop filter output and the VCO. A DCO is also often called a *numerically controlled oscillator* (NCO).

The second type of phase controllers used in DPLLs is implemented by using ASICs or in DSP software. These two types of phase controllers are described below.

DCO as the Phase Controller in a DPLL

A DCO is a VCO with a digital interface. In digital communication applications, a VCO is usually based on an XO and thus called VCXO, or VC-TCXO if it also has

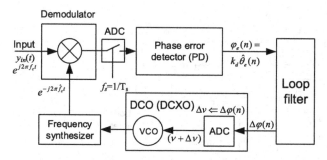

Figure 4.21 DPLL with DCO as phase controller

temperature compensation. The block diagram of such a carrier synchronization block is shown in Figure 4.21.

We first describe the function and characteristics of the DCO. When the input to the DCO is at a nominal value v, it generates a nominal frequency, e.g. 19.2 MHz, which is denoted as f_{DCO} below. When there is a change in the input, e.g., from v to $v + \Delta v$, the generated frequency changes from f_{DCO} to $f_{DCO} + \Delta f_{DCO}$. This DCO's parameter DCO gain, g_{DCO}, is defined by the relative frequency change at its output due to the change in its input, i.e.,

$$g_{DCO} = \frac{\Delta f_{DCO}/f_{DCO}}{\Delta v} \tag{4.81}$$

In receivers, a DCO is often used to synthesize a frequency, which is nominally equal to the received signal's carrier frequency f_c, for performing frequency down-conversion. When there is a change Δv in the input of the DCO, the output of the synthesizer changes by an amount of $\Delta f = g_{DCO} f_c \Delta v$.

Any change Δf of the synthesized frequency at sample $n-1$ results in a phase change of $\Delta \theta = 2\pi \Delta f T_s$ in $\theta(n)$. Thus, the DCO can be used as the phase controller in the DPLL. Assuming that the ADC converts $\Delta \varphi$ to Δv such that $\Delta v = k_v \Delta \varphi$, we express the DCO's gain k_c as

$$k_{c,DCO} = \frac{\Delta \theta}{\Delta \varphi} = \frac{2\pi \Delta f T_s}{\Delta \varphi} = \frac{2\pi g_{DCO} f_c \Delta v T_s}{\Delta \varphi} = 2\pi g_{DCO} k_v f_c T_s \tag{4.82}$$

This is the expression of the phase controller gain k_c in the transfer function of the DPLL with DCO as the phase controller. Because the DCO behaves as an integrator of the phase when used in a phase-locked loop, in Figure 4.21 the integrator in the first-order section is not explicitly shown.

The DCO gain may vary due to part variations, environment changes, and aging. As a result, DPLL designers must take such variations into account and select the DPLL loop parameters such that the DPLL operates normally under the worst-case conditions.

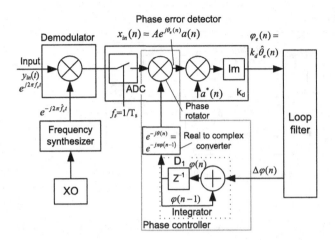

Figure 4.22 DPLL with software/ASIC phase controller implementation

Phase Control Using All-Digital/Software PLL

In an all-digital PLL implemented in ASICs or DSP software, the phase controller can be implemented in an all-digital form in the discrete-time domain. A block diagram of such a DPLL is shown in Figure 4.22 as an example for illustration and analysis. The PD is the same as the one described in Section 4.5.1.1.

The local XO generates a fixed-frequency clock. The clock's frequency is then converted to the nominal carrier frequency $\hat{f}_c$ by the frequency synthesizer as shown in the figure. Given that the frequency of the clock generated by the XO is not adjustable, the phase error given by (4.77) is compensated by the digital phase rotator inserted after the ADC. The angle of the phase rotation is determined by the content of the register D_1, i.e., $\varphi(n - 1)$, in the integrator.

The phase increment $\Delta\varphi(n)$ from the loop filter is accumulated in the register D_1. The value $\varphi(n-1)$ in the register D_1 is interpreted as a fixed-point fractional number in the range of $[-1, 1)$. It corresponds to a phase $\theta(n) = \pi\varphi(n - 1)$ in the range of $[-\pi, \pi)$ and is used to generate a complex number $e^{-j\theta(n)} = \cos\theta(n) - j\sin\theta(n)$ for performing phase rotation of the ADC's nth output sample by the phase rotator.

The phase controller's gain k_c is equal to the ratio of the phase rotation performed on the ADC samples divided by the value of D_1. Hence, $k_c = \theta(n)/\varphi(n - 1) = \pi$, which has a unit of radian.

Because the phase of the samples is periodic with a period of 2π, $\varphi(n)$ is also periodic. When $\varphi(n)$ reaches its limits of -1 or $+1$ during accumulation, it simply wraps around in accordance to the periodicity of the phase. The values $\cos(\pi\varphi(n - 1))$ and $\sin(\pi\varphi(n - 1))$ for performing phase rotation are usually generated by table lookup. If the number that represents $\varphi(n-1)$ is M_1 bits long, a table with 2^{M_1-2} entries is needed.[3] Thus, the word length M_1 should be long enough to ensure the accuracy of the phase $\theta(n)$ that $\varphi(n-1)$ represents. However, it should be short enough so that the size of the table will not be too large, as the table size doubles with a 1-bit increase in the word length of $\varphi(n)$.

Even if the selection of M_1 can ensure the required accuracy of the phase $\theta(n)$, the estimated frequency error may increase due to the underflow that occurred during the accumulation of the phase error $\Delta\varphi(n)$. This underflow issue can be addressed by the same method described previously for the frequency offset accumulator D_2.

Specifically, the number contained in D_1 is N_1 bits long but partitioned into two parts with M_1 bits and N_1-M_1 bits. The M_1 MSBs are used for performing phase rotations and the N_1-M_1 LSBs are used for phase error accumulation. The word length of D_1, i.e., N_1, should be long enough so that no underflow would occur when accumulating $\Delta\varphi(n)$ from the loop filter. Even though some of the LSBs in $\Delta\varphi(n)$ do not impact the value of $\varphi(n)$ immediately, they would accumulate and propagate into the top M_1 bits after multiple updates over time. Such an implementation of D_1 is the same as that shown in Figure 4.20.

4.5.2 An Example of an All-Digital/Software DPLL Implementation

A complete implementation of a DSP/ASIC software-based second-order DPLL is shown in Figure 4.23. The implementation of each of the DPLL's components has been discussed in Section 4.5.1. In this section, we provide a design example. In particular, we will describe the scaling factor selection of the components.

Let us consider a DPLL for the application to carrier synchronization in a digital receiver. To begin with, we outline the system design parameters as follows.

The receiver employs an XO with a fixed frequency as the frequency reference. The accuracy of the XO is better than ± 2.5 ppm, i.e. $<\pm 2.5\times 10^{-6}$. The carrier frequency f_c is equal to 1.9 GHz. Thus, the maximum frequency offset is ± 4750 Hz. Note that this DPLL should be able to operate under such a range of frequency offset, but it is not necessary to track a step change in frequency of this magnitude. This is because a significant portion of the frequency offset should already be estimated and compensated for during factory calibration and receiver initialization.

The received signal is sampled by an ADC at $f_s = 15,000$ samples/second, i.e., 15 kS/s. The ADC word length is equal to 6 bits. The average received signal power is 12 dB below the ADC peak. The decimal point of this fixed-point number is at the middle of its 6 bits, so that the average magnitude A of samples $x_{in}(n)$ at the ADC output is equal to 1 as controlled by the AGC. The received signal for DPLL training and updating are QPSK symbols with the magnitude equal to one. The DPLL is designed to operate at 15 kS/s, the same as the ADC sampling rate.

The DPLL is designed to be trained in two stages. In the first stage, first- and second-order loop gains, k_1 and k_2, are set to 0.1 and 0.0025. In the second stage

---

[3] Normally the table would have 2^{M+1} entries, 2^M each for sine and cosine. However, the number of entries can be reduced by a factor of 8 by taking advantage of the symmetry and the $\pi/2$ offset between sine and cosine values. As a result, only the sine values from 0 to $\pi/2$ need to be stored.

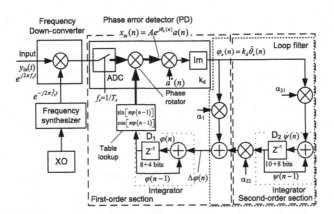

Figure 4.23 An example of second-order DPLL

and in steady state, k_1 and k_2 change to 0.01 and 0.000025. The loop is critically damped in both stages.

Using the phase error expression given by (4.78) and the number formats specified above, $\varphi_e(n) = A \sin(\theta_e(n)) \approx A\theta_e(n)$ and k_d is equal to 1 when $\theta_e(n) \approx 0$. Following the design shown in Section 4.5.1.3, the phase controller's gain k_c is equal to π. Based on these parameters we compute that α_1 and α_2 are equal to 0.032 and 0.0008 during the training stage and equal to 0.0032 and 0.000008 in steady state, respectively.

Let us first consider the phase controller design. As discussed in Section 4.5.1.3, the M_1 bits of the upper portion of register D_1 contains the phase of the input signal samples. The range of D_1 is from -1 to 1 corresponding to a phase range from $-\pi$ to π. By choosing $M_1 = 8$ and using a proper rounding method, the accuracy of the phase is $2^{-8} \approx 0.4$ percent, which should be sufficient for most applications. These 8 bits are used to generate the sine and cosine values by table lookup for performing phase rotation of the samples at the ADC output. The table size is equal to 64 with 8-bit entries. To avoid the loss of precision during accumulation, we have the word length of D_1 as N_1 bits and $N_1 > M_1$. The requirement for N_1 will be determined below. The value of the first-order loop coefficient α_1 as determined above is less than 1. It is usually implemented as a multiplier and a right shifter or just a shifter alone by rounding α_1 to the nearest number in the form of power of 2.

Second, we consider the design of loop filter components. The register D_2 contains the offset frequency that is represented by a fixed-point number in the range of $[-1, 1)$. If the maximum frequency offset is equal to ± 4750Hz as assumed above and the largest magnitude of D_2 is equal to 1, the value of α_{22} can be selected as

$$\alpha_{22} \geq \frac{2\pi \Delta f_{a,\,max} T_c}{A k_c} = 2\Delta f_{a,\,max} T_c = 2 \times 4750/15000 \approx 0.633 \qquad (4.83)$$

according to (4.80) with $A = 1$ and $k_c = \pi$. To simplify implementation and to add some margin, we let α_{22} be equal to 1.0. As a result, the maximum frequency it can handle is about 7500 Hz. In order for the steady-state frequency error to be no greater than ± 10 Hz, we let D_2's upper portion $M_2 = 10$ bits, as $2^{10} > 7500/10$.

To avoid loss of precision during accumulation, the register D_2 has N_2 bits, and $N_2 > M_2$. In the case that $\alpha_2 = \alpha_{21} = 0.000008$, to avoid underflow, N_2 should be equal to, or greater than, $-\log_2(0.000008) \approx 17$ bits. Namely, there are $N_2 - M_2 = 7$ bits for accumulation. Note that 17 bits of N_2 are the minimum number of bits that are needed. It only allows the integer portion of $\varphi_e(n)$ to be accumulated in D_2. It would be desirable to increase the word length of D_2 at least by 1 or 2 bits to ensure more robust operation.

Finally, we determine the total word length of the register D_1. To avoid underflow, we need to ensure that it is adequate for the ranges of both inputs, i.e., $\alpha_1 \varphi_e(n)$ and $\alpha_{22} \psi(n-1)$, to the accumulator. The smallest value of $\alpha_1 \varphi_e(n)$ is equal to 0.0032 if only the integer portion of $\varphi_e(n)$ is considered. In this case, N_1 needs to be 9 bits. If we would also like the fractional portion of $\varphi_e(n)$ to be accumulated in D_1, 11 to 12 bits would be more appropriate. For the input from the second-order loop, because $\alpha_{21} = 1$, the top portion of $\psi(n-1)$ with word length of $M_2 = 10$ bits is directly added to D_1. N_1 should be at least equal to M_2, i.e., 10 bits. Thus, we choose that $N_1 = 12$ to satisfy both requirements for $\alpha_1 \varphi_e(n)$ and $\alpha_{22} \psi(n-1)$.

Based on the given conditions, therefore, the register D_1 should have 12 bits. The top 8 bits are used for table lookup to generate the sine and cosine values used by the phase rotator. The remaining 4 bits is for the accumulation of phase increments $\alpha_1 \varphi_e(n)$. The word length of the register D_2 is set to be 18 bits. The top 10 bits cover the range of frequency offset with sufficient accuracy, and the remaining 8 bits are for the accumulation of $\alpha_{22} \varphi_e(n)$ to track the slow frequency change.

It should be noted that the design discussed above is based on the assumptions for this specific example. The design and values should be changed accordingly if the assumptions change. In particular, above we had assumed that the ADC and the loop filter both operate at the sampling rate $f_s = 15$kS/s. Practically, the ADC could operate at a much higher rate. Thus, the update rate of DPLL's loop filter and sampling rate will be different. Namely, the DPLL will operate in a multirate mode. The implementation of such multirate DPLLs will be considered in the next section.

4.5.3 Topics on System Design and Parameter Selection

So far in this section, we have discussed the implementation of the components of the DPLL. An example of an ASIC/DSP software implementation of a second-order DPLL for carrier synchronization was presented. Below we consider some aspects of DPLL design and implementation from an overall system point of view. The topics to be discussed include the DPLL's operation rate, or update rate, selection, its behavior, and the considerations of the training of the DPLL. In the last part, we consider the characteristics of the DPLL when an additional delay is introduced in the loop.

4.5.3.1   Loop Filter Sampling Rate Selection

In most of the discussions above, we assumed that sampling rate, or update rate, of the loop filter in a DPLL is equal to the receiver signal's sampling rate f_s that, in turn, is equal to the transmitter channel symbol rate $1/T$. This is not the case in most receiver implementations. For example, the loop filter of a DPLL in a wireless receiver usually operates at the rate of the time slots of the received signals. This rate is usually much lower than the channel symbol rate.

Requirement on Loop Filter's Sampling Rate
According to the Nyquist sampling theorem, the minimum sampling rate needs to be higher than twice the maximum frequency of real-valued signals. For complex-valued signals, the minimum sampling rate is equal to the total signal bandwidth. Such a minimum sampling rate is called the *Nyquist sampling rate*. In principle, the same is true for DPLLs. A DPLL's characteristics are determined by its loop filter. Thus, the sampling rate of the loop filter should be at least higher than twice the rate of the time variations that the received signal experiences. These time variations include frequency offset and Doppler fading, which occur when the transmitted signal passes through time-varying channels.

However, the loop filter's sampling rate is usually chosen to be much higher than the minimum requirement. As a rule of thumb, the loop filter's sampling rate should be 10 to 50 times higher than the minimum requirement if possible. The higher rate is necessary because DPLLs are first-order and second-order linear systems. Their roll-off slopes in the frequency domain are not very steep. Hence, a DPLL is not likely to achieve the desired characteristics if the sampling frequency of its loop filter is close to the Nyquist rate of the signal time variation.

Output Phase Jitter
Another factor that needs to be considered is noise aliasing. The additive white noise that corrupts the received signal usually has a bandwidth equal to the received signal's sampling rate. Down sampling the signal to a rate that is lower than the noise bandwidth causes noise aliasing and increases the phase jitter at the DPLL output. For example, if the DPLL's loop filter and PD operate at $1/T_s$, and the DPLL's noise bandwidth is equal to 0.001, the variance of the phase jitter in data samples would be reduced by a factor of 1000 at the DPLL output. If the sampling rate of the loop filter PD is reduced by a factor of N, it may still be capable of tracking the phase's time variation. However, for the tracking characteristic of the DPLL in analog time to remain unchanged after down sampling, the DPLL's loop bandwidth must increase by a factor N.

Since there are N input signal samples between two loop filter updates, N phase error estimates can be generated. By averaging N such estimates to generate a combined phase error estimate, the jitter variance of the combined estimate becomes N times smaller than that of the individual estimates. If the averaged phase error estimates are used as the input to the loop filter with a sampling rate of $1/NT_s$, the variance of DPLL's

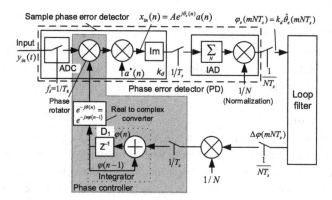

Figure 4.24 A multirate all-digital DPLL

output phase jitter will be the same as the variance when the entire DPLL operates at the rate of $1/T_s$.

Implementation of Multirate DPLLs

A DPLL that has different signal sampling rate and loop filter sampling rate is called a *multirate DPLL*. If a DCO is used as the phase controller, it can be shown that the DPLL's operation is not affected by the loop filter's sampling rate. No change needs to be made in the implementation with a properly designed loop filter. However, as shown below, minor modifications are needed in the design and implementation of the DPLLs with digital rotator-based phase controllers described in Section 4.5.2.

The block diagram of a multirate DPLL is shown in Figure 4.24. The input signal $y(t)$ is sampled by the ADC at the rate of $1/T_s$. The phases of the digital samples at the ADC output are adjusted by a digital phase rotator. The phase errors of the phase-rotated samples are detected by a *sample* phase error detector, which is the same as the one previously described. N-detected phase errors from the samples are added together in a block that performs the *integrate-and-dump* (IAD) function. The generated result is normalized, i.e., divided by $1/N$, to generate an average detected phase error $\varphi_e(mNT_s)$ every NT_s. The sample phase error detector, the IAD and normalization function together form the PD of the multirate DPLL. Due to normalization, the gain k_d of this PD is the same as the gain of the sample phase error detector.

The average detected phase error is processed by the loop filter operated at the rate of $1/NT_s$. The loop filter generates a required phase correction $\Delta\varphi(mNT_s)$ from mNT_s to $(m+1)NT_s$.

The estimated phase correction is integrated in the register D_1. The contents of D_1 are applied to the ADC samples by the digital phase rotator in the phase controller. The integrator and phase-rotator in the shaded area in Figure 4.24 are operated at the rate of $1/T_s$. Thus, the phase correction of $\pi \times \Delta\varphi(mNT_s)$ is distributed over N samples. To achieve the correct per-sample phase-rotation, the input to the integrator is equal to $\Delta\varphi(mNT_s)$ divided by N. The value of D_1 increases by $\Delta\varphi(mNT_s)/N$ every time an integration is done. As a result, the phases of the samples are increased by $\pi \times \Delta\varphi(mNT_s)/N$ from one to the next.

With such an implementation, the behavior of the multirate DPLL will be the same as the single-rate DPLL described in Section 4.5.2, provided they have the same loop filters with the same sampling rates.

4.5.3.2   Multistage Training

When a DPLL is used for synchronization in digital receivers, a multistage training process is usually implemented. This is because when the training starts, the receiver does not have prior information about the frequency and phase of the signal. It is desirable for the receiver to learn such information as quickly as possible, even at the cost of higher estimation error. Once the phase error and/or frequency offset are acquired, it would be necessary to generate accurate estimates of these quantities. However, these two requirements cannot be satisfied by using the same set of DPLL parameters.

To achieve fast learning, a DPLL should have a short time constant, i.e., a wide bandwidth. At the same time, in order to attain low variance of the estimates, the DPLL should have a narrow noise bandwidth to reject most of the noise in the received signal. Because of the conflict between these two requirements, different sets of the parameters, i.e., the loop gain k_1 of the first-order DPLLs or the loop gains k_1 and k_2 of second-order DPLLs, must be used to satisfy these different requirements.

Let us use the first-order DPLL as an example. As shown by (4.48) its time constant is approximately equal to T_s/k_1, i.e., inversely proportional to k_1. Meanwhile, as shown by (4.43), the noise bandwidth of the DPLL is approximately equal to $k_1/2$, i.e., proportional to k_1. The situations of second-order PLLs are more complex than those of first-order PLLs. Nevertheless, the nature of the problem is still the same.

As shown in Figure 4.13, the loop bandwidth is mainly determined by the first-order loop gain k_1 with influence from the damping factor ζ_d, which is determined by the first- and second-order loop gains k_1 and k_2. Moreover, the curves shown in Figure 4.14 and Figure 4.15 indicate that the convergence time is mainly determined by the first-order loop gain k_1 of the second-order PLLs. The difference between the first- and second-order DPLLs is their ability to track phase ramps due to the frequency offset. There is not much difference between them when dealing with step-phase changes.

Better performance is achieved by using a common method in modem receiver design and implementation, i.e., *multistage training*, or "gear shifting."

During multistage training, a DPLL's operations are divided into two or more stages. In the first stage, the DPLL has a wide loop bandwidth, i.e., a short time constant, in order to reach the steady state quickly. However, at the end of this stage, its output phase variance remains high. The DPLL then enters a second training stage with a longer time constant and a narrower loop bandwidth. Given that the majority of phase and frequency errors have already been corrected in the first stage, the objective of the second training stage is to reduce the output phase variance and the remaining phase and frequency errors. More training stages can be employed if necessary.

Changing the loop time constant and bandwidth is straightforward for the first-order DPLL. There is only one parameter, k_1, that we need to determine. Increasing the time constant by a factor of K can be done by simply reducing k_1 to k_1/K. Consequently, the loop bandwidth reduces by a factor of K. However, changing the time constant and loop bandwidth of the second-order loop is more difficult because both loop gains k_1 and k_2 need to be selected correctly. Scaling both by the same factor changes not only the loop time constant but also the damping factor. Such a change could cause undesirable effects if not done carefully.

As shown by (4.61) in Section 4.4.3.3, the damping factor ζ_d is equal to $k_1/2\sqrt{k_2}$. Thus, if k_1 is changed by a factor of $1/K$, it is desirable to change k_2 by a factor of $1/K^2$. Doing so makes the time constant of the loop K times longer, and the damping factor remains unchanged. This conclusion is useful for designing and implementing multistage training.

One question about multistage training often encountered by modem designers is how to decide one stage of training has ended and the next stage could begin. A simple method is to let the DPLL stay in the first stage for a duration of three or four times its time constant. If the time constant is known, this approach can be very effective. However, the time constant of a DPLL is determined by the gains k_d and k_c of its PD and phase controller and these parameters may not be known.

The gain k_d depends on a number of factors such as the received signal magnitude and/or signal to noise ratio of the signal. The gain k_c would be affected by the DCO's gain if the DCO was used in the design. Caution must be taken to make sure that these parameters do not depart much from the design. The adverse effects can also be reduced by setting the duration of the first training stage long enough to ensure that it has converged in the case when the DPLL's time constant is longer than expected.

The transition point from the one stage training to the next could also be determined by measuring the phase variance at the DPLL's output. This method can be effective if all of the conditions are as expected. However, the accuracy of the switching point so determined is often affected by the noise variance and other impairments in the received signal. As a result, the decision could be erroneous. This approach should be thoroughly evaluated under various conditions and used with caution.

4.5.3.3 DPLL with Additional Loop Delay

In some DPLL applications, the estimated phase error $\theta_e(n)$ or $\varphi_e(n)$ may not be available immediately after the input signal sample $x_{in}(n)$ is received. For example, when the DPLL is used for carrier and/or for timing synchronization in receivers, the PDs often use the recovered transmitted data symbols as phase references. In channel-coded transmission, reliable symbol decisions can only be obtained after decoding. With such a design, the phase reference is not available until the decoding is completed.

Another example is timing and/or frequency synchronization in data networks. To simplify the implementation and to reduce overhead, it is attractive to send information needed for achieving synchronization between different entities of the network through packet transmission. However, delays and their variations may be introduced by the packet network during the transmission of such information.

In such cases, it is necessary to introduce delay into the DPLL for its operation. The impact of the additional loop delay on the behavior of the DPLL is considered below.

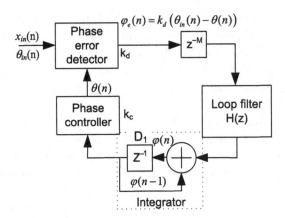

Figure 4.25 Block diagram of a DPLL with loop delay

A simplified block diagram of a DPLL with a loop delay of M samples is shown in Figure 4.25. The system transfer function of such a DPLL can be expressed as

$$\frac{\Phi(z)}{\Phi_{in}(z)} = \frac{H(z)z^{-M}}{z - 1 + H(z)z^{-M}} = \frac{H(z)}{z^M(z - 1) + H(z)} \tag{4.84}$$

For a first-order DPLL, $H(z)$ is a constant k_1, and the transfer function is given by

$$\frac{\Phi(z)}{\Phi_{in}(z)} = \frac{k_1}{z^M(z - 1) + k_1} \tag{4.85}$$

As shown in [14], the loop is stable, provided

$$0 < k_1 < 2\sin\left(\frac{\pi}{2(2M + 1)}\right) \tag{4.86}$$

As a sanity check, if $M = 0$, i.e., there is no delay, we have $0 < k_1 < 2\sin(\pi/2) = 2$, and this result agrees with the stability criterion given in 4.4.2.1. Moreover, the stability criterion given by (4.86) is a special case of the delayed LMS (DLMS) algorithm analyzed in [15, 16].

In addition to stability, the delay introduced in the DPLL also affects its noise bandwidth. By applying the results of the DLMS algorithm given in [16], we can show that the noise bandwidth B_L increases from $k_1/(2 - k_1)$ to $k_1/[2 - (1 + 2M)k_1]$, provided $k_1 \ll 1$ and $(1 + 2M)k_1 < 1$.

When first-order DPLLs are used for synchronization, k_1 is commonly chosen to be much smaller than one. As a result, the stability is usually not a problem as long as M is not too large. If there is a concern, the loop gain of the DPLL should be chosen such that (4.86) is satisfied and simulations should be performed to evaluate the system's behavior.

The stability of second-order DPLLs with loop delays was studied in [14]. The results are much more complex than those for the first-order loop. One useful conclusion given there is that the second-order DPLL is stable if k_2 is less than $1/(2M+1)^2$ and $\sqrt{k_2} \leq k_1 \leq 2.6\sqrt{k_2}$, which corresponds to the damping factor $0.5 \leq \zeta_d \leq 1.3$. This result is useful when introducing delays into second-order DPLLs becomes necessary.

4.6 Applications of Phase-Locked Loops

PLLs play an important role in digital communication systems, in particular, for carrier and timing synchronization. As these applications are the subjects of the next two chapters, only a general view of the PLLs' roles in these functions is given in this section.

PLLs and related techniques are also used to perform non-synchronization functions in transmitters and receivers. For example, even though the automatic gain control is usually not viewed as a type of PLL, the operations of an AGC loop are based on the same principles. In this book, we will not discuss in detail the operations and functions of the AGC. Interested readers can find information about an AGC in the literature, including [17, 18, 19].

Later in this section, we will provide a brief introduction of the PLL's application to frequency synthesis, which is widely used and closely related to synchronization in digital communication systems.

4.6.1 Application of PLLs to Synchronization in Digital Communications

4.6.1.1 PLL for Carrier Synchronization

The PLL is one of the key components in carrier synchronization. Except for feedforward realizations, most realizations of carrier synchronization in digital receivers for communications over static and nearly static channels are implemented as a type of PLL. The differences of carrier synchronization in receivers between different types of systems lie in how the phase error detection is performed. A comprehensive description of second-order PLLs in carrier synchronization for wireline communications will be presented in Section 5.6. What is discussed there is also applicable to digital communications over other nearly static channels, such as microwave and satellite communication channels.

4.6.1.2 Frequency-Locked Loop for Carrier Synchronization

Second-order PLLs can achieve both phase and frequency synchronization if the phase and/or frequency variations introduced by channels are relatively slow. However, if these quantities, especially frequency, change rapidly, a second-order PLL would not be very effective. This is because the second-order section of a second-order PLL usually has a much smaller loop coefficient than the first-order coefficient to attain a proper damping factor. As a result, the time constant of the second-order section is much longer than that of the first-order section. Therefore, a second-order PLL cannot effectively track fast frequency variations. For communication over fast time-varying channels, such as wireless communications over mobile channels, it would be advantageous to track frequency and phase variations separately.

For mobile wireless communications, carrier synchronization is usually divided into two parts: frequency tracking and phase tracking. Frequency tracking is commonly performed by a frequency-locked loop, also called automatic frequency control (AFC), while phase tracking is usually performed through channel estimation.

An FLL is similar to a PLL except it has a frequency error detector (FD) and a frequency controller (FC), while the latter has the phase error detector and phase controller. The operation, implementation, and performance of the FLL will be studied in Section 5.4.7.

4.6.1.3   PLL for Timing Synchronization

In almost all communication systems, timing synchronization in receivers is implemented in the form of first- or second-order PLLs. The changes in timing phase and frequency are usually much slower than the changes in carrier phase and frequency. As a result, a second-order PLL, in this case usually called a *timing locked loop*, *timing control loop*, *timing tracking loop* (TTL), or *delay locked loop* (DLL), can effectively track and maintain receiver sample timing to be synchronous to the remote transmitter timing.

The implementation and analysis of TLLs are very similar to those for PLLs given above. The main differences are in the design and realization of the timing phase error detector (TPD) and the timing phase controller (TPC) in different types of communication systems as will be presented in Chapter 6.

In recent years, the TPCs are implemented by using digital resampling techniques in most digital receivers. This topic is the subject of Chapter 7.

4.6.2   An Introduction to Applications of PLL in Frequency Synthesis

Frequency synthesis is one of the major application areas of PLLs. In this section, we provide the basics of frequency synthesis related to the applications of PLLs. Note that the frequency synthesis technologies are very well developed and involve many technical disciplines. Many details and issues must be addressed and solved in order to build a practical working frequency synthesizer. The involved disciplines and the solutions of such issues, especially RF-related issues, are beyond the scope of this book. Relevant information can be found in the literature such as [6, 20, 21]. In this section, we will only provide some examples to illustrate the roles of the PLL in frequency synthesis.

4.6.2.1   A Basic Frequency Synthesizer

As has been mentioned in Sections 4.5.1.1 and 4.5.1.3, to demodulate the received signal with carrier frequency f_c from a transmitter, the receiver needs to synthesize a local clock, or waveform, with a nominal frequency equal to f_c. As shown in Figure 4.21 and Figure 4.22, the function of a frequency synthesis block is to convert the frequency of the clock generated by a LO, such as an XO or a VCXO/DCXO, to a clock at the carrier frequency. Since the LO's frequency is usually lower than f_c, the conversion

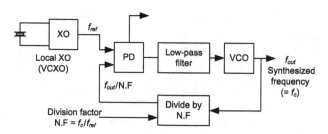

Figure 4.26 A conceptual frequency synthesizer

performed is called frequency *up-conversion*. Figure 4.26 shows the high-level block diagram of a typical frequency synthesizer that performs frequency up-conversion.

As shown in the figure, the output clock with frequency f_{out} is generated by a VCO, which is usually implemented by using either passive LC components or ceramic resonators. The frequency generated is equal to, or higher than, the carrier frequency f_c. In the latter case, f_{out} is divided down to the carrier frequency. Below, we consider the case that f_{out} is equal to f_c.

At the beginning, the output clock with frequency f_{out} generated by the VCO is not exactly equal to f_c. The frequency f_{out} is divided by a precomputed ratio of $N.F$, where N is the integer part, and F is the fractional part. The PD compares the divided-down output clock to the reference clock, which is generated by a local XO with the frequency f_{ref}. The objective of the synthesizer is to lock the phase of the divided-down clock to the phase of the reference clock. Once the phases are locked, the output is at the frequency $f_{out} = N.F \times f_{ref}$ and is phase-locked to the reference. Thus, by choosing $N.F = f_c/f_{ref}$, the output of the frequency synthesizer is equal to f_c and can be used for frequency down-conversion of the received signal.

In order to facilitate the operation of the divider, the VCO output is usually first converted to a rectangular waveform. Dividing down such waveform by an integer, i.e., $F = 0$, can be done easily by using digital logic circuits. If the ratio $N.F$ is a rational number, a hardware divider can be implemented. However, the rectangular clock waveform at the output of a rational-number divider is usually not uniformly distributed. As a result, the phase difference detected would be jittery. Low-pass filtering is used to smooth out the jitter in the PD output before the PD output is used to control the VCO.

If the ratio is not a rational number or the implementation of such a required rational divider is not convenient, it is possible to use the so-called *dithering* technique [22]. Namely, the divider is implemented as a combination of a divided-by-N divider and a divided-by-$(N+1)$ divider. If the ratio of the time spent on using N-divider and $(N+1)$-divider is equal to α/β, the actual division ratio is equal to $N + \beta/(\alpha + \beta)$. This approach can also be used to approximate any irrational ratio. The problem with this approach is that it is likely to generate higher interference in the form of frequency spurs. It has been shown that the spur effect can be reduced by using sigma-delta PLL architectures [6, 22].

The output of the PD is the phase-difference between the divided-down VCO output clock and the reference clock from the local XO output. Since these two inputs are analog waveforms, the PD is usually implemented by using analog multipliers or digital logic circuits, e.g., XOR gates or flip-flop, as commonly used in APLLs [1, 2, 6].

The PD's output is used to control the VCO to form a negative feedback loop. A low-pass filter is inserted between the PD and the VCO to reduce the jitter and interference and to ensure the stable output of the VCO. Because the low-pass filter's bandwidth is usually much wider than the bandwidth of the synthesizer, the filter behaves more like a delay element in the loop than a part of the loop filter. Thus, the entire frequency synthesizer behaves like a first-order mixed-signal PLL.

4.6.2.2 Mobile/Wireless Device Transmitter Frequency Synthesis

In wireless communication systems, the carrier frequencies generated by the BTS are usually very accurate. This is because BTSs can afford to use high-precision (yet high-cost) oscillators. Moreover, BTSs often synchronize their clocks to the clock of GPS or other similar systems to achieve high-frequency accuracy.

However, to save cost and power, mobile device manufacturers prefer to use low-cost XOs. When free running, such XOs can generate only frequency references with low precision, e.g., with an inaccuracy of 2 ppm to 10 ppm or worse. However, due to standard and local regulatory requirements, mobile devices must maintain necessary precision of the carrier frequency of their transmitted signals. During a normal communication session, a device receiver acquires the carrier frequency of the remote BTS through carrier synchronization. The acquired carrier frequency can be used as the local frequency reference by the device. Therefore, the synthesized local reference frequency in the device can approach the accuracy of the remote transmitter's carrier frequency.

The carrier synchronization of receivers is achieved through a phase-locked loop as discussed in Section 4.5.2 and as will be discussed in the next chapter. Carrier synchronization of mobile and wireless devices may also be implemented by using FLLs, which has been mentioned in Section 4.6.1.2 and will be described in Section 5.4.7. If the phase or frequency controller in the PLL or FLL is implemented by using a type of VCXOs, such as VC-TCXO, the output of the local oscillator is synchronous to the remote transmitter carrier after carrier synchronization is achieved. Thus, using the output of the VC-TCXO as the reference for the frequency synthesizer can ensure that the carrier frequency of the device's transmitter attain the required accuracy.

Another possibility is that the receiver's PLL or FLL for carrier synchronization is implemented by using a free-running XO together with a software phase controller implemented in DSP or ASIC. While the frequency of the XO's output is not synchronous to the remote carrier frequency, accurate local transmitter's carrier frequency can be generated based on knowledge of the estimated frequency offset from the PLL or FLL. For example, the local transmitter carrier frequency can be generated as illustrated in Figure 4.27.

As shown in the figure, the output of the XO is used as the reference input to the synthesizer as before. However, the division ratio $N.F$ is modified based on the estimated frequency error of the actual output frequency of the XO as described below.

The receiver synthesizes a frequency $\hat{f}_c$, which is equal to the XO's output frequency scaled up by a factor of $N_c.F_c$. The synthesized frequency $\hat{f}_c$ is nominally equal to the

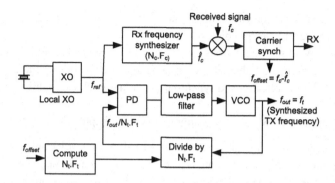

Figure 4.27 Device transmitter frequency synthesizer (carrier recovery DPLL with software phase controller)

carrier frequency f_c and used for demodulation of the received signal. As will be shown in Chapter 5, the carrier synchronization block of the receiver generates an estimate of the frequency offset $\Delta f = f_c - \hat{f}_c$. Thus, the actual XO's output frequency f_{XO} can be computed as $f_c - \Delta f$ scaled down by the factor $N_c.F_c$, i.e., $f_{XO} = (f_c - \Delta f)/H_c.F_c$. The division ratio $N_t.F_t$ to synthesize the required local transmitter carrier frequency $f_{out} = f_t$ is equal to

$$N_t.F_t = f_t/f_{XO} = N_c.F_c \times f_t/(f_c - \Delta f) \qquad (4.87)$$

The operation of the synthesizer is the same as that described above with the understanding that the division ratio is computed in real time, and it could change during operation. When the division ratio is close to an integer N_t, three dividers with the divided ratios of N_t-1, N_t, and N_t+1, may be needed for implementing the dithering method.

Since the offset frequency obtained in the carrier synchronization block may be affected by channel fading, additional averaging of the estimated offset frequency would be necessary to mitigate this effect. Note, however, that the averaging time constant should be less than the XO frequency variation caused by temperature changes and other factors.

4.7   Summary and Remarks

In its history of over 80 years, phase-locked loops have always been closely related to telecommunications. The PLL was developed initially for performing demodulation and is still being widely used in many telecommunication systems today.

For a long time, PLLs were mainly implemented by analog components. As a result, analog APLLs were very well studied, and there are many references in the literature. However, due to the advances in computer and DSP technologies, PLLs used for digital communication applications today are mainly digital DPLLs.

The PLL plays a central role in the development and implementation of synchronization in digital communication systems. It is not an exaggeration to say that it is impossible to study synchronization in digital communications without having a good understanding of the PLL technology.

In this chapter, we discussed various aspects of PLLs, especially DPLLs, and related techniques. In addition to the theoretical fundamentals, details of the implementations and examples of their applications to synchronization in digital communication systems were also given.

The remarks given below serve as a summary of this chapter.

- The most widely used analog and digital PLLs are first-order and second-order PLLs. In most cases, the linear analysis of PLLs is sufficient for an understanding of the properties and behaviors of PLLs in their applications in digital communications.

- The analysis of PLLs is based on the theory of low-order linear systems and, thus, is simple and within the reach for anyone with the understanding of basic linear system theory. Despite its simplicity, such analysis provides useful insights into the understanding of the characteristics and the implementation considerations of PLLs.

- Both the APLL and the DPLL each have three basic components: a phase error detector, a loop filter, and a phase controller. In this chapter, the variations of these components in different applications were presented and their practical implementations were considered and described.

- The key parameters of a PLL, either analog or digital, are the noise bandwidth and the convergence time. The noise bandwidth characterizes the PLL's ability to reject input phase jitter. The convergence time determines how fast the input phase or frequency change can be completely compensated during the transition period after the occurrence of the change. Since a PLL with a narrow noise bandwidth must have a long convergence time, it is only fair to compare different PLLs with one of these two parameters fixed.

- Another important parameter of a second-order PLL is the damping factor, which characterizes its transitional behavior. The DPLL's transition behavior is often studied by mapping it to the corresponding APLL behaviors that are thoroughly investigated and understood. However, this mapping approach has limitations.

- As shown in Section 4.4, DPLLs' transition characteristics can be directly derived in the discrete-time domain in closed-form expressions, similar to those given in [9]. Using these discrete-time expressions directly is convenient and accurate for evaluating DPLL's performance.

- The expression of discrete-time transition behavior shows that the damping factor of a second-order DPLL is related to its loop gains by $\zeta_d = k_1/2\sqrt{k_2}$. This relationship is useful for changing the loop bandwidth and time constant of the DPLL in real time to maintain a constant damping factor such as for multistage training.

- During the multistage training, a short time constant is employed at the beginning to speed up the convergence of the PLL. The residual errors and large jitter variance are reduced in later stages if the time constant is increased and, thus, the noise bandwidth is reduced.

- Additional delays in a DPLL loop often occur due to various system design constraints. Such delays will introduce degradation of the DPLL performance. Nevertheless, the impact of the delay would not be too serious if the amount of delay is relatively short compared with the loop time constant.

- PLLs used for synchronization in wireless devices have been mainly implemented in digital form for the last 30 years. For many years, however, the phase controller in the DPLLs had been implemented in a partly analog form such as a DCO. To reduce the device cost and due to the advances in digital processing in recent years, manufacturers have often chosen all-digital phase controllers based on free running XOs for the design of mobile devices.

- Such a change also presents challenges to meeting the system requirements while saving cost. A good understanding of the fundamentals of PLLs and careful considerations of designs are essential for the system designs to succeed. The design and implementation of the all-digital implementations of DPLLs and the examples given in this chapter could be useful to the readers who are interested in related projects.

References

[1] F. M. Gardner, *Phaselock Techniques*, 1st, 2nd and 3rd edns, New York: John Wiley and Sons, 1966, 1979, 2005.

[2] R. E. Best, *Phase-Locked Loops: Design, Simulation, and Applications*, 5th edn, New York: McGraw-Hill, 2003.

[3] W. F. Egan, *Phase-Lock Basics*, 2nd edn, Hoboken, NJ: John Wiley and Sons, 2007.

[4] W. C. Lindsey and C. M. Chie, Eds., *Phase-Looked Loops*, New York: IEEE Press, 1986.

[5] H. de Bellescize, "La réception synchrone," *L'onde électrique*, vol. 11, pp. 225–40, May 1932.

[6] V. F. Kroupa, *Phase Lock Loops and Frequency Synthesis*, New York: John Wiley and Sons, 2003.

[7] L. A. Hoffman, *Receiver Design and the Phase-Locked Loop*, El Segundo, CA: Aerospace Corp., 1963.

[8] J. G. Proakis and D. G. Manolakis, *Introduction to Digital Signal Processing*, New York: Macmillan Publishing Company, 1988.

[9] W. C. Lindsey and C. M. Chie, "A Survey of Digital Phase-Locked Loops," *Proceedings of IEEE*, vol. 69, no. 4, pp. 410–31, April 1981.

[10] A. V. Oppenheim and R. W. Schafer, *Digital Signal Processing*, Englewood Cliffs, NJ: Prentice Hall, 1975.

[11] A. H. Sayed, "Performance of Sign-Error LMS," in *Adaptive Filters Performance of Sign-Error LMS*, Hoboken, NJ: John Wiley and Sons, 2008.

[12] F. Ling and J. G. Proakis, "Numerical Accuracy and Stability: Two Problems of Adaptive Estimation Algorithms Caused by Round-Off Error," in IEEE International Conference on ICASSP '84 Acoustics, Speech, and Signal Processing, San Diego, CA, 1984.

[13] C. Maxfield and A. Brown. [Online]. Available: www.clivemaxfield.com/diycalculator/popup-m-round.shtml#A1.

[14] J. W. M. Bergman, "Effect of Loop Delay on Stability of Discrete-Time PLL," *IEEE Trans. on Circuits and Systems – I*, vol. 42, no. 4, pp. 229–31, April 1995.

[15] P. Kabal, "The Stability of Adaptive Minimum Mean Square Error Equalizer Using Delayed Adjustment," *IEEE Transactions on Communications*, vols. COM-31, no. 3, pp. 430–32, March 1983.

[16] G. Long, F. Ling, and J. G. Proakis, "Corrections to 'The LMS algorithm with delayed coefficient adaptation'," *IEEE Transactions on Signal Processing*, vol. 40, no. 1, pp. 30–32, January 1992.

[17] J. P. Alegre Perez, S. C. Pueyo, and B. C. Looez, *Automatic Gain Control, Techniques and Architectures for RF Receivers*, Berlin: Springer, 2011.

[18] M. Vucic and M. Butorac, "All-Digital High-Dynamic Automatic Gain Control," in IEEE International Symposium on Circuits and Systems, 2009. ISCAS 2009, Taipei, Taiwan, 2009.

[19] A. Liu, J. An, and A. Wang, "Design of a Digital Automatic Gain Control with Backward Difference Transformation," in Sixth International Conference on Wireless Communications Networking and Mobile Computing (WiCOM), Chungdu, 2010.

[20] V. Manassewitsch, *Frequency Synthesizers: Theory and Design*, Hoboken, NJ: Wiley-Interscience, 1987.

[21] W. F. Egan, *Frequency Synthesis by Phase Lock*, 2nd edn, Hoboken, NJ: John Wiley and Sons, 1999.

[22] A. E. Hussein and M. I. Elmasry, "A Fractional-N Frequency Synthesizer for Wireless Communications," in 2002 IEEE International Symposium on Circuits and Systems, Phoenix-Scottsdale, AZ, 2002.

5 Carrier Synchronization

5.1 Introduction

Carrier synchronization is one of the key components in digital communication systems. Briefly, carrier synchronization concerns the estimation and compensation of the carrier frequency and phase differences between the transmitted signal and the corresponding received signal. Although carrier synchronization is relatively simple in principle, it directly affects the overall system performance including capacity and the robustness under adverse conditions. In this chapter, we discuss various aspects of the theory and realizations of carrier synchronization. Examples of designs and practical implementations will also be presented. This chapter is organized as follows.

In Section 5.2, we provide an overview of carrier synchronization in digital communication systems. Classic analog processing-based passband carrier synchronization techniques are presented in Section 5.3. Aspects of carrier synchronization based on digital signal processing of baseband signal samples are presented in the subsequent sections. Specifically, Sections 5.4 and 5.5 focus on basic techniques of carrier synchronization for single-carrier and multicarrier digital communications. Section 5.6 considers the implementations and properties of carrier synchronization for wireline communications. These aspects of wireless communications are presented in Section 5.7. Section 5.8 outlines the implementations and practical considerations for four popular wireless communication systems in deployment today. Finally, Section 5.9 provides a summary and remarks on what we discussed in this chapter.

5.2 Overview of Carrier Synchronization in Digital Communication Systems

To facilitate the description of carrier synchronization in communication systems, Figure 5.1 depicts a simplified block diagram of a typical single-carrier digital communication system. For clarity, only the blocks pertinent to carrier synchronization are shown.

As shown in the figure, in the transmitter, coded bits are mapped to transmitter modulation symbols for various constellations, e.g., PSK, PAM, and QAM. The modulation symbols are represented by complex numbers, generated at the Tx symbol rate $1/T$, where T is the transmitter symbol interval, and filtered by a *transmitter filter*. The output digital samples from the filter are generated at a rate higher than, but synchronous to, the transmitter symbol rate, e.g., at m/T, where m is an integer. The

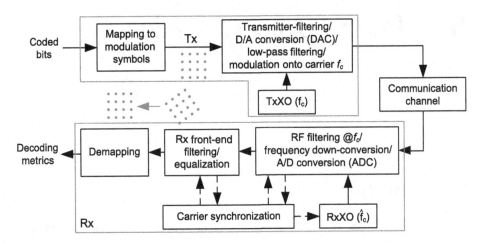

Figure 5.1 A simplified block diagram of a communication system

digital samples from the transmitter filter are *baseband* transmitter signal samples that have a spectrum determined by the transmitter filter and suitable for transmission. Thus, the transmitter filter is also called a *spectrum-shaping filter*, or *pulse-shaping filter*.

The baseband transmitter signal samples are converted to analog form by a DAC operated at the rate of m/T and low-pass filtered. The analog baseband signal is first modulated onto carrier frequency f_c. The modulated signal is transmitted over a communication channel and received by the receiver.

In the receiver, the received signals from the communication channel at the carrier frequency of the remote transmitter are down-converted in frequency to baseband. The *down-conversion frequency* $\hat{f}_c$, sometimes also called *demodulation frequency*, is generated locally and is nominally equal to f_c. The generated baseband signal is filtered by a receiver front-end filter and converted to digital samples by an ADC. To achieve timing synchronization, the Rx sampling rate should be synchronous to the transmitter symbol rate $1/T$. However, since the receiver sampling clock is generated locally in the receiver, this is usually not true. How to achieve timing synchronization is the subject of the next chapter. In this chapter, when dealing with carrier synchronization, we will assume that the transmitter and receiver sampling clocks are already synchronized.

The generated digital samples are processed to generate the estimates of the transmitted modulation symbols, which, in turn, generate the decoding metrics of the coded bits. To coherently receive the original transmitted data, the generated symbol estimates should be as close to the original transmitted symbols, maybe up to a scale factor, as possible. This requires the time response of the *overall channel* to be approximately an impulse. Equivalently, the aliased frequency spectrum of the overall channel should satisfy the Nyquist criterion [1, 2]. However, even if this condition is satisfied, there may still be differences between the phases of the transmitted symbols and their estimates if carrier synchronization is not achieved.

Let us look at the example shown in Figure 5.2 for 16QAM symbols. The small circles are the constellation points of the transmitted modulation symbols, and the gray

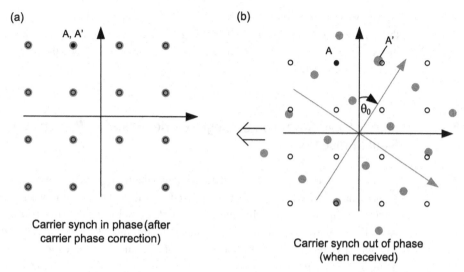

(a)

A, A'

Carrier synch in phase (after
carrier phase correction)

(b)

A   A'

θ_0

Carrier synch out of phase
(when received)

Figure 5.2 Transmitted and received modulation symbols

balls are their estimates generated by the receiver when there is no noise. Assume that
the symbol at constellation point A (the black dot) is transmitted. When the phase of
the transmitted symbol is close to the phase of its estimate, the constellation point A′
is received and is on top of A, as shown in Figure 5.2(a). Thus, the receiver can
determine with high probability from A′ that A is transmitted even if there are other
impairments.

However, if there is an offset between the transmitter and receiver carrier phases,
for example, the estimate A′ is rotated by an angle θ_0 as shown in Figure 5.2(b), then
it may not be possible to reliably recover the original bits that generate the modulation symbol from the estimate of the symbol. The objective of carrier synchronization is to correct such phase errors by bringing the phase of the estimate to align
with the phase of the originally transmitted symbol from (b) to (a) as shown in the
figure.

If noncoherent modulations, such as DPSK or orthogonal modulations were
employed, the phase offset described above would not be a problem for noncoherent
reception. However, noncoherent receivers usually have inferior performance compared
to coherent receivers. Thus, coherent communications are preferred for system designs
when possible.

Phase differences between the Tx modulation symbols and their estimates are mainly
caused by the following two factors. First, the overall channel, including the transmitter
filter, the communication channel, and the receiver filter/equalizer, may introduce a
fixed or slow varying nonzero phase offset θ. Second, the down-conversion frequency
$\hat{f}_c$ generated in the receiver is usually not exactly equal to the transmitter carrier
frequency f_c. Because the phase is equal to the integration of the frequency over time,
the phase differences between the transmitted modulation symbols and their estimates

change with time according to $\theta(t) = \theta + 2\pi\Delta f t$ (radian), where $\Delta f = f_c - \hat{f}_c$ is called the *carrier frequency offset*. Thus, as long as Δf is not exactly equal to zero, the phase difference will always become large enough over time to degrade the receiver performance if not corrected.

5.3   Carrier Synchronization of Single-Carrier Systems: Passband Analog Signal Processing

In order to perform effective carrier synchronization, it is necessary to detect the carrier phase and frequency of the received signal. In this section, we present classic carrier synchronization techniques that employ analog processing of passband signals. These analog signal processing-based techniques were widely used in wireline and wireless receivers in the early days of digital communications. They are still being used in some of today's digital receivers. Carrier synchronization based on digital processing of baseband signal samples, which are used in most of today's single- and multicarrier communication systems, will be discussed in the sections that follow.

5.3.1   Passband Carrier Phase and Frequency Synchronization: Non-Data-Assisted Analog Processing

In digital communication systems, the passband signal arriving at the receiver front end after passing through the channel can be expressed in the general form of

$$y_c(t) = \mathrm{Re}\left[x(t)e^{j(2\pi f_c t + \theta)}\right] + z_c(t) \tag{5.1}$$

where $x(t)$ is the received baseband signal, f_c is the carrier frequency, θ is the fixed channel phase, and $z_c(t)$ is the additive noise. Both θ and $z_c(t)$ are introduced during transmission. As discussed in Section 1.3, a large class of received baseband signals can be expressed as

$$x(t) \cong x(t, \tau) = \sum_{k=-\infty}^{\infty} a_k g_c(t - kT - \tau) \tag{5.2}$$

where a_k's are the channel modulation symbols, which may take different forms of constellations such as AM, PSK, or QAM; $g_c(t-kT-\tau)$ is the impulse response of the *composite baseband channel*, or simply *composite channel*, including the transmitter filter and the channel; and τ is an unknown, but fixed, time delay introduced during transmission.

In Section 4.2, we considered APLLs with sinusoidal inputs. For carrier synchronization, the carrier phase information is contained in the data-modulated received signal. In most systems, carrier phase and frequency information cannot be extracted from the received signal by linear processing as shown below.

For example, the mean of $y_c(t)$ given by (5.1) is

$$E[y_c(t)] = \mathrm{Re}\left[E\left[x(t)e^{j(2\pi f_c t+\theta)}\right] + z_c(t)\right] = \mathrm{Re}\left[\bar{x}e^{j(2\pi f_c t+\theta)}\right] = \bar{x}\cos\left(2\pi f_c t + \theta\right) \quad (5.3)$$

If the mean $\bar{x}$ of $x(t)$ is not equal to zero, the carrier information can be extracted by averaging $y_c(t)$. However, transmission with $\bar{x} \neq 0$ is not efficient, because a portion of the transmission power is dedicated to the carrier instead of the information carrying signal $x(t)$. In fact, *suppressed-carrier* (SC) modulation is usually used so that no power is allocated to the transmission of the carrier, i.e., $\bar{x} = 0$. Though the carrier information cannot be obtained by averaging $y_c(t)$, we can still extract the phase information from $y_c(t)$ by taking advantage of its *cyclostationarity* through nonlinear processing [2, 3, 4] as shown below.

5.3.2 Squaring Loop

Let us consider the transmission with double-sideband/suppressed-carrier (DSB/SC) modulation, such as pulse amplitude modulation and binary phase shift keying, of real $x(t)$. The received passband signal can be expressed as

$$y_c(t) = x(t)\cos\left(2\pi f_c t + \theta\right) + z_c(t) \quad (5.4)$$

The phase information can be extracted by squaring the signal $y_c(t)$ to generate a second harmonic component at $2f_c$. By ignoring the noise term, from (5.4) we have

$$E[y_c^2(t)] = E\left\{[x(t)\cos\left(2\pi f_c t + \theta\right)]^2\right\} = \frac{1}{2}E[x^2(t)][1 + \cos\left(4\pi f_c t + 2\theta\right)] \quad (5.5)$$

In (5.5), note that $E[x^2(t)] > 0$ and $E[y_c^2(t)]$ contains the carrier frequency and phase information. As shown in [3], the carrier frequency and phase information can be recovered by using a PLL with its output at twice the carrier frequency. A block diagram of such a carrier recovery circuit is shown in Figure 5.3.

The figure shows that the squarer output given by (5.5) is multiplied by $\sin\left(4\pi f_c t + 2\hat{\theta}\right)$, which is generated by a VCO with a natural frequency equal to twice the carrier frequency f_c. The multiplier's output has three terms at $2f_c$, $4f_c$, and zero frequencies. After passing through a low-pass filter with frequency response H_{LP}, the two higher frequency components are removed. The remaining term is proportional to twice the phase difference between the input signal and the VCO output. The multiplier's output is filtered by the loop filter H_{LF} and sent back to control the VCO to close the loop. This loop filter has a pole at zero to form a second-order PLL so that the average phase error $\theta - \hat{\theta}$ is equal to zero at steady state as discussed in Section 4.2.

The output of PLL is locked to the squarer output with a phase shift of $\pi/2$. Its frequency is divided down by a factor of two to yield the recovered carrier of the received signal with a phase offset of $\pi/2$.

It should be pointed out that the squaring operation removes the sign information of $x(t)$. Thus, the output of the PLL has a 180-degree ambiguity relative to the carrier phase in $y_c(t)$. Namely, a modulation symbol a_k could be mistaken as $-a_k$. This ambiguity can be overcome by using differential encoding or 180-degree phase-invariant channel coding.

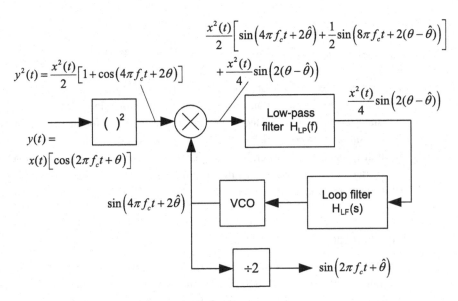

Figure 5.3 Squarer-based carrier recovery circuit

The squaring loop shown above follows the design described in [3]. There are other variations existing in the literature. The low-pass filter H_{LP} in Figure 5.3 can be removed, and the high-frequency components could be rejected by the loop filter [4]. Alternatively, a band-pass filter can be placed at the output of the squarer and the output of the filter is multiplied with the VCO output [5, 2]. These variations have their own merits and shortcomings. The choice should be based on the requirements of the application and implementation.

Above we have demonstrated that the squaring loop can be used to recover the carrier sinusoid for PAM and BPSK modulations. To simplify the description, we ignored the noise term in the signal. In reality, the squaring operation introduces additional noise at the input of the PLL.

By using the expression of $y_c(t)$ as in (5.1) with the noise term $z_c(t)$ included and defining $s(t) \cong x(t) \cos (2\pi f_c t + \theta)$ to simplify notation, we express $y_c^2(t)$ as

$$y_c^2(t) = [s(t) + z_c(t)]^2 = s^2(t) + 2s(t)z_c(t) + z_c^2(t) \tag{5.6}$$

The last two terms on the right side of (5.6) are due to noise $z_c(t)$. Due to these two terms, the variance of the phase jitter at the output of a squaring PLL is larger than that at the output of a PLL with a sinusoidal input at the same SNR as $y_c(t)$. The increase of the noise variance is characterized by the *squaring loss factor* S_L [5], which is defined by

$$\sigma_{\theta,\ squaring\ loop}^2 = \sigma_{\theta,\ PLL}^2 / S_L \tag{5.7}$$

where S_L is always less than one. The exact value of S_L depends on the design and implementation of the squaring loop. From the results given in [2, 4, 5], we can express S_L by the following general form:

$$S_L = \left(1 + a/\gamma_{y_c}\right)^{-1} \tag{5.8}$$

where a is a constant depending on the implementation of the squaring loop and is most likely in the range of 0.25 to 2. From (5.8), we conclude that the increase in the jitter variance is inversely proportional to the SNR of the input signal. Therefore, the performance of the squaring loop asymptotically approaches the performance of PLLs with sinusoidal inputs when the SNR of the input signal goes to infinity.

5.3.3 Costas Loop

A popular alternative to the squaring loop for carrier recovery of DSB/SC signals is the *Costas loop*, which was developed in the 1950s by John Costas [6] at General Electric Company. A block diagram of the Costas loop is shown in Figure 5.4. Its operations are described below.

The natural frequency of the VCO in the Costas loop is the same as the received signal carrier frequency f_c. The VCO generates a sinusoidal waveform $\sin(2\pi f_c t + \hat{\theta})$, which is sent to the multiplier at the lower branch of the loop. Another sinusoidal waveform $\cos(2\pi f_c t + \hat{\theta})$ is generated by a $\pi/2$ phase shift of $\sin(2\pi f_c t + \hat{\theta})$ and is sent to the multiplier in the upper branch. The input signal $y_c(t) = x(t)\cos(2\pi f_c t + \theta) + z(t)$ is multiplied by these two sinusoidal waveforms. Without the noise terms, the outputs of the two multipliers have the following expressions:

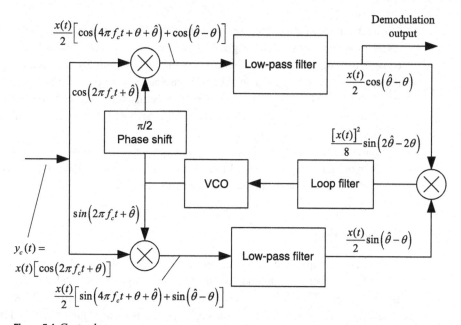

Figure 5.4 Costas loop

$$x(t) \cos\left(2\pi f_c t + \theta\right) \cos\left(2\pi f_c t + \hat{\theta}\right) = \frac{x(t)}{2} \cos\left(4\pi f_c t + \theta + \hat{\theta}\right) + \frac{x(t)}{2} \cos\left(\hat{\theta} - \theta\right)$$

$$(5.9)$$

and

$$x(t) \cos\left(2\pi f_c t + \theta\right) \sin\left(2\pi f_c t + \hat{\theta}\right) = \frac{x(t)}{2} \sin\left(4\pi f_c t + \theta + \hat{\theta}\right) + \frac{x(t)}{2} \sin\left(\hat{\theta} - \theta\right)$$

$$(5.10)$$

As shown in the figure, the terms at twice the carrier frequency at the output of multipliers are removed by the low-pass filtering. A third multiplier of the loop multiplies the two remaining terms, i.e., $[x(t)/2] \cos\left(\hat{\theta} - \theta\right)$ and $[x(t)/2] \sin\left(\hat{\theta} - \theta\right)$, to generate

$$\frac{x(t)}{2} \sin\left(\hat{\theta} - \theta\right) \frac{x(t)}{2} \cos\left(\hat{\theta} - \theta\right) = \frac{x^2(t)}{8}\left[\sin\left(2\hat{\theta} - 2\theta\right)\right] \qquad (5.11)$$

The phase of the output of the third multiplier given by (5.11) is approximately equal to twice the phase difference between the phases of the input signal and the output of the VCO if the difference is small. It is then filtered by the loop filter of the Costas loop. The output of the loop filter controls the frequency of the VCO to form a feedback loop. With a properly designed loop filter, the Costas loop drives the phase difference to zero. Thus, the VCO generates the recovered carrier of $y_c(t)$ with a $\pi/2$ phase shift.

With the additive noise in the input signal $y_c(t)$, we must consider the noise components at the input of the loop filter and thus the phase jitter in the VCO output. It can be shown that the variance of the phase jitter is the same as that generated from a squaring loop. Actually, the squaring loop and the Costas loop have equivalent performances as shown in [7].

An advantage of the Costas loop is that its VCO operates at the same frequency as the carrier frequency of the input signal. Thus, it can be used to perform frequency down-conversion without additional circuitry. As shown in Figure 5.4, the multiplier output in the upper branch of the loop contains the baseband signal once the loop converges, i.e. $E\left[\hat{\theta} - \theta\right] = 0$. Moreover, if the low-pass filter is matched to the channel and sampled at the right time, the filter generates samples that are optimal for estimation of the data symbols and for decoding by the receiver.

Finally, while the Costas loop was developed for carrier recovery of DSB signals, it can be modified to operate with MPSK signals [7]. However, due to the complexity of the modified Costas loop, it has rarely been used in practice.

5.3.4   Mth-Power Loop

The squaring loop and the corresponding Costas loop described above are appropriate for carrier recovery and demodulation of DSB/SC signals with real $x(t)$, e.g., PAM and BPSK signals. They do not work if $x(t)$ is complex, such as for MPSK and QAM modulation. To show why this is the case, we consider the signal modulated by generic complex symbols.

The received signal $y_c(t)$ with complex modulation symbols can be expressed as

$$y_c(t) = \text{Re}\left[x(t)e^{j(2\pi f_c t + \theta)}\right] + z(t) = a(t)\cos(2\pi f_c t + \theta) - b(t)\sin(2\pi f_c t + \theta) + z(t)$$

(5.12)

where both $a(t)$ and $b(t)$ are real. By ignoring the noise term as done earlier, we can express the expectation of $y_c^2(t)$ by

$$E\left[y_c^2(t)\right] = E[a^2(t)]\cos^2(2\pi f_c t + \theta) + E[b^2(t)]\sin^2(2\pi f_c t + \theta)$$
$$- 2E[a(t)b(t)]\cos(2\pi f_c t + \theta)\sin(2\pi f_c t + \theta)$$

(5.13)

Assuming that $E[a^2(t)] = E[b^2(t)] = A^2$ and $E[a(t)] = E[b(t)] = E[a(t)b(t)] = 0$, we can show that

$$E\left[y_c^2(t)\right] = A^2\left[\cos^2(2\pi f_c t + \theta) + \sin^2(2\pi f_c t + \theta)\right] = A^2$$

(5.14)

Thus, there is no frequency and phase information of the carrier in $E[y^2(t)]$. Since the assumptions that lead to (5.14) are true for many types of complex modulation symbols, including MPSK and QAM, the squaring loop is not usable for carrier recovery of such signals.

For signals transmitted using MPSK modulation, the squaring loop can be generalized for performing their carrier recovery. MPSK modulation symbols are represented by complex numbers in the form of $e^{j2\pi m/M}$, $m = 0, \ldots, M-1$, and the received MPSK signal can be expressed as

$$y_c(t) = \text{Re}\left[Ae^{j2\pi m/M}e^{j(2\pi f_c t + \theta)}\right] + z_c(t) = A\cos\left(2\pi f_c t + 2\pi\frac{m}{M} + \theta\right) + z_c(t) \quad (5.15)$$

The value of m determines the phase of the MPSK symbol, which carries the transmitted information. This value changes from one data symbol to the next and is unknown to the receiver. To recover the carrier frequency and phase information, we must remove this unknown phase by using, not the squaring device, but an Mth-power-law device. At the output of this device, there is an Mth harmonic of the carrier given by

$$\cos\left[M\left(2\pi f_c t + 2\pi\frac{m}{M} + \theta\right)\right] = \cos(2\pi Mf_c t + 2m\pi + M\theta) = \cos(2\pi Mf_c t + M\theta)$$

(5.16)

Due to the 2π-periodicity of the cosine function, the unknown value of m is removed from the Mth harmonic. The right side of (5.16) contains the carrier phase and frequency information, which can be extracted by using a VCO with a natural frequency of Mf_c.

For this purpose, we modify the squaring loop shown in Figure 5.3. In a carrier recovery loop for MPSK signaling, the squarer is replaced by the Mth-power device and the VCO is tuned to Mf_c instead of $2f_c$. The output of the VCO is divided by a factor of M in frequency to generate the recovered carrier signal for performing frequency down-conversion.

Due to the Mth-power operation, the recovered carrier has a phase ambiguity of $2\pi/M$. This issue can be addressed by using differential encoding in the transmitter and corresponding differential decoding in the receiver.

Another issue of the Mth-power loop is its noise enhancement. In general, a larger M always associates with a larger phase jitter at the output of VCO. The cases of $M = 4$ and 8 were evaluated in [5].

5.3.5   Carrier Recovery of QAM by a Fourth-Power Loop

Previously, we have shown that the Mth-power device can be used for carrier recovery of signals with M-PSK modulation. One special case is to use fourth-power loops for performing carrier recovery of QAM signals. This case is of special interest because QAM modulations are widely used in many digital communication systems.

Symbols with QAM constellations are two-dimensional modulation symbols. A 2^MQAM symbol represents M binary bits and with 2^M constellation points. If M is an even number, e.g., $2^M = 16$, 64, or 256, 2^MQAM symbols have a perfect square constellation, such as the 16QAM constellation shown in Figure 5.2. If M is an odd number, the 2^MQAM constellation can be constructed from a square constellation by taking off some corner points. A fourth-power loop works for QAM signals and is especially effective for QAM signals with square constellations.

QAM symbols can be represented by complex numbers in the form of $a(t)+jb(t)$. The fourth-power of $y_c(t)$ given by (5.12) can be shown to have a DC term, second harmonic ($2f_c$) terms. and fourth harmonic ($4f_c$) terms. After simplification and taking averaging, the second harmonic terms are zero and only the fourth-harmonic terms contain the phase information, which can be expressed as

$$E\left[y_c^4(t)\right] = \frac{1}{8}\left(\overline{[a(t)]^4} - 6\overline{[a(t)]^2[b(t)]^2} + \overline{[b(t)]^4}\right)\cos\left(8\pi f_c t + 4\theta\right)$$
$$- 4\left(\overline{[a(t)]^3[b(t)]} - \overline{[a(t)][b(t)]^3}\right)\sin\left(8\pi f_c t + 4\theta\right) + ... \tag{5.17}$$

Furthermore, the coefficient of $\sin\left(8\pi f_c t + 4\theta\right)$ is zero since both $a(t)$ and $b(t)$ have zero means. It can also be shown that the coefficient of $\cos\left(8\pi f_c t + 4\theta\right)$ is not zero for QAM constellations. Thus, $y^4(n)$ can be used for carrier recovery of QAM signals.

A fourth-power loop for carrier recovery of QAM signals is shown in Figure 5.5. The received signal $y_c(t)$ is first processed by a fourth-power device. Its output is filtered by a band-pass filter centered at $4f_c$ to remove the $2f_c$ and DC terms. The output of the band-pass filter is multiplied by a sinusoidal waveform at a frequency of $4f_c$ generated by the VCO in the loop. The multiplier output has a component at $8f_c$ and a DC component, which contains the information of the phase difference between the VCO output and the input signal. The $8f_c$ term is removed by the loop filter. The phase difference of the loop filter output controls the VCO to form a feedback loop.

The output of the VCO can be used as the recovered carrier after being divided by 4 in frequency. However, it should be noted that, due to the fourth-power operation, there is a 90-degree phase ambiguity of the recovered carrier. Differential encoding or 90-degree phase invariant channel coding should be used to overcome such a phase ambiguity.

While the fourth-power loop can be used for carrier recovery of QAM signals, its behavior depends on the mean of the second- and fourth-power terms in (5.17). Moreover,

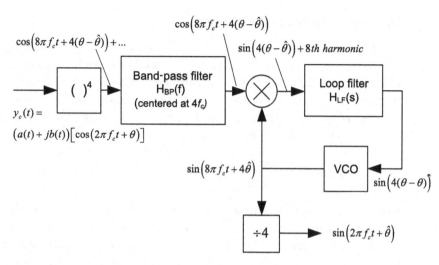

Figure 5.5 Fourth-power loop for QAM carrier recovery

its phase jitter could be quite large due to the interference introduced by the off-diagonal elements in the QAM constellation. As a result, the performance of the fourth-power loop for carrier recovery of QAM signals is worse than the performance when it is used for the QPSK signal.

5.4 Carrier Synchronization of Single-Carrier Systems: Baseband Digital Signal Processing

In today's digital communication systems, receivers are most likely implemented by using digital processing techniques. Starting from this section, we will concentrate on these techniques and their implementations.

Detection of carrier phase and frequency using discrete-time digital signal processing of baseband signal samples is used in most modern digital communication systems. Hence, it has special theoretical and practical importance. In this section, we discuss a few of the basic detection techniques in single-carrier digital communications. Both data-assisted and non-data-assisted techniques for detecting the carrier phase and frequency from the received signal samples will be discussed. The frequency locked loop for carrier frequency synchronization will be specifically described and analyzed as it is widely used in mobile wireless receivers. The methods appropriate for synchronization in OFDM systems will be discussed separately in Section 5.5.

When dealing with baseband signal samples, the carrier phase is reflected in the phase of the baseband equivalent channel, and the carrier frequency error is reflected in the frequency offset generated during the frequency down-conversion of the received signal. Therefore, in the discussion below, we will consider the detection and correction of the channel phase and frequency offset. To begin, we describe maximum likelihood carrier phase estimation.

5.4.1   Maximum Likelihood Phase Estimation

As discussed in Section 5.2, the expression for the passband received signal at the carrier frequency f_c is given by (5.1) and its equivalent baseband signal has the form of (5.2). The passband signal at the carrier frequency is down-converted to baseband by a frequency down-converter that uses a locally generated down-conversion frequency $\hat{f}_c$, which is nominally equal to f_c. The baseband signal $r(t)$ at the output of the down-convertor can be expressed as

$$r(t) = e^{j(2\pi\Delta ft+\theta)} \sum_{k=-\infty}^{\infty} a_k g_c(t - kT - \tau) + z(t) \cong e^{j\theta_t} x(t,\tau) + z(t) \qquad (5.18)$$

where $z(t)$ is an AWGN process; θ and τ are the constant phase offset and time delay introduced during the transmission; Δf is the frequency offset, which is equal to $f_c - \hat{f}_c$ and is generated during frequency down-conversion; a_k's are the transmitter channel modulation symbols; and $x(t,\tau)$ is the equivalent baseband signal. The impulse response $g_c(t)$ of the composite channel consisting of the transmitter filter and the channel has a finite support, i.e., $g_c(t) = 0$, if $t < -T_1$ or $t > T_2$.

The objective of carrier phase detection is to determine the total phase error θ_t introduced by the channel and during frequency down-conversion. Since Δf is usually small, we can assume that $\theta_t = (2\pi\Delta ft + \theta)|_{\text{mod } 2\pi}$ is a constant between $-\pi$ and π during the phase detection time near t. To make the analysis tractable, we assume that $z(t)$ is AWGN with zero mean and noise density N_0. Moreover, we assume perfect timing for the known delay $\tau = \tau_0$.

If the *receiver filter*, which performs the overall receiver baseband filtering, has an impulse response of $g_c^*(-t)$, it is the matched filter of the received signal $r(t)$. As shown in Sections 2.2.2 and 3.2.1, the log-likelihood function of the carrier phase based on the received signal $r(t)$, which contains the data symbols a_k's, $n - L < k \leq n$, can be expressed by

$$l(\theta_n) = \text{Re}\left[e^{-j\theta_n} \sum_{k=n-L+1}^{n} a_k^* y(kT + \tau_0)\right] \qquad (5.19)$$

where θ_n is the carrier phase at nT, which is a constant during the detection time $(n - L + 1)T \leq t \leq nT$, a_k's are the known transmitted modulation symbols as defined in Section 3.2.1, and $y(kT + \tau_0)$ is the receiver filter output sampled at $kT + \tau_0$. It can be shown that for $r(t)$ given by (5.18), $y(kT + \tau_0)$ is the MF output that maximizes the signal energy of the symbol a_k in the sample. Since the data symbols a_k's and the delay τ_0 are assumed to be known to the receiver, only the phase θ_n needs to be estimated.

The summation in (5.19) can be expressed as

$$\sum_{k=n-L+1}^{n} a_k^* y(kT + \tau_0) \cong \sum_{k=n-L+1}^{n} a_k^* y_k = e^{j\hat{\theta}_n} \left|\sum_{k=n-L+1}^{n} a_k^* y_k\right| \qquad (5.20)$$

where, by definition, $y_k \cong y(kT + \tau_0)$ and $\hat{\theta}_n = \text{Arg}\left|\sum_{k=n-L+1}^{n} a_k^* y_k\right|$. Substituting (5.20) into (5.19), we obtain

$$l(\theta_n) = \text{Re}\left[e^{j(\hat{\theta}_n - \theta_n)} \left| \sum_{k=n-L+1}^{n} a_k^* y_k \right|\right] = \cos\left(\hat{\theta}_n - \theta_n\right) \left| \sum_{k=n-L+1}^{n} a_k^* y_k \right| \qquad (5.21)$$

It is clear from (5.21) that $l(\theta_n)$ reaches the maximum if $\hat{\theta}_n = \theta_n$. Thus, if the noise term $z(t)$ in (5.18) is AWGN, it can be shown that $\hat{\theta}_n$ is the ML estimate of θ_n.

In short, if the receiver filter is the MF of the received signal, the ML estimate of θ_n is

$$\hat{\theta}_{n,ML} = \text{Arg}\left[\sum_{k=n-L+1}^{n} a_k^* y_k\right] \cong \text{Arg}[u_{n,L}] = \arctan\left(\frac{\text{Im}[u_{n,L}]}{\text{Re}[u_{n,L}]}\right) \qquad (5.22)$$

where $u_{n,L}$ is a complex scalar that is equal to $\sum_{k=n-L+1}^{n} a_k^* y_k$.

5.4.2 Data-Assisted Carrier Phase Detection

As shown above, the ML estimate of the carrier phase is the angle of $u_{n,L}$, which is the correlation of the matched filter output sample sequence $\{y_k\}$ and the corresponding transmitted data symbol sequence $\{a_k\}$, $n - L < k \leq n$. It should be noted that matched filtering of the received signal also constitutes the optimal receiver front end of other receiver functions. For example, the optimal front end of an equalizer consists of a matched filter and a T-spaced sampler [8, 9]. Initial acquisition described in Chapter 3 also uses similar preprocessing. Thus, this conclusion leads to a unified architecture of the receiver front end. Because the data-assisted carrier phase estimation techniques are based on the scalar quantity $u_{n,L}$ generated from the correlation given by (5.22), it is important to have a good understanding of the properties of $u_{n,L}$.

Assume that the receiver filter is the matched filter of the received signal with an output rate of $1/T$. The ML estimate of θ_n can be generated from the output samples of the filter near nT, i.e., y_k's, $n - L < k \leq n$, as shown by (5.22). Moreover, if the overall channel including $g_c(t)$ and the receiver filter satisfies the Nyquist criterion at the sampling time τ_0, y_k can be expressed as

$$y_k = g_n a_k + z_k \cong e^{j\theta_n}|g_n|a_k + z_k \qquad (5.23)$$

where $g_n \cong e^{j\theta_n}|g_n|$ is the single tap of the *overall channel impulse response* (CIR) at nT. The term z_k contains noise and interference, including ISI, with a variance of σ_z^2. To simplify the analysis, we assume that z_k is Gaussian, i.e., normal, distributed with zero mean. The SNR of y_k, including y_n, is equal to

$$\gamma_y = A^2|g_n|^2/\sigma_z^2 \qquad (5.24)$$

where $A^2 = E\left[|a_k|^2\right]$.

The correlation $u_{n,L}$ defined in (5.22) can be expressed as

$$u_{n,L} \cong e^{j\theta_n}|u_{n,L}| = \sum_{k=0}^{L-1} a_{n-k}^* y_{n-k} = g_n \sum_{k=0}^{L-1} a_{n-k}^2 + z_n' \qquad (5.25)$$

As can be seen from (5.25), $u_{n,L}$ is an estimate of g_n scaled by $\sum_{k=0}^{L-1} a_{n-k}^2$, and $E[u_{n,L}] = LA^2 g_n$. The variance $\sigma_{z'}^2$ of the additive noise z_n' is equal to $LA^2\sigma_z^2$. Thus, the SNR of $u_{n,L}$ can be expressed by

$$\gamma_u = |E[u_{n,L}]|^2/\sigma_{z'}^2 = A^4 L^2 |g_n|^2 / (LA^2\sigma_z^2) = LA^2 |g_n|^2/\sigma_z^2 = L\gamma_y \qquad (5.26)$$

Namely, it is L times higher than the SNR of the sample y_k.

In the discussion below, we call the correlation-based phase estimate given by (5.22) the *ML phase estimate* of θ_n. It should be noted that *it is a true ML estimate only if the receiver filter is a matched filter of* $r(t)$. Otherwise, it can be regarded as an approximate ML estimate.

Moreover, we call a phase detector that implements (5.22) the maximum likelihood phase detector (MLPD) with the understanding that it produces an ML phase estimate only if the receiver filter is the MF of $r(t)$ and it otherwise generates an approximate ML estimate.

Optimality of the Correlation-Based Phase Estimation

When the receiver filter is the MF of the received signal, the phase estimate derived above is optimal under the ML criterion. This is the case for carrier phase detection using $u_{n,L}$ based on (5.22). As it was shown in [10], when the overall channel satisfies the Nyquist criterion, i.e., there's no ISI at the receiver filter output, the variance of the estimated phase satisfies the Cramer–Rao bound and thus the estimated phase is optimal. However, when the channel does not satisfy the Nyquist criterion, i.e., there is ISI at the receiver filter output, the phase detection performance can be improved if the ISI is suppressed. Therefore, in receivers, especially in wireline receivers, phase detections during the data mode are often performed at the output of the equalizer rather than at the output of the receiver filter.

The optimality of the phase detection based on the received signal samples y_k's depends on the strength of ISI. Such phase detection is close to optimal if ISI in y_k is much weaker than other interference such as noise and inter-user interference. This situation is typical in the CDMA systems for voice communications. In such an environment, the correlation-based phase detector in a CDMA RAKE receiver is nearly optimal.

Combining Multiple Estimates

The ML phase detection is derived based on a sequence of data symbols. In the summation in (5.25), each term can be viewed as an estimate of the channel. These terms are combined to yield an overall estimate with a reduced phase error. If the carrier phase remains constant, the combining is optimal in the ML sense and the sum is still an ML estimate. As shown above, the SNR of $u_{n,L}$ given by (5.26) is L times higher than the SNR of y_n. Therefore, a longer data symbol sequence could be used to obtain a better phase estimate.

The same conclusion is true for the phase noise, if z_k is complex Gaussian, i.e., it is circularly symmetric. The variance of the noise component perpendicular to $u_{n,L}$, i.e., along the phase direction, is equal to a half of the total noise variance. As a result, the ratio of the signal power to the variance of $u_{n,L}$'s phase noise component is equal to $2\gamma_u$. The variance of the phase estimate $\hat{\theta}_n$ in (5.22) is equal to the variance of $u_{n,L}$'s phase noise normalized, i.e., divided, by $|E[u_{n,L}]|^2$ in a unit of $(radian)^2$. From (5.26), we have,

$$\sigma_{\hat{\theta}_n}^2 = \frac{\sigma_{z'}^2}{2|E[u_{n,L}]|^2} = \frac{\sigma_{z'}^2}{2(LA^2|g_n|)^2} = \frac{1}{2\gamma_u} = \frac{1}{2L\gamma_y} \tag{5.27}$$

This conclusion is especially important to wideband communication systems. Since the symbol rates are proportional to the signal bandwidth, the channel phase does not change significantly during many symbol time intervals. Multiple estimates can be combined before they are further processed by PLLs or by FLLs. Such an implementation can reduce the computation requirement, as the PLL or FLL can operate at a rate much lower than the data symbol rate.

To reduce the jitter variance, an alternative method is to average the multiple phase estimates, each of which is generated from the individual $a_k^* y_k$. However, this approach is not optimal when compared to the ML phase estimation that uses the average of multiple $a_k^* y_k$ before extracting the phase using (5.25) and (5.22). This is because the reliability of the phase estimate of each $a_k^* y_k$ depends on its magnitude. When summation is performed on the phases, the magnitude information is removed, and the combining process becomes less effective than the ML estimation. However, this conclusion is only valid when the phase errors are due to additive Gaussian noise. If the phase errors are the result of the phase noise caused by phase jitter, averaging the phases computed from each product term would be effective. These conclusions are important when signals with multilevel constellations, e.g., QAM, are considered.

5.4.3 Simplified Phase Detector: The Quadrature Phase Detector

The phase estimate given by (5.22) is the optimal ML estimate if the receiver filter is the MF of the received signal. Specifically, (5.22) yields an unbiased phase estimate with minimal variance under the given conditions shown in [10]. The mean of the phase estimate is equal to the true phase θ_n. Thus, it can be used directly to correct the modulation symbol estimates' phase error introduced during transmission.

For practical implementations, the realization of (5.22) could be computationally complex because it needs to perform a division and to evaluate the value of the arctangent function. It would be desirable if the detector can be simplified. Simplifications are possible when the detector is used as the PD of a PLL.

Because a PLL is a feedback control system, in order for the PLL to generate unbiased phase estimates, it is sufficient that the actual phase error and the expectation of its estimate generated by the PD have the same sign. From (5.22), we observe that the numerator of the quotient, i.e., the imaginary part of $u_{n,L}$, $Im[u_{n,L}]$, satisfies this condition. Thus, it can be used as the input to a PLL instead of the phase estimate $\hat{\theta}_n$. We call this simplified form of phase detector a *quadrature phase detector* (QPD) because the imaginary part of a complex number is usually called its *quadrature* component. Some of the characteristics of the QPD are described below.

5.4.3.1 Phase Characteristics

From (5.22), it can be shown that

$$\text{Im}[u_{n,L}] = |u_{n,L}| \sin \hat{\theta}_n \tag{5.28}$$

Thus, the imaginary part of the summation has the same sign of $\hat{\theta}_n$ in the range between $-\pi$ and π. Moreover, when $\hat{\theta}_n$ is close to zero, $\hat{\theta}_n \approx \sin \hat{\theta}_n$ and $\text{Im}[u_{n,L}] \approx |u_{n,L}|\hat{\theta}_n$. Therefore, for the carrier phase or frequency offset estimation $\text{Im}[u_{n,L}]$ can be used instead of the ML carrier phase estimate $\hat{\theta}_n$ as the input to a PLL. The characteristic of the QPD output is shown in Figure 5.6 where the expectation of $\text{Im}[u_{n,L}]$ is plotted as a function of θ_n. As can be seen from the figure, the mean of the QPD output is almost equal to the carrier phase θ_n multiplied by $|u_{n,L}|$ if θ_n is less than $\pi/4$. It reaches the maximum at $\pi/2$ and then starts to decline. It becomes zero when the carrier phase reaches π.

In the literature, the curve shown in Figure 5.6 is referred to as the *S-curve* of the phase detector. The corresponding curve of an MLPD is a straight line from $(-\pi, -\pi)$ to (π, π) and periodically repeats with a period of 2π.

5.4.3.2 Detector Gain

Another important parameter of the PD in a PLL is its gain k_d. The gain of the QPD is a function of the detected phase θ_n and is equal to

$$k_{d,QPD} = \frac{E\left[|u_{n,L}| \sin \hat{\theta}_n\right]}{\theta_n} = LA^2 |g_n| \text{sinc}(\theta_n/\pi) \qquad \left(-\pi \le \hat{\theta}_n < \pi\right) \tag{5.29}$$

as is depicted in Figure 5.7. In comparison, k_d of the MLPD is a constant of one from $-\pi$ to π.

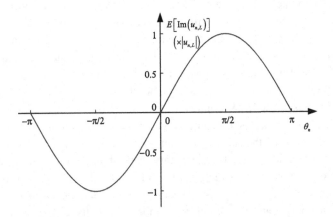

Figure 5.6 S-curve of the correlator-based QPD

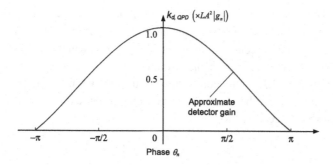

Figure 5.7 Gains of correlator-based approximate and ideal phase detectors

5.4.3.3   Noise Characteristics

Recall from Section 5.4.2 that the noise variance of $u_{n,L}$ is equal to $\sigma_{z'}^2$ and the variance at the QPD output, i.e., the noise variance of $\mathrm{Im}[u_{n,L}]$, is equal to $\sigma_{z'}^2/2$. It can be shown that the variance of the QPD output is equal to $(LA^2|g_n|)^2 \sigma_{\theta_n}^2$, where $\sigma_{\theta_n}^2$ is the phase variance at the MLPD output given by (5.27). Thus, it is $(LA^2|g_n|)^2$ times larger than the variance at the MLPD output. However, by taking into account their phase detector gains, the variances of the output phase jitter of a DPLL using QPD and MLPD in steady state are the same, as will be shown in Section 5.4.4.2.

By comparing the characteristics of a QPD and an MLPD, we can conclude the following. When a PLL has converged, i.e., the carrier-phase error is near zero, the PLLs driven by the QPD and MLPD have the same steady-state behavior. A PLL with a QPD has a smaller tracking range than the one with an MLPD because of the smaller loop gain when the phase offset increases. Therefore, it is less capable of tracking rapidly increasing phase errors caused by a large frequency offset.

5.4.4   Applications of Carrier Phase Detection to Coherent Demodulation

At the beginning of this section, we mentioned that one key factor in achieving optimal coherent detection of data symbols is to remove the carrier-phase error introduced during transmission. There are two basic approaches to using the detected carrier phase for this purpose: the feed-forward and feedback methods. Below, we describe these two approaches used in receivers. Their characteristics and examples will also be given.

5.4.4.1   Feed-Forward Carrier-Phase Correction

As shown above, the carrier phase can be derived from the average of the correlation between the expected data symbol and the receiver filter output. Namely,

$$\hat{\theta}_n = \mathrm{Arg}\left[\sum_{k=0}^{L-1} a_{n-k}^* y_{n-k}\right] \cong \mathrm{Arg}[u_{n,L}] \tag{5.30}$$

as long as the carrier phase does not change in the samples that are involved in the summation. The phase estimate given by (5.30) is an ML estimate if the receiver filter is the matched filter of the received signal. Even when the receiver filter is not exactly a matched filter, $\hat{\theta}_n$ will still be a good estimate if the sampling delay τ is chosen to maximize the energy at the receiver filter output. The estimate can be used to correct the carrier-phase error of the received signal samples containing the data symbols for recovering the transmitted data bits. For example, if the receiver filter output y_n contains the data symbols a_n with a scaling factor of A and a carrier phase of θ_n, a scaled estimate of a_n can be generated according to

$$e^{-j\hat{\theta}_n}y_n = e^{j(\theta_n - \hat{\theta}_n)}Aa_n + e^{-j\hat{\theta}_n}z'(n) = A\hat{a}_n \tag{5.31}$$

If the mean of $\hat{\theta}_n$ is equal to θ_n, $\hat{a}_n$ is an unbiased estimate of a_n. This method is called *feedforward carrier-phase correction*, or *feedforward carrier-phase recovery*.

One good example of the application of the feedforward phase correction is the phase-error correction performed in the RAKE fingers in DS-CDMA receivers described in Section 1.5.

However, it should be noted that in most communication systems with channel coding, the phase angle values are usually not computed and not explicitly used in feedforward phase correction. The conjugate of the correlator outputs $u_{n,L}$ in (5.25) is directly multiplied with the data samples, i.e.,

$$u_{n,L}^* y_n = Ce^{j(\theta_n - \hat{\theta}_n)}a_n + u_{n,L}^* z'(n) \tag{5.32}$$

to generate a scaled, or weighted, estimate of the data symbol a_n. As can be seen from (5.25), the weight is proportional to the channel gain, which determines the reliability of the estimate. Thus, $u_{n,L}^* y_n$ is particularly suitable for generating decoding metrics of the coded bits. As an additional benefit, the computational complexity is also reduced.

5.4.4.2 Feedback Method

Another way to use detected carrier phases to perform data symbols recovery is through a PLL. In the literature, this approach is referred to as the *feedback carrier-phase correction* and data symbol recovery. To illustrate the feedback method, we show the block diagram of an exemplary implementation in Figure 5.8. The system shown here is essentially a DPLL, which has been described in Chapter 4. A few of its features are discussed below.

In a DPLL, the detected phase errors can be either unbiased or biased. The sufficient requirement of the PD is that the expectation of its output has the same sign as the true phase error that the output represents. Thus, the QPD shown in Section 5.4.3 is suitable for this purpose. The average phase errors are reduced in the feedback process and eventually become zero after the DPLL converges. Because the input samples are phase corrected during the same time, the data samples can be directly used for data symbol estimation with no additional operation. The feedback method is also robust to quantization errors. To reduce complexity, L samples are first accumulated before phase detection, which is shown by (5.22), is performed. Thus, the loop filter can operate at a rate of $1/LT$, which is L times lower than the modulation symbol rate $1/T$. Nonetheless, the integrator and the phase rotator must operate at the symbol rate as discussed in Section 4.5.3.1.

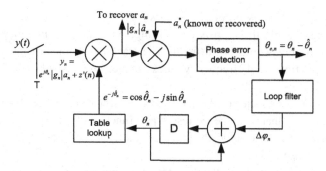

Figure 5.8 Data-assisted phase-error estimation and correction by using a DPLL

For the system shown in the figure, we assume that the correlation for phase detection is performed based on the decisions of the detected data symbol in the data mode. The accuracy of the detected phase depends on the reliability of such decisions. If the detected symbol a_n^* is a hard decision directly obtained from $\hat{a}_n$, this method is more suitable for relatively undistorted channels, or it is being placed at the output of an equalizer. If a_n^* is obtained after decoding, it is not immediately available for performing the correlation with the phase-corrected samples y_n. Hence, the PLL coefficient update needs to be delayed. The impact of the delayed update is usually not serious if the loop gain is small, as discussed in Section 4.5.3.3.

Another advantage of the feedback method is that it can also perform carrier frequency synchronization if a second-order PLL is employed. As a result, no stand-alone frequency-tracking loop is needed. Such designs are widely used in wireline data modems and have been adapted for other applications as will be described in Section 5.6.

DPLL with QPD

The QPD presented in the last section is particularly suitable for DPLL implementation. Below we consider its convergence and steady-state phase-error behaviors. To simplify the analysis, we consider a first-order DPLL. The results can be extended to higher-order DPLLs.

A first-order DPLL with a QPD is similar to the one shown in Figure 5.8. The phase-error detection block in the figure is implemented by using the imaginary part of the correlation of L samples and their corresponding data symbols given by (5.25) and (5.28). The loop filter of the DPLL could be operated at the rate of $1/LT$.

As has been discussed in Chapter 4, in response to a step-phase change, this first-order DPLL converges exponentially to the new phase with a time constant of $k_1/(2 - k_1) \times NT$ (seconds), where k_1 is the overall loop gain. The gain k_d of the QPD given by (5.29) and shown in Figure 5.7 is a function of the phase error angle detected by the QPD and the gain varies during the convergence of the DPLL.

Because the loop gain k_1 is proportional to k_d, the convergence time constant of the DPLL changes while the loop is converging. An analytical treatment of such a time-varying system would be difficult. The tracking characteristics, including the convergence time and frequency lock range, are usually evaluated by simulation.

Another important parameter of the DPLL is the variance of its steady-state phase jitter. Below, we show that it can be expressed as a function of the input signal's SNR. The expression is derived analytically based on the characteristics of the first-order DPLL studied in Chapter 4.

As was shown in Section 4.3.2, the variance of the output phase jitter of a DPLL is determined by its noise bandwidth B_n. For a first-order DPLL, the variance of the output phase jitter is equal to the variance of the *input* phase jitter multiplied by B_n. In Section 5.4.3, it was shown that the variance at the QPD *output* due to the noise in the input signal is equal to $\sigma_z^2/2$. In order to apply the general result, we need to convert the phase jitter variance at the output of the QPD to its input as shown in the block diagram in Figure 5.9.

It is shown in the figure that the phase variance at the output of the phase detector can be moved to its input after being divided by k_d^2. In steady state, k_d of the QPD is equal to $LA^2|g_n|$ when θ_n is close to zero. Hence, the variance of the DPLL's *equivalent input noise* is equal to $\sigma_z^2/(2L^2A^4|g_n|^2)$. From (5.26) and (5.27), it can be shown that the equivalent noise variance is equal to $1/2\gamma_u$ and is the same as that of the DPLL with an ML phase detector. The phase variance at the output of a DPLL is equal to the input noise variance multiplied by the noise-bandwidth. For the first-order DPLL with a QPD, it is equal to $k_1/[(2-k_1)2\gamma_u]$.

The above analysis also leads to an important conclusion: *For a fair comparison of the output noise characteristics of DPLLs that employ different types of PDs, it is sufficient to compare the variances of their equivalent input noise, which is equal to PDs' output noise variances divided by the respective squared detector gains.* We will apply this conclusion when comparing non-data-assisted phase-detection techniques.

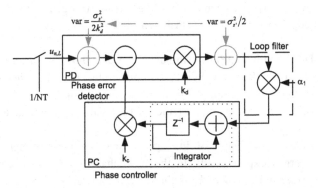

Figure 5.9 Block diagram of first-order DPLL for output phase-error analysis

5.4.5   Data-Assisted Frequency-Offset Detection

In the previous subsections, we described data-assisted carrier-phase estimation based on baseband received signal samples. In that discussion, we assumed that the carrier phase does not change significantly during detection. In this section, we consider how to perform carrier-frequency offset detection from the detected phases. Digital frequency-locked loops (DFLLs), which are widely used in wireless mobile communications, will be presented in Section 5.4.7.

Recall from Section 5.2 that the carrier phase offset between a transmitter and a receiver normally consists of two components. One is approximately constant and the other changes linearly with time. The first one is equal to the phase of the overall communication channel. The other component is caused by the frequency offset Δf due to the difference between the transmitter modulation frequency f_c and the receiver down-conversion frequency $\hat{f}_c$. Such a frequency offset results in a linear phase change with time. As a result, the sampled output of the receiver filter has the form of (5.18) and the carrier phase offset can be expressed as

$$\theta_t = 2\pi\left(f_c - \hat{f}_c\right)t + \theta_c \cong 2\pi\Delta ft + \theta \tag{5.33}$$

As has been analyzed previously, if the time support of the overall CIR $g(t)$ is significantly shorter than $1/\Delta f$, the phase of $a_n^* y_n$ is an unbiased estimate of $\theta_n \cong \theta_{nT}$.

5.4.5.1   Frequency Offset Detector Based on ML Phase Estimates

Let us denote the phase of $a_n^* y_n$ by $\hat{\theta}_n$, which is an estimate of $\theta(nT)$. From (5.33), we observe that if $NT < 1/(2\Delta f)$, $\left(\hat{\theta}_n - \hat{\theta}_{n-N}\right)\Big|_{\text{mod}2\pi}/(2\pi NT)$ is an estimate of Δf, namely,

$$\widehat{\Delta f} = \left(\text{Arg}\left[a_n^* y_n\right] - \text{Arg}\left[a_{n-N}^* y_n\right]\right)\Big|_{\text{mod}2\pi}/(2\pi NT) \tag{5.34}$$

In (5.34), the range of the difference between the phase angles in the numerator is defined to be between $-\pi$ and π, and the detection range of $\widehat{\Delta f}$ is between $-1/(2NT)$ and $1/(2NT)$. Below we will assume that this is always the case with the modulo operation not showing explicitly.

The variance of the estimate of Δf can be reduced by using the sum of multiple $a_n^* y_n'$ s, i.e., $u_{n,L}$, given by (5.25). Note that the integration length L and the delay between the adjacent estimates N can be chosen independently. In most cases, L is less than or equal to N. To simplify notation, we will not show L in the subscript of u_n in the text below. Moreover, we assume that the estimates are generated once every N samples and define $n \cong mN$, $u(m) \cong u_{mN}$, $g(m) \cong g_{mN}$ and $\hat{\theta}_u(m) \cong \arg\left[u(m)\right]$. Applying these notations, we can express the estimate of Δf as

$$\widehat{\Delta f}(m) = \left(\text{Arg}[u(m)] - \text{Arg}[u(m-1)]\right)/(2\pi NT) \cong \left(\hat{\theta}_u(m) - \hat{\theta}_u(m-1)\right)/(2\pi NT)$$

$$\tag{5.35}$$

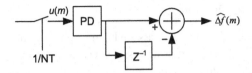

Figure 5.10 ML phase estimate based frequency detector

The *frequency error detector* based on ML phase estimates is shown in Figure 5.10. We call such an FD the *MLFD* below. The output of the MLFD, denoted as $f_e(m)$, is given by

$$f_e(m) = \mathrm{Arg}[u(m)] - \mathrm{Arg}[u(m-1)] \tag{5.36}$$

It is a scaled estimate of Δf. The gain of the MLFD is equal to

$$k_d(m) = E[f_e(m)]/\Delta f = 2\pi NT \tag{5.37}$$

which is a constant independent of the frequency offset.

As was shown in Section 5.4.2, the variance of the phase estimate $\theta_u(m)$ is equal to $\gamma_u^{-1}/2$. When the estimate of Δf is used as the input to the DFLL with the update interval equal to NT, the z-transform of (5.35) is

$$\hat{\Delta f}(z) = \left(1 - z^{-1}\right)\hat{\theta}_u(z)/(2\pi NT) \tag{5.38}$$

Note that the noise in the estimate of Δf is equal to the noise component in $\hat{\theta}_u(m)$ passing through a filter with transfer function of $1 - z^{-1}$. Because such a filter is a high-pass filter with a zero at $f = 0$, it removes a large portion of the noise near zero frequency. When this characteristic of the noise in $\hat{\Delta f}(m)$ is combined with the narrow-band low-pass characteristic of a DFLL, the variance of the steady-state frequency error at the DFLL output is significantly less than the output variance of the DFLL with white input noise that has the same variance. We will analyze this property when discussing DFLL in Section 5.4.7.

5.4.5.2   Quadri-Correlator Frequency Detector

Similar to the QPD for phase detection discussed in Section 5.4.3, a scaled estimate of the frequency error can be obtained as

$$\hat{\Delta f}_{QC}(m) = \mathrm{Im}[u(m)u^*(m-1)] = u_i(m)u_r(m-1) - u_r(m)u_i(m-1) \tag{5.39}$$

where $u_r(m)$ and $u_i(m)$ are the real and imaginary components of $u(m)$, respectively. This form of frequency detector is called *quadri-correlator frequency detector* (QCFD) in [11]. Assuming that $E[|u(m-1)|] = E[|u(m)|] = A^2|g|$, by either choosing $L = 1$ or normalizing $u(m)$ by $1/L$, the expectation of the QCFD output is

$$E\left[\hat{\Delta f}_{QC}(m)\right] = A^4|g|^2 \sin(2\pi NT\Delta f) \tag{5.40}$$

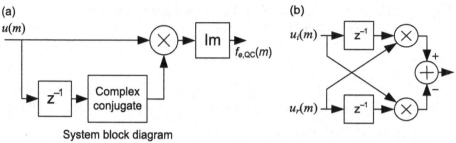

Figure 5.11 Block diagrams of quadri-correlator frequency detector

It can be used as the FD in a DFLL and has been studied for carrier frequency synchronization [11, 12, 13]. Figure 5.11(a) depicts the system block diagram of a QCFD. Its actual implementation is shown in Figure 5.11(b).

The gain of the QCFD as a function of Δf is given by

$$k_{d,QC}(m) = \frac{E\left[\hat{\Delta}_{fQC}(m)\right]}{\Delta f} = \frac{A^4|g|^2 \sin{(2\pi NT\Delta f)}}{\Delta f} = 2\pi NTA^4|g|^2\mathrm{sinc}(2\pi NT\Delta f) \quad (5.41)$$

When $\Delta f \approx 0$, the detector's gain can be expressed as

$$k_{d,QC}(m)\big|_{\Delta f\approx0} = E\left[f_{e,QC}(m)\right]/\Delta f\big|_{\Delta f\approx0} \approx A^4|g|^2 2\pi NT \quad (5.42)$$

Therefore, it is equal to the gain of the MLFD multiplied by a constant $A^4|g|^2$.

QCFDs are widely used in DFLLs' implementations of carrier frequency synchronization in digital mobile receivers. Similar to the MLFD described above, the output noise of a QCFD also has the high-pass characteristic. As a result, the variance of the steady-state frequency error at the DFLL output is significantly less than the variance of the DFLL's output frequency error if the noise at the output of the QCFD is white with the same variance. A detailed analysis of the characteristics of the DPLL with a QCFD is given in Section 5.4.7.2.

5.4.6 Non-Data-Assisted Baseband Carrier Phase and Frequency Detection

In this section, we consider non-data-assisted carrier phase and frequency detection using baseband digital signal samples at the output of the receiver filter given by (5.23) for $L=1$ and $k = n$. Given that the data symbols a_n's are random with zero means, the mean of y_n is also zero. As a result, it is not possible to recover the carrier phase θ_n by linear processing of the samples y_n. However, similar to carrier recovery of passband signal considered in Section 5.3.1, in many cases, it is possible to recover the carrier phase by nonlinear processing. Below we present some of these nonlinear processing techniques and their characteristics.

5.4.6.1   Carrier Phase Detection Based on Squaring Operations

By squaring y_n in (5.23), we obtain

$$y_{s,n} \cong y_n^2 = e^{j2\theta_n}|g_n|^2 a_n^2 + 2e^{j\theta_n}|g_n|a_n z_n + z_n^2 = e^{j2\theta_n}|g_n|^2 a_n^2 + z_{s,n} \tag{5.43}$$

where $z_{s,n} = 2e^{j\theta_n}|g_n|a_n z_n + z_n^2$ is white with zero mean but not Gaussian.

For transmission with real modulation symbols, such as BPSK and PAM, by defining $A = \sqrt{E[a_n^2]}$, we have the mean of $y_{s,n}$ as equal to

$$E[y_{s,n}] \cong E[y_n^2] = e^{j2\theta_n} A^2 |g_n|^2 \tag{5.44}$$

Thus, the carrier phase information is available from $y_{s,n}$ Such a carrier-phase error detector is referred to as a *squaring PD*. The variance of $z_{s,n}$ can be computed as

$$\sigma_{z_s}^2 = E\left[|2e^{j\theta_n}|g_n|a_n z_n + z_n^2|^2\right] = 4|g_n|^2 E\left[|a_n|^2|z_n|^2\right] + E\left[|z_n^2|^2\right] \tag{5.45}$$

Denoting $z_n = z_{n,r} + j z_{n,i}$, $z_n^2 = z_{n,r}^2 - z_{n,i}^2 + 2j z_{n,r} z_{n,i}$ and assuming that $z_{n,r}$ and $z_{n,i}$ are Gaussian and independent, since $E[z_{n,r}^2] = E[z_{n,i}^2] = \sigma_z^2/2$ we have $E[z_{n,r}^4] = E[z_{n,i}^4] = 3\sigma_z^4/4$,

$$E\left[|z_n^2|^2\right] = E[z_{n,r}^4] + E[z_{n,i}^4] - E[2z_{n,r}^2 z_{n,i}^2] + E[4z_{n,r}^2 z_{n,i}^2] = \frac{3}{4}\sigma_z^4 + \frac{3}{4}\sigma_z^4 + 2\frac{\sigma_z^2}{2}\frac{\sigma_z^2}{2} = 2\sigma_z^4$$

and

$$4|g_n|^2 E\left[|a_n|^2|z_n|^2\right] = 4|g_n|^2 E\left[|a_n|^2\right]\left(E[z_{n,r}^2] + E[z_{n,i}^2]\right) = 4A^2|g_n|^2\sigma_z^2 \tag{5.46}$$

Therefore,

$$\sigma_{z_s}^2 = 4A^2|g_n|^2\sigma_z^2 + 2\sigma_z^4 \tag{5.47}$$

The SNR of $y_{s,n}$ can be expressed by

$$\gamma_{y_{s,n}} = \frac{E\left[|y_{s,n}|^2\right]}{\sigma_{z_s}^2} = \frac{|g_n|^4 A^4}{4|g_n|^2 A^2\sigma_z^2 + 2\left(\sigma_z^2\right)^2} = \frac{1}{4\gamma_y^{-1} + 2\gamma_y^{-2}} \tag{5.48}$$

where $\gamma_y = |g_n|^2 A^2/\sigma_z^2$, as given by (5.24).

To reduce the variance of the estimated carrier phase, multiple squarer outputs can be added together before being used to compute the phase estimate. Namely, we generate an averaged value, $u_{s,n}$, one per NT, by forming the sum of L squarer outputs, such that

$$u_{s,n} = \frac{1}{L}\sum_{k=n-L+1}^{n} y_k^2 \approx e^{j2\theta_n}|g_n|^2 A^2 + 2e^{j\theta_n}|g_n|A\tilde{z}'_n + \tilde{z}_n'^2 \cong e^{j2\theta_n}|g_n|^2 A^2 + z'_{s,n} \tag{5.49}$$

where $z'_{s,n} = \frac{1}{L}\sum_{k=n-L+1}^{n}\left(2e^{j\theta_n}|g_n|a_k z_k + z_k^2\right)$ is an additive white noise term. It is approximately Gaussian if L is large according to the central limit theorem. From (5.47) it can be shown that the variance of $z'_{s,n}$ is equal to

$$\sigma_{z'_s}^2 = \frac{1}{L}\left(4A^2|g_n|^2\sigma_z^2 + 2\sigma_z^4\right) \tag{5.50}$$

If the carrier phase does not change significantly in the L values in the summation in (5.49), the estimated carrier phase computed from $u_{s,n}$ is the same as that from $y_{s,n}$. However, its noise variance is L times smaller.

Similar to (5.22), the carrier phase estimate can be computed as

$$\hat{\theta}_{n,s} = \mathrm{Arg}[u_{s,n}]/2 \tag{5.51}$$

The variance of the noise component of $u_{s,n}$ along the phase direction, i.e., perpendicular to $u_{s,n}$, is equal to half of $\sigma_{z'_s}^2$. Thus, similar to (5.27), the variance at the squaring phase detector output given by (5.51) can be shown to be

$$\sigma_{\hat{\theta}_{n,s}}^2 = \frac{0.5\sigma_{z'_s}^2/2^2}{|E[u_{s,n}]|^2} = \frac{\sigma_{z'_s}^2}{8|g_n|^4 A^4} = \frac{4A^2|g_n|^2\sigma_z^2 + 2\sigma_z^4}{8|g_n|^4 A^4 L} = \frac{\gamma_y^{-1}}{2L} + \frac{\gamma_y^{-2}}{4L} \tag{5.52}$$

Comparing the variances given by (5.52) with (5.27), it can be seen that they only differ by a term proportional to γ_y^{-2}. Hence, the phase variance of the squaring PD approaches that of the data-assisted MLPD when the SNR increases.

The carrier phase estimate can be used to correct the phase of data symbols directly or through a PLL in the feedback form. It can also be used for the estimation of the offset carrier frequency through a frequency locked loop as will be discussed in Section 5.4.7.

From (5.43) it can be seen that the carrier phases θ and $\theta + \pi$ of a symbol a_n result in the same $y_{s,n}$. In other words, the squaring operation results in a π, or 180-degree, ambiguity in the estimated carrier phase. This ambiguity of the estimated phase can be overcome by employing differential encoding or 180-degree invariant channel coding, as has been studied previously, e.g., in [14]. Moreover, due to this phase ambiguity, the phase detection range of the squaring PD is reduced to $[-\pi/2, \pi/2)$, i.e., half of the detection range of a data-assisted MLPD.

Squaring QPD

When used as the PD in a DPLL, a simplified form of the squaring PD can be derived by using the imaginary part of $y_{s,n}$, instead of its angle $\mathrm{Arg}[y_{s,n}]$. This simplified form is the counterpart of the data-assisted QPD studied in Section 5.4.3. Below, we consider two of the properties of the squaring QPD: the PD gain and the phase variance at the DPLL output.

Its PD gain is a function of the carrier phase θ_n and can be expressed by

$$k_{d,s} = \frac{E[\mathrm{Im}[y_{s,n}]]}{\theta_n} = \frac{A^2|g_n|^2\sin(2\theta_n)}{\theta_n} \tag{5.53}$$

The maximum of the detector gain is attained when $\theta_n \approx 0$ and is equal to $2A^2|g_n|^2$.

The variance of $\text{Im}[y_{s,n}]$ is half of the variance of $y_{s,n}$ given by (5.47), i.e., $2A^2|g_n|^2\sigma_z^2 + \sigma_z^4$. As shown in Section 5.4.4.2, to compare squaring QPD's output noise with other types of PDs, we can simply normalize the output noises by their detector gains. When the phase error is near zero, the normalized noise variance is equal to

$$\frac{2A^2|g_n|^2\sigma_z^2 + \sigma_z^4}{\left(2A^2|g_n|^2\right)^2} = \frac{\sigma_z^2}{2A^2|g_n|^2} + \frac{1}{4}\left(\frac{\sigma_z^2}{A^2|g_n|^2}\right)^2 = \frac{\gamma_y^{-1}}{2} + \frac{\gamma_y^{-2}}{4} \tag{5.54}$$

where γ_y is as defined by (5.24). This is the same result given by (5.52). Thus, we conclude that at high SNR, the output noise variance of the squaring QPD asymptotically approaches that of the data-assisted PD given by (5.27) with $L = 1$. The variance of the output noise is reduced by a factor of L if the squaring QPD uses the average of L squared samples as its input.

Thus, at high SNR, the term related to γ_y^{-2} can be ignored and the squaring QPD has a phase variance similar to the variance of its data-assisted counterpart. The penalty paid is that the detection range of phase errors by the squaring QPD is reduced by a factor of two relative to the latter. At low-signal SNRs, the output phase variance of the squaring QPD increases due to the term related to γ_y^{-2}.

5.4.6.2   Carrier-Phase Detection by Using an *M*th-Power Phase Detector

The MPSK modulation symbols are usually defined by $Ae^{j\pi(1+2k)/M}$, $k = 0, ..., M - 1$. By taking the *M*th power of y_n, we obtain

$$y_{M,n} \cong y_n^M = e^{jM\theta_n}A^Me^{j\pi(1+2k)} + Mz_nA^{M-1}e^{j\pi(1+2k)(M-1)/M} + \text{high order terms of } z_n \tag{5.55}$$

Since $e^{j\pi(1+2k)} = e^{j\pi} = -1$, we have

$$E[y_{M,n}] = -A^Me^{jM\theta_n} \tag{5.56}$$

Therefore, the estimate of θ_n can be computed by

$$\hat{\theta}_n = \text{Arg}[-y_{M,n}]/M \tag{5.57}$$

Alternatively, the MPSK modulation symbols can be defined by $Ae^{j\pi 2k/M}$, $k = 0, ..., M - 1$. In this case, the estimate of θ_n can be computed by $\hat{\theta}_n = \text{Arg}[y_{M,n}]/M$.

Similar to the squaring PD and data-assisted PD, the *M*th-power PD can be used for carrier-phase correction for detecting data symbols. An *M*th-power QPD based on the imaginary part of $-y_{M,n}$ can also be used as the PDs in DPLLs. In addition, it can be used to detect a frequency offset and used as the FD in a DFLL. Because the applications and the derivations of the characteristics of the *M*th-power PDs are similar to those of other phase detectors, we will not elaborate here.

Due to the *M*th-power operation, it can be shown that there is a $2\pi/M$ phase ambiguity in the detected phase of the *M*th-power PD. As a result, the detectable phase angle is limited to $\pm\pi/M$. This range is $1/M$ of the range of the data-assisted PDs. When used to form frequency detectors in DFLLs, the usable range of frequency control is also reduced by a factor of M of the range of their data-assisted counterparts.

Finally, similar to what is discussed in Section 5.3.5, it is possible to use fourth-power PDs for carrier phase and frequency synchronization in systems with QAM signaling. However, compared to QPSK, QAM data symbols cause additional phase noise at the output of the detector. Thus, it is only useful if the channel is relatively static so that a DPLL with long time constants, which suppress such phase noise, can be used.

5.4.6.3   Non-Data-Assisted Detection of Baseband Carrier Frequency Offset

In the same way as with the data-assisted frequency offset detection, the non-data-assisted PD can also be used to detect baseband frequency offset. In this section, we will only consider the squaring PD. Mth-power PDs can be used in a similar manner.

Squaring FD

The squaring PD described in Section 5.4.6.1 can be used for carrier frequency offset detection in the same way as the data-assisted MLPD. From (5.51) and (5.34) the estimate of the frequency offset can be computed by using the estimated phases at mNT and $(m-1)NT$. Using the averaged squared output defined by (5.49) and defining

$$n = mN, \quad u_s(m) = u_{s,mN}, \quad z'_s(m) = z_{s,mN} \text{ and } \hat{\theta}_{s,u}(m) = \text{Arg}[u_s(m)] \qquad (5.58)$$

we have

$$\Delta f_{e,s}(m) = (\text{Arg}[u_s(m)] - \text{Arg}[u_s(m-1)])/4\pi NT \qquad (5.59)$$

If we construct an FD, which computes $(\text{Arg}[u_s(m)] - \text{Arg}[u_s(m-1)])/2$ as the estimate of the frequency offset, the mean of its output is equal to

$$E[f_{e,s}(m)] = E\left[\frac{\text{Arg}[u_s(m)] - \text{Arg}[u_s(m-1)]}{2}\right] = 2\pi NT\Delta f \qquad (5.60)$$

The gain of such a squaring PD-based FD, called a *squaring FD* below, is given by

$$k_{d,s}(m) = E[f_{e,s}(m)]/\Delta f = 2\pi NT \qquad (5.61)$$

The squaring FD is equivalent to the FD based on data-assisted MLPD given by (5.37). However, the phase detection range of the squaring PD is equal to $[-\pi/2, \pi/2)$ instead of $[-\pi, \pi)$ of the MLPD. Consequently, the maximum frequency offset that the squaring FD can handle is equal to $\pm(4NT)^{-1}$, which is half of that a data-assisted MLPD-based FD can detect.

Similar to the data-assisted MLPD-based FD, the noise at the output of the squaring FD can be treated as the output noise of the squaring PD passing through a $1-z^{-1}$ filter, which has a null at the zero frequency.

Squaring QCFD

Similar to the QCFD described in the last section, it is possible to construct a squaring QCFD, which can generate an estimate of the frequency offset as

$$\hat{\Delta}_{fs,QC}(m) = \text{Im}\left[u_s(m)u_s^*(m-1)\right] = u_{s,i}(m)u_{s,r}(m-1) - u_{s,r}(m)u_{s,i}(m-1) \quad (5.62)$$

Its properties are very similar to that of the data-assisted QCFD. We will not go through the derivation again but state only the main results below.

The gain of squaring QCFD is a function of Δf given by

$$k_{d,s,QC}(m) = \frac{E\left[f_{e,s,QC}(m)\right]}{\Delta f} = \frac{A^4|g|^4 \sin(4\pi NT\Delta f)}{\Delta f} = 4\pi NTA^4|g|^4 \text{sinc}(4\pi NT\Delta f)$$

$$(5.63)$$

When $\Delta f \approx 0$, the detector's gain can be expressed as

$$k_{d,s,QC}(m)\big|_{\Delta f \approx 0} = E\left[f_{e,s,QC}(m)\right]/\Delta f\big|_{\Delta f \approx 0} \approx A^4|g|^4 4\pi NT \quad (5.64)$$

The maximal frequency offset that the squaring QCFD can handle is equal to $\pm(4NT)^{-1}$. However, to ensure that the convergence time does not increase too much, the frequency offset should be limited to 0.6 to 0.75 of its maximum value.

The output noise of the squaring QCFD also exhibits a high-pass characteristic, which results in reduced output noise variance when used in a DFLL. This property will be analyzed in Section 5.4.7.3.

The squaring FDs, including the regular form and QCFD form, can be used as the FD of DFLL in the same way the corresponding data-assisted FDs can. As can be shown by normalizing their detector gains, at high SNRs the variances of their output phase jitter are similar if the higher-order noise terms can be ignored. As a result, the variances of the steady-state output phase-jitter of DFLLs employing the squaring FDs and their data-assisted counterparts are the same at high SNR. The penalty introduced by the squaring operation without the known data symbols is that the DFLL with a squaring FD can compensate for only half of the maximum frequency offset that a DFLL with the corresponding data-assisted FD can.

Mth-power FDs can be constructed similarly. Their characteristics are similar to those of the squaring FD with the frequency tracking range reduced to $1/M$ of the corresponding data-assisted FDs. Moreover, the SNRs that are required by the Mth-power FDs to achieve similar performances of the data-assisted FDs are even higher.

5.4.7   Digital Frequency-Locked Loop

The frequency-locked loop, also called an *automatic frequency control* loop or frequency tracking loop (FTL), is a variation of a PLL. Its objective is to achieve carrier frequency synchronization rather than carrier-phase synchronization. As it is unlikely that the carrier would have a consistent frequency ramp, at least not consistently for a long time, practical FLLs are usually implemented as first-order loops.

While PLLs have been thoroughly studied and documented in the literature, references on FLLs are relatively scarce. This is mainly because the performance of a receiver is directly impacted by carrier-phase synchronization. If the channel is relatively static, PLLs are suitable for achieving both phase and frequency synchronization at the same time as we have shown in Chapter 4. As a result, historically, PLLs were widely used for achieving carrier synchronization in communication systems, especially for communication over wireline and other types of static and quasi-static channels.

However, FLLs have become popular since the early 1990s because of the widespread deployment of wireless mobile communication systems. In such systems, the communication channels usually experience fast time-variations caused by channel fading. Due to the wireless channel conditions and the system design, it is more convenient and effective to track and control the carrier phase and frequency separately. The phase change due to fast channel fading is commonly tracked by channel estimation based on periodically transmitted reference, or pilot, symbols. FLLs are used to learn and track the average frequency offset.

Due to the current popularity of DFLLs in practice, it is important to have a good understanding of DFLLs' properties analytically and of their implementations in detail. In this section, the properties and characteristics of DFLLs will be analyzed, and details of their implementation considerations will be discussed.

5.4.7.1 Basic Structure and Characteristics of DFLLs

Figure 5.12 depicts the block diagram of a typical first-order DFLL with an MLPD-based FD.

Similar to a DPLL, a DFLL consists of three components: A *frequency error detector*, or *frequency detector*, a *loop filter* and a *frequency controller*. DFLLs operate at a rate of $1/NT$, where T is the data symbol interval and $N \geq 1$.

The input to the FD of a DFLL is the correlations $u(k)$, $k = \ldots, m-1, m, m+1, \ldots$, at the rate of $1/NT$, as defined in Section 5.4.5.1. The FD shown is based on an MLPD as given by (5.36). The MLPD detects the phase $\theta_u(m)$. From the detected phase, the frequency error is computed as $f_e(m) = \theta_u(m) - \theta_u(m-1)$ with the FD gain k_d equal to $2\pi NT$. The generated frequency error is scaled by the loop coefficient α_2 and accumulated in a perfect integrator to generate an estimate of the frequency offset.

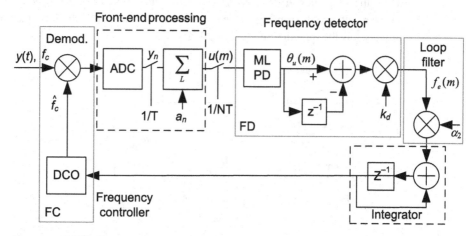

Figure 5.12 Block diagram of a typical DFLL

The estimate is sent to the FC to correct the frequency offset of the input signal. The FC may be implemented by a DCO as shown or, alternatively, by a digital phase rotator. The properties of these two types of FCs will be described in Section 5.7 when we discuss frequency synchronization in wireless communication receivers.

Transfer Function and Convergence Characteristics

The first-order DFLL shown in Figure 5.12 has more resemblance to a second-order PLL than to a first-order PLL. For example, since the DCO behaves as an integrator as discussed in Chapter 4, the first-order DFLL contains two integrators, the same as a second-order PLL. However, the FD's transfer function $1-z^{-1}$ introduces a zero at $(1,0)$ and cancels one of the two poles created by the two integrators. Therefore, its transfer function effectively has only one pole. This is also true if a digital phase controller with an integrator as described in Section 4.5.1.3 is used instead of the DCO.

The first-order DFLL can be represented by the equivalent discrete-time system block diagram shown in Figure 5.13. Due to its similarity to the second-order PLL, the notations used there are consistent with those of a corresponding second-order PLL.

By using the results of the first-order DPLL given in Section 4.3.2, the system transfer function of the DFLL shown in Figure 5.13 can be expressed by

$$\frac{F(z)}{F_{in}(z)} = \frac{k_c k_d \alpha_2 z^{-1}}{1 - (1 - k_c k_d \alpha_2)z^{-1}} \cong \frac{k_2 z^{-1}}{1 - (1 - k_2)z^{-1}} \tag{5.65}$$

where $k_2 = k_d k_c \alpha_2$ is the loop gain of the DFLL. The time response of the first-order DFLL to a step change of frequency, $f(0) \times U(m)$, can be shown to be

$$f(m) = [1 - (1 - k_2)^m]f(0) \times U(m) \tag{5.66}$$

Its frequency error at the input of the FD can be expressed as

$$\Delta f(m) = f(0) - f(m) = (1 - k_2)^m f(0) \times U(m) = f_e(m)/k_d \tag{5.67}$$

Thus, the frequency error converges to zero with a time constant of NT/k_2 (seconds). These results are the same as those of a generic first-order PLL with frequency, instead of phase, as the quantity being controlled.

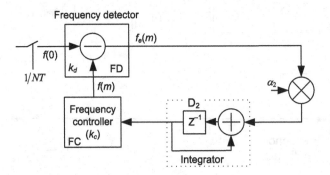

Figure 5.13 System diagram of first-order DFLL

Steady-State Frequency Error

Once the DFLL has converged, the mean of the residual frequency error is equal to zero. However, due to the noise/interference in the input sample $u(m)$, there is always a random steady-state frequency error. The expression of its variance is derived below.

In Figure 5.12, the sample $u(m)$ at the input of the PD has a variance of $\sigma_{z_i}^2$ and the SNR of $u(m)$ is γ_u. As can be seen in the figure, the noise of the phase component, which has a variance of $\gamma_u^{-1}/2$, is filtered by a $1-z^{-1}$ filter and scaled by k_d, which is equal to $2\pi NT$ as discussed in Section 5.4.5.

Recall from Section 4.4 that the variance of the output noise of a DPLL can be computed by integrating its input noise density in frequency over its loop frequency response. However, the analysis given in Section 5.4.5 only gave us the variance of the noise at the FD's output, not at its input. As it is for DPLLs, for a DFLL, the equivalent noise at the FD's input is equal to the FD's output noise divided by k_d. This procedure is shown in Figure 5.14.

From (5.65) and the expression of input noise given in Figure 5.14, the variance of the frequency error at the output of the DFLL can be computed as

$$\sigma_{f,out}^2 = \frac{\sigma_{z_\theta}^2}{2\pi} \int_{-\pi}^{\pi} \left| \frac{k_2 z^{-1}(1-z^{-1})}{1-(1-k_1)z^{-1}} \frac{1}{k_d} \right|^2_{z=e^{j\omega}} d\omega \tag{5.68}$$

where $\sigma_{z_\theta}^2$ is the variance of $z_\theta(m)$, which is equal to $1/(2\gamma_u)$, and $k_d = 2\pi NT$. Moreover, it can be shown that

$$\int_{-\pi}^{\pi} \left| \frac{z^{-1}(1-z^{-1})}{1-(1-k_2)z^{-1}} \right|^2_{z=e^{j\omega}} d\omega = \int_{-\pi}^{\pi} \frac{2(1-\cos\omega)}{(1-k_2)^2 + 1 - 2(1-k_2)\cos\omega} d\omega = \frac{4\pi}{2-k_2}$$

$$\tag{5.69}$$

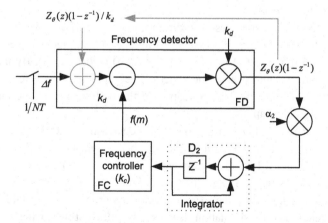

Figure 5.14 Equivalent block diagram of DFLL for frequency error analysis

Substituting this result into (5.68), we obtain the following expression:

$$\sigma_{f,out}^2 = \frac{k_2^2}{4\pi\gamma_u k_d^2} \frac{4\pi}{2-k_2} = \frac{k_2^2}{4\pi^2 N^2 T^2 \gamma_u (2-k_2)} \tag{5.70}$$

It is interesting to note that the variance of the frequency error at the DFLL's output is proportional to k_2^2 with the nonwhite noise at the FD output due to the $1-z^{-1}$ filter. If we had assumed that the noise were white, it could have been shown from the result of first-order PLL that the frequency error would be equal to $k_2/\left[8\pi^2 N^2 T^2 \gamma_u (2-k_2)\right]$, which is about $0.5/k_2$ times larger than the actual value. As k_2 is usually a small number, the variance of the output frequency error would be overestimated if the $1-z^{-1}$ filter were not considered.

The analysis is based on the assumption that the noise at the input of the frequency detector is stationary. Each input $u(m)$ is the sum of multiple received signal samples multiplied by the complex conjugate of the transmitter data symbols as given by (5.25). If the data symbols have the same magnitude, the stationary noise assumption is valid, and the analysis is accurate. However, if the amplitudes of the symbols change from one to the next, e.g., for the symbol from the 16QAM constellation, there is an additional time-varying component in the additive noise. Thus, the analysis given above would be less accurate. However, if the number L of the samples involved in averaging is large, the variation would be small, so is the error introduced into the analysis. These conclusions are also applicable to DFLLs employing other types of data-assisted and non-data-assisted FDs.

5.4.7.2   Characteristics of DFLLs with QCFD

QCFD-based DFLLs are widely used in practical receiver implementations. The DFLL has almost the same block diagram shown in Figure 5.12, with the only difference being that the QCFD shown in Figure 5.11 is used instead of the MLPD-based FD in Figure 5.10. The analysis presented above on DFLLs can be applied in principle to the QCFD-based DFLLs with only a few exceptions as discussed below.

Convergence Characteristics

A DFLL with the QCFD converges according (5.66) and (5.67). However, since the detector gain of the QCFD given by (5.41) is a function of the frequency error that it observes, k_d in (5.66) and (5.67) is a function of $\Delta f(m) = f(0) - f(m)$. By taking this factor into account, we can evaluate the exact convergence characteristic of a DFLL through simulation. To simplify the evaluation of the performance of the DPLL, we make the following observations.

First, the gain of the QCFD is a monotonically decreasing function of $\Delta f(m)$. The maximum value $k_{d,max}$ is attained at $\Delta f(m) = 0$. This occurs when the frequency offset is fully corrected, i.e., the DFLL is in steady state. Otherwise, $k_d(m)$ decreases as $\Delta f(m)$ increases. Second, since the DFLL is a first-order loop, it is always stable when $k_d(m)$ becomes smaller. Finally, while the exact convergence property of the DFLL with a QCFD can be determined by simulation, it is usually sufficient in practice to consider the worst-case convergence time.

For example, the shortest time constant of the DFLL, denoted by T_{min}, is attained when $\Delta f(m) = 0$ and $k_d = A^2|g_n|^2$. Assume that the DPLL converges in $4T_{min}$ seconds with such a time constant. If we would like the DFLL to converge always within $8T_{min}$ seconds, we can select the system parameters such that the worst-case frequency error is no greater than $0.3/NT$. As can be seen from (5.41) and (5.42), if this condition is met, the detector gain $k_d(m)$ is always larger than $0.5k_{d,max}$ and the convergence time of DFLL will be within twice the converging time of a DFLL with $k_d = k_{d,max}$.

Steady-State Frequency Error

The steady-state error of a DFLL with the QCFD can be analyzed in the same way a DFLL with an MLPD-based FD is analyzed. The input to the QCFD is $u(m)$, which is equal to the channel estimate $g(m)$ plus a noise/interference component $z'(m) \cong z'_{mN}$, which has a variance of $\sigma_{z'}^2$. From Figure 5.11, the QCFD output can be expressed by

$$
\begin{aligned}
\mathrm{Im}[u(m)u^*(m-1)] &= \mathrm{Im}\big[(A^2g(m) + z'(m))(A^2g^*(m-1) + z'^*(m-1))\big] \\
&= \mathrm{Im}\big[A^4g(m)g^*(m-1)\big] + \mathrm{Im}\big[A^2z'(m)g^*(m-1) + A^2g(m)z'^*(m-1)\big] \\
&\quad + \mathrm{Im}\big[z'(m)z'^*(m-1)\big]
\end{aligned}
\tag{5.71}
$$

The first term on the right side of (5.71) is a scaled estimate of the frequency offset. The other two terms are the noise components in the estimate. The last term, denoted by $z_{d,2}(m)$, is the imaginary part of the product of two independent Gaussian noises. It is not Gaussian distributed but is white with zero mean, and its variance can be shown to be equal to $(\sigma_{z'}^2)^2/2$ since $z'(m)$ and $z'(m-1)$ are independent. The middle term, denoted by $z_{d,1}(m)$, is the imaginary part of the sum of two terms, each of which is white and Gaussian distributed with a variance of $A^4|g|^2\sigma_{z'}^2$, where $|g| = |g(m)| = |g(m-1)|$. However, the sum is no longer white. In steady state, Δf is equal to zero on average and

$$
g(m) = g(m-1) = |g|e^{j\theta_g}
\tag{5.72}
$$

When (5.72) is true, the middle terms on the right side of (5.71) can be expressed as

$$
\begin{aligned}
z_{d,1}(m) &\cong A^2\mathrm{Im}\Big[z'(m)h^*(m-1) + h(m)z'^*(m-1)\Big] \\
&= A^2\mathrm{Im}\Big[|g|e^{-j\theta_g}z'(m) + |g|e^{j\theta_g}z'^*(m-1)\Big]
\end{aligned}
\tag{5.73}
$$

By defining $z''(m) \cong e^{-j\theta_g}z'(m)$, which is AWGN and has the same variance as $z'(m)$, i.e., $\sigma_{z'}^2/2$, (5.73) becomes

$$
z_{d,1}(m) = A^2\mathrm{Im}\Big[|g|z''(m) + |g|z''^*(m-1)\Big] = A^2|g|\big(z''_i(m) - z''_i(m-1)\big)
\tag{5.74}
$$

where $z''_i(m)$ and $z''_i(m-1)$ are the imaginary parts of $z''(m)$ and $z''(m-1)$, respectively.

As shown by (5.74), in steady state, the noise term $z_{d,1}(m)$ at the QCFD output can be modeled as white noise that passes through a $1 - z^{-1}$ filter. This is similar to the case of using $\hat{\theta}_u(m) = \mathrm{Arg}[u(m)]$ as the phase estimate discussed in the last subsection. As a

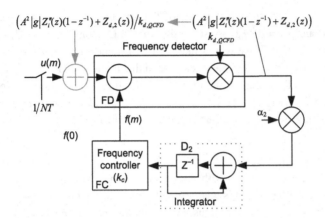

Figure 5.15 Equivalent block diagram of the DFLL with a QCFD for output error analysis

result, the variance of the steady-state frequency error at the DFLL output is significantly less than the DFLL output variance if the FD output noise is white with the same variance [13].

To facilitate the analysis, a modified DFLL system block diagram is given in Figure 5.15.

As shown in the figure, the equivalent input noise of the QCFD is equal to its output noise divided by the gain $k_{d,QCFD}$. Using the results obtained for a first-order DPLL, the variance of the noise component in $f(m)$ due to $z_{d,2}$ is attenuated by a factor of $k_2/(2-k_2)$. Namely, the variance is equal to

$$\sigma_{f,2}^2 = \frac{k_2}{2-k_2}\frac{\sigma_{z_{d,2}}^2}{k_{d,QCPD}^2}\bigg|_{\Delta f=0} = \frac{k_2}{2-k_2}\frac{(\sigma_z^2)^2/2}{\left(A^4|g|^2 2\pi\ NT\right)^2} = \frac{k_2}{2-k_2}\frac{1}{8\pi^2 N^2 T^2 (\gamma_u)^2} \quad (5.75)$$

where $\gamma_u = A^4|h|^2/\sigma_z^2$ as defined in (5.26).

Using the same derivation that led to (5.70), we express the variance of the noise component in $f(m)$ due to $z_{d,1}$ as

$$\sigma_{f,1}^2 = \frac{k_2^2|h|^2\sigma_z^2/2}{2\pi\ k_{d,QCPD}^2}\frac{4\pi}{2-k_1} = \frac{k_2^2 A^4|g|^2\sigma_z^2}{4\pi\left(A^4|g|^2 2\pi NT\right)^2}\frac{4\pi}{2-k_2} = \frac{k_2^2}{4\pi^2 N^2 T^2 \gamma_u(2-k_2)} \quad (5.76)$$

Thus, the total noise variance at the DFLL output is equal to

$$\sigma_f^2 = \sigma_{f,1}^2 + \sigma_{f,2}^2 = \frac{k_2^2}{4\pi^2 N^2 T^2 (2-k_2)\gamma_u}\left(1+\frac{1}{2k_2\gamma_u}\right). \quad (5.77)$$

From (5.77), it can be seen that if $2k_2\gamma_u \ll 1$, i.e., if k_2 is small and/or the SNR γ_u is relatively low, we can ignore the first term "1" in the parenthesis. In other words, the frequency variance is dominated by the noise product term. The variance of the frequency estimate at the DFLL output can be approximated by

$$\sigma_f^2 \approx \frac{k_2}{(4\pi N T \gamma_u)^2} \tag{5.78}$$

Thus, the variance is proportional to k_2 and inversely proportional to the square of $N T \gamma_u$. Conversely, for high SNR the first-order noise term dominates. The variance of the frequency estimate is approximately equal to

$$\sigma_f^2 \approx \frac{k_2^2}{(2\pi N T)^2 (2 - k_2) \gamma_u} \tag{5.79}$$

It is proportional to k_2^2 but inversely proportional to the SNR. This is the same result for the MLPD-based DFLL given by (5.70). Hence, the steady-state output error of the QCFD-based DPLL is close to the error of the MLPD-based DFLL for high SNR input signals.

The results given above are the same as those given in [13] with different expressions of the quantities involved. It was shown in [13] that simulation results agreed very well with the results obtained from the analysis.

5.4.7.3   DFLL with Squaring Frequency Detector

Squaring FDs can be used in exactly the same way in DFLLs as in the data-assisted FDs shown in Figure 5.11. Below, we briefly discuss how to apply the results of the data-assisted DFLLs to their squaring counterparts.

The block diagram of a DFLL with a squaring frequency detector is essentially the same as the data-assisted DFLL given in Figure 5.12 with the MLPD-based FD. With the squaring frequency detector specified by (5.60), the FD gain is the same as the MLPD-based FD. One difference between them is that the FDs based on a squaring PD and the MLPD have different output variances. As shown by (5.27) and (5.52), the variance of the squaring PD output error has an additional second-order term compared to the output variance of the MLPD. Otherwise, the analysis of the data-assisted DFLL can be directly applied to the squaring DFLL. Therefore, their convergence behaviors are similar. In addition, the steady-state output variance of the squaring DFLL approaches that of the data-assisted DFLL when the SNR of the input signal increases. Without the assistance of data symbols, the penalty paid is that the frequency offset detection ability of the squaring DFLL is reduced by a factor of 2.

The characteristics of the DFLL with a squaring QCFD discussed in 5.4.6.3 can be analyzed similarly as the data-assisted QCFD-based DFLL. Its convergence analysis is the same as that of the data-assisted QCFD. Below we outline the analysis of its output frequency error in steady state.

The input to the squaring QCFD is $u_s(m)$, which is equal to the scaled channel estimate $A^2 g^2(m)$ plus a noise/interference component $z_s'(m)$ defined in (5.49) and (5.58). The variance of $z_s'(m)$ is equal to $\sigma_{z_s'}^2$ given by (5.50). According to Figure 5.11, the squaring QCFD output can be expressed as

$$\begin{aligned}
\mathrm{Im}\left[u_s(m)u_s^*(m-1)\right] &= \mathrm{Im}\left[A^4 g^2(m)g^{2*}(m-1)\right] \\
&\quad + \mathrm{Im}\left[A^2 g^{2*}(m-1)z_s'(m) + A^2 g^2(m)z_s'^*(m-1)\right] + \mathrm{Im}\left[z_s'(m)z_s'^*(m-1)\right]
\end{aligned} \tag{5.80}$$

Following an analysis similar to (5.71), we can show that the last term on the right side of (5.80), denoted by $z'_{s,2}(m)$, is white and with zero-mean, and has a variance of $\sigma^2_{z'_{s,2}} = \left(\sigma^2_{z'_s}\right)^2/2$. In steady state when Δf is near zero on average, we have

$$g^2(m) = g^2(m-1) = |g|^2 e^{j2\theta_g} \tag{5.81}$$

Similar to what is seen in (5.74), the middle terms, denoted by $z'_{s,1}(m)$, can be modeled as a white noise with a variance of $\sigma^2_{z'_s}/2$ passing through a $1 - z^{-1}$ filter [13]. Using the derivation leading to (5.76), we can express the variance $z_{s,1}$ as

$$\sigma^2_{s,f,1} = \frac{k_2^2 A^4 |g|^4 \sigma^2_{z'_s}/2}{2\pi k^2_{s,d,QC}} \frac{4\pi}{2-k_2} = \frac{k_2^2}{4\pi^2 N^2 T^2 (2-k_1)L}\left(\gamma_y^{-1} + \gamma_y^{-2}/2\right) \tag{5.82}$$

where $\gamma_y = A^2|g|^2/\sigma^2_z$ was defined in (5.26).

The variance of the noise component $z_{s,2}$ is equal to

$$\sigma^2_{s,f,2} = \frac{k_2}{2-k_2} \frac{\sigma^2_{z'_{s,2}}}{k^2_{s,d,QC}\Big|_{\Delta f=0}} = \frac{k_2}{2-k_2} \frac{\left(\sigma^2_{z'_s}\right)^2/2}{\left(A^4|g|^4 4\pi NT\right)^2} = \frac{k_2}{2-k_2} \frac{\left(2\gamma_y^{-1} + \gamma_y^{-2}\right)^2}{4\pi^2 N^2 T^2 L^2} \tag{5.83}$$

Thus, the total noise variance at the DFLL output is equal to

$$\sigma^2_{s,f} = \sigma^2_{s,f,1} + \sigma^2_{s,f,2} = \frac{1}{4\pi^2 N^2 T^2 (2-k_2)}\left(\frac{k_2^2\left(\gamma_y^{-1} + \gamma_y^{-2}/2\right)}{L} + \frac{k_2\left(2\gamma_y^{-1} + \gamma_y^{-2}\right)^2}{L^2}\right) \tag{5.84}$$

According to (5.84), in high SNR cases, by ignoring the high order terms and using the relationship $\gamma_u = L\gamma_y$, we have

$$\sigma^2_{s,f} \approx \frac{k_2^2 \gamma_y^{-1}}{4\pi^2 N^2 T^2 (2-k_2)L} = \frac{k_2^2 (L\gamma_y)^{-1}}{4\pi^2 N^2 T^2 (2-k_2)} = \frac{k_2^2 \gamma_u^{-1}}{4\pi^2 N^2 T^2 (2-k_2)} \tag{5.85}$$

This is the same as the result obtained from data-assisted QCFD-based DFLL given by (5.79). Thus, the performance of the squaring QCFD-based DFLL approaches that of the data-assisted QCFD at high SNRs. The penalty paid is that its frequency tracking range is only half of the latter.

5.5   Carrier Phase and Frequency Synchronization: OFDM Systems

So far in this chapter, we have described carrier phase and frequency synchronization in single-carrier systems. In this section, we consider carrier synchronization in OFDM systems.

As shown in Section 1.6, an OFDM communication link can be viewed as a number of parallel single-tap, i.e., single-path, subchannels, each of which is on a subcarrier.

When communicating over frequency selective, i.e., multipath, channels, the phases of the subcarriers are different from each other. Thus, unlike in single-carrier communications, there is not a single-carrier phase to recover. In OFDM communications, all of the phases of the subcarriers need to be estimated and corrected in order to recover the data symbols modulated on them. The most efficient way to perform OFDM carrier-phase synchronization is to estimate all of the subchannel phases at the same time, usually by *frequency-domain channel estimation*.

In contrast, OFDM carrier frequency synchronization is done on the entire received signal stream. Therefore, it is possible to use the total energy of the received signal for frequency synchronization to achieve better accuracy. It is important to understand the characteristics and requirements of OFDM carrier frequency synchronization analytically in addition to using simulation tools.

Moreover, in order to effectively reduce ISI, the bandwidth of an OFDM subcarrier is chosen to be narrower than the coherence bandwidth of the channel. Due to the narrow subcarrier bandwidth, the time duration of OFDM symbols is quite large. Hence, it is commonly believed that OFDM communication is more sensitive to carrier frequency offset than single-carrier communication. However, this conclusion must be evaluated quantitatively.

In this section, we first present the basic characteristics and techniques of pilot-assisted channel estimation for obtaining subcarrier phase estimates required for data symbol recovery. Second, the impact of frequency offset on the receiver performance will be analyzed. Techniques of frequency offset estimation and compensation are then presented. These analytical results and the techniques will be applied to carrier synchronization of practical OFDM systems to be discussed later.

5.5.1 Pilot-Assisted Channel Estimation for Carrier Phase Estimation

As discussed in Section 1.6, the demodulation of transmitted data symbols in an OFDM receiver is performed in a DFT block. It can be described as follows.

The timing circuitry of the receiver determines the proper position of an N_{DFT}-long *DFT window* in the received signal sample stream as will be shown in Section 6.6. The samples inside the window form an N_{DFT} long vector, which is processed by a DFT processor. Each element of the DFT output is a data symbol multiplied by the channel response sampled at the subcarrier frequency, on which the symbol is modulated, plus an additive noise term. Thus, the data symbol can be recovered once the phases and magnitude of the subcarrier coefficient are learned through channel estimation. In other words, *subcarrier phase estimation in OFDM communications is performed through channel estimation*.

It should be noted that channel estimation is essential for effective OFDM communications. Due to its importance, various aspects of its analyses and implementations have been thoroughly studied, and can be found in the literature [15, 16, 17, 18, 19, 20]. Due to the scope of this book, only the basics of the channel estimation that are pertinent to synchronization in OFDM systems will be discussed.

5.5.1.1   Fundamentals of OFDM Channel Estimation

Consider a discrete-time channel that consists of $M\,T_s$ spaced paths, where T_s is the OFDM sample time interval, with time-domain coefficients, or taps, $h_0, h_1, \ldots, h_{M-1}$. In addition, it is assumed that the DFT window aligns with the main portion of the DFT symbol through the first arriving path (FAP) h_0. As shown in Section 1.6, the kth element of the DFT output of the nth OFDM symbol, $Y(n,k)$, can be expressed as

$$Y(n, k) = X(n, k) \sum_{m=0}^{M-1} h_m W_{N_{DFT}}^{mk} + z'(n, k) \cong H(n, k)X(n, k) + z'(n, k) \qquad (5.86)$$

where, by definition $W_{N_{DFT}}^{mk} = e^{j2\pi mk/N_{DFT}}$, $X(n,k)$ is the data symbol modulated on the kth subcarrier, $H(n, k) = \sum_{m=0}^{M-1} h_m W_{N_{DFT}}^{mk}$ is the Fourier transform of the T_s-spaced discrete-time filter with coefficients $h_0, \ldots, h_{M-1}$ sampled at the kth-subcarrier frequency f_k, and $z'(n,k)$ is the additive noise on the kth subcarrier. $H(n,k)$ is called the kth-*subcarrier coefficient* of the nth OFDM symbol.

Let us construct a vector $\mathbf{h}_M = (h_0, h_1, \ldots, h_{M-1})^t$ and expand it to an N_{DFT} dimensional vector $\tilde{\mathbf{h}}_{N_{DFT}}$ by post-pending $N_{DFT} - M$ zeros. The subcarrier coefficients, $H(n,k)$, $k = 0, \ldots, N_{DFT} - 1$, are the output of the N_{DFT} point DFT of $\tilde{\mathbf{h}}_{N_{DFT}}$.

If the kth subcarrier coefficient $H(n,k)$ is known or can be estimated, the estimate of the data symbol $X(n,k)$ can be obtained as

$$\hat{X}(n, k) = \hat{H}^*(n, k)Y(n, k)/|\hat{H}(n, k)|^2 + z''(n, k)$$

where $\hat{H}(n, k)$ is the estimate of $H(n,k)$ and $z''(n, k)$ is the additive noise in $\hat{X}(n, k)$.

To estimate the subcarrier coefficients $H(n,k)$'s, known symbols, called FDM pilots, or reference symbols, are usually embedded in the data symbol stream, modulated onto selected subcarriers and transmitted. If the symbol modulated on the kth subcarrier of the nth OFDM symbol is the pilot symbol $X_p(n,k)$, the kth subcarrier is called a *pilot subcarrier*. Similarly, subcarriers modulated with data symbols are called *data subcarriers*. The estimate of the pilot subcarrier coefficient $H_p(n,k)$ can be computed as

$$\hat{H}_p(n, k) \cong \left[X_p^*(n, k)/|X_p(n, k)|^2 \right] Y(n, k) \qquad (5.87)$$

To recover the data symbols in an OFDM symbol, all of the coefficients of the subcarriers that are modulated by data symbols must be estimated. The simplest brute force method is to periodically send OFDM symbols with known data symbols, i.e., pilots, modulated on all of their subcarriers. Such OFDM symbols that are completely known to the receiver are called *TDM pilot symbols*, or *TDM pilots*. All of the subcarrier coefficients of a TDM pilot can be computed according to (5.87). If the channel changes slowly relative to the OFDM symbol rate, the subcarrier coefficients estimated from the TDM pilot can be used to demodulate the OFDM symbols that are close to it in time. The frequency of sending TDM pilots should be determined by the rate of channel's time-variation. According to the Nyquist sampling theorem, the TDM pilot frequency should be at least twice the highest channel fading frequency.

TDM pilots are often sent at the beginning of transmission for channel estimation during initial acquisition to establish the communication link. Examples of the initial acquisition have been given in Chapter 3 for 802.11a/g Wi-Fi and LTE OFDM communication systems. However, because the entire TDM pilot symbol is known to the receiver, it carries no data information. This approach is not efficient for data transmission over channels that have coherence bandwidth narrower than the OFDM signal bandwidth. For such channels, *FDM pilots* are often used.

5.5.1.2   FDM Pilot Symbol-Assisted Channel Estimation

In OFDM communication systems, data packets may be transmitted in short bursts or continuously. In the first case, if the channel does not change much during each burst, the channel estimate generated at the beginning of the burst is sufficient for recovery of the entire data packet. This case often occurs in Wi-Fi OFDM systems. However, if the transmission duration of a data packet is longer than the channel coherence time, it is essential for the OFDM signal design to ensure that the channel estimation can track the channel time-variation.

For channels with a coherence bandwidth narrower than the OFDM signal band-width, pilot symbols are usually modulated onto some of the subcarriers, i.e., pilot subcarriers, of the OFDM symbols, which also carry data on the other subcarriers. Since the pilot subcarriers are distributed in the frequency domain, they are called *FDM pilots*. OFDM symbols with FDM pilots are transmitted periodically in time for tracking channel time-variations.

Design Criteria of FDM Pilot Parameters
Given that FDM pilots are distributed in both the frequency and time domains, they are best expressed as nodes in a two-dimensional time-frequency grid. Figure 5.16 shows two examples of such FDM pilot patterns.

The FDM pilots shown in Figure 5.16(a) are distributed on a rectangular grid. The shortest distance between the two pilots in the frequency domain in an OFDM symbol is $L_p = 6$ subcarrier spacings. In the time domain, the shortest distance between two OFDM symbols that contain FDM pilots is $N_p = 3$ OFDM symbols. According to the time-domain sampling theorem, the maximum frequency of the channel variation should be lower than the OFDM symbol rate divided by $2N_p$.

The required spacing of the FDM pilots in the frequency domain can be determined based on the following two criteria. Both of them are related to the time-domain characteristics of the channel.

First, let us assume that the channel has an actual time span of $M_c T_s$ and that J_c out of the $M_c T_s$-spaced taps are not zero. Since these J_c taps span a J_c-dimensional space, at least J_c FDM pilots in an OFDM symbol are needed to uniquely determine the channel frequency response.

Second, since the channel has a time span of $M_c T_s$, the estimated subcarrier coefficients $\hat{H}(n, k)$ can be viewed as the samples of the channel frequency response. Thus, the problem can be viewed as signal sampling in the frequency domain. To avoid aliasing in the time domain, the sampling rate in frequency must be at least equal to

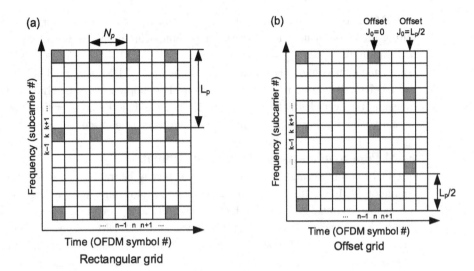

Figure 5.16 Examples of FDM pilot distributions

$1/(M_c T_s)$. Since the frequency bandwidth at the DFT output is equal to $1/T_s$, at least M_c pilot subcarriers are needed to cover the entire OFDM signal bandwidth.

When pilot subcarriers also exist in the guard subcarrier regions (called virtual subcarriers in some references) and the M_c channel taps considered in the first case are all nonzero, i.e., $J_c = M_c$, these two criteria are equivalent. Namely, at least M_c pilot subcarriers per OFDM symbol are needed. In systems with guard subcarriers, both can be viewed as approximate designing guidelines. The same conclusions are given in [18].

The conditions discussed above are applicable to both pilot patterns shown in Figure 5.16(a) and (b). The pattern in Figure 5.16(b) shows the FDM pilots in an OFDM symbol with one out of two subcarrier index offsets. Such a *staggered* pilot pattern provides additional flexibility as will be discussed later in this section. Below we first describe how to generate estimates of data subcarrier coefficients using FDM pilots for data symbol recovery. In the description, we will not consider the guard carrier effects and will assume that the pilots are evenly distributed in the entire DFT bandwidth of $1/T_s$. We also assume that there are M pilot subcarriers and that MT_s is greater than or equal to the channel time span $M_c T_s$.

Implementation of Channel Estimation Using FDM Pilots

The pilot subcarriers coefficients can be estimated according to (5.86) and (5.87). With the limited channel time span to satisfy the conditions given above, the coefficients of the nearby subcarriers are correlated. Thus, interpolation can be used to generate the coefficients of the data subcarriers that are between the pilot subcarriers. Since the subcarriers in OFDM symbols can be viewed as in a time-frequency two-dimensional plane, the optimal interpolation can be achieved using a two-dimensional Wiener filter based on an MMSE criterion [18, 19]. In practice, since the frequency-domain and time-domain variations are independent, interpolation can be optimized in these two domains independently. Below we first consider channel coefficient interpolation in the frequency domain.

As has been studied extensively in the field of digital signal processing [21], interpolation between time samples can be achieved by using low-pass filtering. In principle, this is also true for interpolation in the frequency domain. The commonly used interpolation methods are first-order linear interpolation, multitap linear interpolation filter, MMSE-Wiener filter, least squares (LS) estimator, and robust filter/interpolator.

Among these interpolators, the simplest one is the first-order linear interpolator. It performs linear interpolation between two known channel estimates on adjacent pilot subcarriers. For example, if we have two estimates $\hat{H}(n,k)$ and $\hat{H}(n,k+L_p)$, the estimate of the subcarrier coefficient at $k+j$, $0 < j < L_p$, can be computed by

$$\hat{H}(n,k+j) = \left(1 - j/L_p\right) \times \hat{H}(n,k) + \left(j/L_p\right) \times \hat{H}(n,k+L_p).$$

The first-order linear interpolator is the simplest to implement, but it could introduce severe distortion that would cause degradation in receiver performance. A multitap low-pass FIR filter, which can provide better noise suppressing and thus less error in the estimates, can be a better interpolator. The performance of such a linear interpolation filter can be optimized based on different criteria. For example, it can be optimized by using the MMSE criterion, if the statistics of the channel taps h_m in (5.86) are known [19].

In practice, the statistics of the channel taps are usually not known. One possible alternative to the *MMSE estimator* is the so-called *robust estimator* [22]. A robust estimator provides the best estimates under the worst conditions. The time response of a robust frequency-domain estimator has a rectangular shape with a width equal to the channel time span. Thus, the frequency-domain response of the robust estimator is a sync function.

The digital linear interpolators can be implemented in different forms. More details of the design and implementation of such interpolators will be presented in Chapter 7 of this book for time-domain sample interpolation. Below, we describe an LS estimator/interpolator based on FFT/iFFT operations, which is widely used for OFDM channel estimation due to a number of its attractive properties.

FFT/iFFT-Based Least Squares Estimator/Interpolator
Consider the pilot patterns shown in Figure 5.16, with a total of M evenly distributed pilots in each OFDM symbol. The positions of the FDM pilots with an offset j_0 can be expressed as $j_0+L_p j$, where $0 \le j_0 < L_p$ and $j = 0,\ldots, M-1$. Moreover, to simplify the description and analysis, we assume that $M \times L_p = N_{DFT}$, i.e., N_{DFT} is divisible by L_p and there are M pilot subcarriers. In addition, we denote the time-domain channel coefficient vector by $\mathbf{h}_M$, which has M elements. The actual channel span $M_c T_s$ is less than or equal to MT_s. If $M_c < M$, the elements, $h_{M_c}, \ldots, h_{M-1}$, in $\mathbf{h}_M$ are zeros.

From the definition given by (5.86) and with the assumptions listed above, the estimates of the pilot subcarrier coefficients given by (5.87) can be expressed by

$$\hat{H}\left(n, j_0 + L_p j\right) = \sum_{m=0}^{M-1} h_m W_{N_{DFT}}^{m\left(j_0 + L_p j\right)} + z_j'', \quad j = 0, 1, \ldots J - 1 \tag{5.88}$$

where $W_N = e^{-j2\pi/N}$, z_j'' is the additive noise in $\hat{H}\left(n, j_0 + L_p j\right)$, and $h_m = 0$ for $m \geq M_c$. The equation set (5.88) can be expressed in matrix form as

$$
\begin{pmatrix}
\hat{H}(j_0) \\
\hat{H}(j_0 + L_p) \\
\hat{H}(j_0 + 2L_p) \\
\vdots \\
\hat{H}(j_0 + (M-1)L_p)
\end{pmatrix}
=
\begin{pmatrix}
1 & W_{N_{DFT}}^{j_0} & W_{N_{DFT}}^{2j_0} & \cdots & W_{N_{DFT}}^{(M-1)j_0} \\
1 & W_{N_{DFT}}^{j_0+L_p} & W_{N_{DFT}}^{2(j_0+L_p)} & \cdots & W_{N_{DFT}}^{(M-1)(j_0+L_p)} \\
1 & W_{N_{DFT}}^{j_0+2L_p} & W_{N_{DFT}}^{2(j_0+2L_p)} & \cdots & W_{N_{DFT}}^{(M-1)(j_0+2L_p)} \\
\vdots & \vdots & \vdots & \ddots & \vdots \\
1 & W_{N_{DFT}}^{j_0+(M-1)L_p} & W_{N_{DFT}}^{2[j_0+(M-1)L_p]} & \cdots & W_{N_{DFT}}^{(M-1)[j_0+(M-1)L_p]}
\end{pmatrix}
\begin{pmatrix}
h_0 \\
h_1 \\
h_2 \\
\vdots \\
h_{M-1}
\end{pmatrix}
+
\begin{pmatrix}
z_0'' \\
z_1'' \\
z_2'' \\
\vdots \\
z_{M-1}''
\end{pmatrix}
$$
$$\tag{5.89}$$

To simplify the notation, we omit the OFDM symbol index n in (5.89) and in the text below given that there is no confusion. The FCE $\hat{\mathbf{H}}_{M,j_0}$, i.e., the subcarrier coefficient vector estimate on the left side of (5.89), can be expressed as

$$\hat{\mathbf{H}}_{M,j_0} = \mathbf{W}_{M \times M, j_0} \mathbf{h}_M + \mathbf{z}' \cong \mathbf{W}_M \mathbf{D}_{M_c, j_0} \mathbf{h}_M + \mathbf{z}_M'' \tag{5.90}$$

where $\mathbf{W}_{M \times M, j_0}$ is the $M \times M$ matrix shown on the right side of (5.89), $\mathbf{h}_M = (h_0, h_1, \ldots, h_{M-1})^t$ is the time-domain channel coefficient vector, $\mathbf{D}_{M,j_0}$ is an $M \times M$ diagonal matrix whose ith element is equal to $e^{j2\pi i j_0/N_{DFT}}$, $i = 0, \ldots, M-1$, $\mathbf{W}_M$ is the DFT matrix that performs an M point DFT as defined by (1.3.2) in Section 1.6.3, and $\mathbf{z}_M'' = \left(z_0'', z_1'', \ldots, z_{M-1}''\right)^t$ is an M-dimensional noise vector. The LS estimate of $\mathbf{h}_M$ can be computed by

$$\hat{\mathbf{h}}_M = \mathbf{D}_{M,j_0}^{-1} \left(\mathbf{W}_M^H \mathbf{W}_M\right)^{-1} \mathbf{W}_M^H \hat{\mathbf{H}}_{M,j_0} = \mathbf{D}_{M,j_0}^{-1} \mathbf{W}_M^H \hat{\mathbf{H}}_{M,j_0} \tag{5.91}$$

where we have used the result given by (1.33) that $\mathbf{W}_M^H \mathbf{W}_M$ is an identity matrix and $\mathbf{W}_M^H$ performs iDFT operation.

An N_{DFT}-long FCE $\hat{\mathbf{H}}_{N_{DFT}}$ is generated by performing the DFT of the N_{DFT}-long time-domain coefficient vector estimate $\hat{\mathbf{h}}_{N_{DFT}}$, which is generated by post-pending $N_{DFT}-M$ zeros to $\hat{\mathbf{h}}_M$. The elements of the FCE, $\hat{H}(n, k)$, $k = 0, \ldots, N_{DFT} - 1$, are the interpolated coefficients of all of the subcarriers.

There are a few attractive properties of this iDFT/DFT interpolation approach compared to direct interpolation in the frequency domain.

First, if the CIR of the channel is sparse, i.e., most of its coefficients are zeros, the small elements of the estimated time-domain channel coefficient vector $\hat{\mathbf{h}}_M$ are likely due to noise rather than being the estimates of the true channel paths. When the elements with magnitude below a predetermined threshold are eliminated, the estimated subcarrier coefficients could be less noisy. This approach is especially beneficial for low SNR channels. Conversely, when the frequency-domain direct interpolation is used, there is no good way to eliminate the contributions from the elements in $\mathbf{h}_M$ due to noise.

Second, as a by-product, the estimates of time-domain channel coefficients, h_i, $i = 0,\ldots$, $M-1$, is generated by this iDFT/DFT interpolation method in the process. As will be shown in Chapter 6, timing synchronization of an OFDM receiver is equivalent to determining the proper DFT window position and can be achieved based on the estimated channel coefficients. Thus, the channel coefficient estimates generated from the iDFT/ DFT interpolation can be used for timing synchronization with few additional operations.

Third, if the channel coherence time is longer than the time gap between the two OFDM symbols that contain pilots, the channel estimates from them can be averaged to improve the accuracy of the estimates of the subcarrier coefficients. If the pilots on these two or more OFDM symbols are on the subcarriers with the same index, averaging can be performed on the estimated pilot subcarrier coefficients before interpolation. However, if the pilots in adjacent OFDM symbols are on different subcarriers as shown in Figure 5.16(b), the subcarrier coefficients must be interpolated first before averaging is performed. Averaging can also be performed in the time domain as shown below.

Using the DFT/iDFT method, the LS estimates of the channel time-domain coefficients are generated according to (5.91). For the pilot patterns shown in Figure 5.16(b), j_0's are equal to 0 and $L_p/2$ for the OFDM symbols at n and $n + 3$, respectively. Furthermore, by using the previous assumption that N_{DFT} is divisible by M, we have

$$\hat{\mathbf{h}}_M(n) = \mathbf{W}_M^H \hat{\mathbf{H}}_M(n) \text{ and } \hat{\mathbf{h}}_M(n+3) = \mathbf{D}_{M,3}^{-1} \mathbf{W}_M^H \hat{\mathbf{H}}_M(n+3) \tag{5.92}$$

where $\hat{\mathbf{H}}_M(n)$ and $\hat{\mathbf{H}}_M(n+3)$ are the FCEs generated from the FDM pilots in the OFDM symbols at n and $n + 3$, respectively.

Based on the channel conditions and implementation considerations, various methods can be used to compute the FCEs of the OFDM symbols that do not contain FDM pilots. If the channel does not change from nT_{Sym} to $(n+3)T_{Sym}$, we can perform algebraic averaging of the two coefficient vectors, such that

$$\bar{\mathbf{h}}_M(n, n+3) = \left[\hat{\mathbf{h}}_M(n) + \hat{\mathbf{h}}_M(n+3)\right]/2 \tag{5.93}$$

The averaged coefficient vector given by (5.93) can be used as the estimate of the time-domain coefficient vectors at n, $n + 1$, and $n + 2$. Alternatively, these estimates can be generated from $\hat{\mathbf{h}}_M(n)$ and $\hat{\mathbf{h}}_M(n+3)$ by using other interpolation/extrapolation techniques.

Estimation of a Channel with Long Time Span

If the coherence time is much longer than the time difference between two adjacent OFDM symbols with pilots, the pilot pattern shown in Figure 5.16(b) should be able to work with channels with a time span of up to $2MT_s$. For example, if the channel does not change from nT_s to $(n+3)T_s$, the estimated FCEs, i.e., $\hat{\mathbf{H}}_M(n)$ and $\hat{\mathbf{H}}_M(n + 3)$ in (5.92), can be interleaved and combined to construct a $2M$ long subcarrier coefficient vector, $\hat{\mathbf{H}}_{2M}$. The iDFT of $\hat{\mathbf{H}}_{2M}$ generates a time-domain channel estimate that is twice as long as $\mathbf{h}_M$ in (5.90).

Generating such a $2MT_s$ long time-domain channel estimate is also straightforward when the iDFT/DFT approach is used for FCE interpolation. Similar to (5.91), it can be shown that $\hat{\mathbf{H}}_{2M}$ can be generated by performing the DFT of the estimate of the $2M$-long channel-domain coefficient vector $\hat{\mathbf{h}}_{2M}$, defined by

$$\hat{\mathbf{h}}_{2M} \cong \begin{pmatrix} \hat{\mathbf{h}}_M \\ \hat{\mathbf{h}}_{M,ex} \end{pmatrix} = \begin{pmatrix} \mathbf{h}_M \\ \mathbf{h}_{M,ex} \end{pmatrix} + \mathbf{z}'_{2M} \cong \mathbf{h}_{2M} + \mathbf{z}'_{2M} \qquad (5.94)$$

where $\mathbf{h}_{2M}$ is the time-domain coefficient vector, which is 2M long, and $\mathbf{h}_{M,ex}$ is called its *excess channel portion*.

Note that $\hat{\mathbf{H}}_M(n)$ and $\hat{\mathbf{H}}_M(n+3)$ contain the even and odd elements of $\hat{\mathbf{H}}_{2M}$, respectively. The iDFT outputs of $\hat{\mathbf{H}}_M(n)$ and $\hat{\mathbf{H}}_M(n+3)$ with proper phase correction are $\hat{\mathbf{h}}_M(n)$ and $\hat{\mathbf{h}}_M(n+3)$ given by (5.92). As shown by (5.93), an estimate of $\mathbf{h}_M$ can be obtained by averaging $\hat{\mathbf{h}}_M(n)$ and $\hat{\mathbf{h}}_M(n+3)$. From the basic properties of the iDFT, it can be shown that the estimate of the excess portion of $\mathbf{h}_{2M}$, i.e., $\hat{\mathbf{h}}_{M,ex}$, can be computed by

$$\hat{\mathbf{h}}_{M,ex} = \frac{\hat{\mathbf{h}}_M(n) - \hat{\mathbf{h}}_M(n+3)}{2} \qquad (5.95)$$

The derivation of (5.95) and more details on utilizing the staggered FDM pilots can be found in [23]. The method discussed above can also be extended to more general cases to construct even longer estimates of the time-domain channel with properly staggered pilot patterns.

5.5.2   Carrier Frequency Synchronization

As discussed previously, an OFDM communication link can be viewed as consisting of N parallel subchannels, each of which is carried by a subcarrier. Ideally, these subchannels are orthogonal to each other and there is no interference among them. However, the orthogonality holds only if the *frequency offset* after down-conversion, i.e., $\Delta f = f_c - \hat{f}_c$, is equal to zero. Otherwise, the orthogonality between subcarriers is not preserved and there will be interference, called *intersubcarrier interference*, among them.

In reality, the frequency offset Δf is never exactly equal to zero due to the inaccuracy of the locally generated down-conversion frequency. Such a nonzero frequency offset generates ICI, which directly affects the OFDM receiver's performance. To understand what the receiver's tolerable frequency offset is, we derive below a quantitative relationship between Δf and the resulting ICI.

5.5.2.1   Impact of Carrier Frequency Offset on OFDM Receiver Performance

As described in Section 1.6, an OFDM symbol is the sum of N data symbols modulated on different subcarriers at the frequencies kf_{sc} and k is in the range from 0 to $N_{DFT} - 1$. The received OFDM signal samples are covered by a $1/f_{sc}$ long rectangular window before the DFT operation. Due to the windowing effect, the spectrum of the received OFDM signal samples is the sum of sinc functions, each of which is centered at a different subcarrier frequency.

The orthogonality between the subcarriers without a frequency offset can be seen from the subcarriers' spectra in the form of sync functions shown in Figure 5.17(a). Specifically, with $\Delta f = 0$, the DFT outputs are the sampled peak values of the sinc functions of the subcarriers. For such a sync function, the values of a subcarrier's

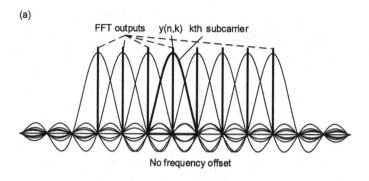

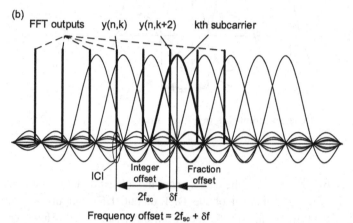

Figure 5.17 Integer and fraction frequency offset

spectrum at the peaks of other subcarriers' spectra are equal to zero. Thus, these values do not interfere with the other DFT outputs.

However, if the frequency offset is not equal to zero, the subcarriers' spectra of the received signal are shifted. The DFT outputs are no longer the sampled peak values of the subcarriers. As a result, there is ICI as shown in Figure 5.17(b). The frequency offset can be larger or smaller than a subcarrier spacing f_{sc}. When it is larger than f_{sc}, the frequency offset can be expressed as an integer part plus a fractional part, called the *integer and fractional frequency offsets*. Namely, $\Delta f = m f_{sc} + \delta f$, where m is an integer and $|\delta f| < f_{sc}/2$ as shown in Figure 5.17(b). The integer part is usually resolved and removed during initial acquisition or training.

After an OFDM receiver enters the data mode, it can be assumed that the remaining frequency offset is less than half of a subcarrier spacing as shown in Figure 5.18. Below we derive the expression of the receiver performance degradation due to such a frequency offset based on basic OFDM signal characteristics. A more rigorous analysis that leads to the same result can be found in [24].

As can be seen from Figure 5.18, if $\delta f = 0$, the kth element of the DFT output is the sampled peak value of the sinc function centered at $k f_{sc}$. It is equal to the value of the data symbol modulated on the subcarrier weighted by the channel frequency response at

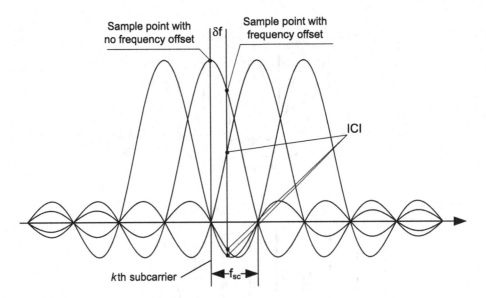

Figure 5.18 OFDM signal spectrum with frequency offset

kf_{sc}, as given by (5.86). Thus, the data symbol modulated on the kth subcarrier can be estimated from the kth element of the DFT output. DFT outputs are corrupted by additive noise but not by the interference from other subcarriers. However, if $\delta f \neq 0$, the sampling points of the DFT output elements deviate from the peaks of the sinc functions. Each DFT output element has a reduced magnitude and it is corrupted by the ICI from the nearby subcarriers.

From the figure, we observe that the magnitude of the kth element at the DFT output is reduced to $\mathrm{sinc}^2(\delta f/f_{sc})$ relative to its peak value. Moreover, there is ICI generated by other data and pilot subcarriers. The kth element of the DFT output can be expressed as

$$Y(n,k) = H(n,k)X(n,k)\mathrm{sinc}(\delta f/f_{sc})$$

$$+ \sum_{\substack{l \in \text{data/pilot subcarries} \\ l \neq k}} H(n,l)X(n,l)\mathrm{sinc}(k-l+\delta f/f_{sc}) + z'(k) \qquad (5.96)$$

The last two terms in (5.96) are due to the ISI and additive noise. Assuming that all of the DFT output elements have the same average power P_{sc} and are independent, the ISI power is

$$P_{ISI} = \sum_{\substack{l \in \text{data/pilot subcarries} \\ l \neq k}} P_{sc} \times \mathrm{sinc}^2(k-l+\delta f/f_{sc}) \qquad (5.97)$$

The power of the kth DFT output is $P_k = P_{sc}\mathrm{sinc}^2(\delta f/f_{sc})$. Thus, the signal-to-ISI ratio of the kth DFT output is

$$\gamma_{k,ISI} = \frac{\text{sinc}^2(\delta f/f_{sc})}{\sum\limits_{\substack{l \in \text{ data/pilot subcarries} \\ l \neq k}} \text{sinc}^2(l - k - \delta f/f_{sc})} \tag{5.98}$$

The total SNR of the kth DFT output is the geometric mean of the signal-to-ISI ratio and the SNR γ_k due to additive noise only, i.e.,

$$\gamma_{k,total} = \left[\left(\gamma_{k,ISI}\right)^{-1} + \gamma_k^{-1}\right]^{-1} \tag{5.99}$$

The formulas given by (5.98) and (5.99) can be used to evaluate the SNR degradation of the DFT's output at the kth subcarrier if the number of subcarriers in an OFDM symbol is known. In [24], an approximate but generic formula is derived for a system of any DFT size and can be expressed by

$$\gamma_{ISI} > \frac{3}{\pi^2(\delta f/f_{sc})^2} \tag{5.100}$$

Due to the various approximations made in the derivation, the expression (5.100) can only be used as a lower bound for rough receiver performance evaluation.

The formulas given above assume that the frequency offset δf is a constant. Nonetheless, they can also be used to evaluate the performance degradation in channels with Doppler fading by integrating δf over the entire Doppler spectrum.

5.5.2.2 Initial Frequency Offset Detection in OFDM Systems

It can be shown from formulas given above that a frequency offset, which may be as small as a quarter of a subcarrier spacing, can cause significant degradation in receiver performance. When an OFDM receiver starts its operation, the initial frequency offset could be as large as a few subcarrier spacings due to the inaccuracy of the local oscillator. However, most of the data-mode frequency detection methods can only correctly detect a frequency offset up to half of a subcarrier spacing. Therefore, an initial frequency offset larger than $0.5f_{sc}$ must be detected and corrected before reliable data demodulation can start.

There are a few different approaches to estimating the initial frequency offset in OFDM systems. In most packet communication systems, TDM pilots are usually sent at the beginning of a packet transmission to facilitate receiver performing system acquisition. These OFDM symbols can also be used for initial frequency estimation. We have discussed this approach to some degree in Chapter 3 in the context of the initial acquisition of OFDM systems. In broadcasting systems such as DVB-T for terrestrial TV broadcasting, signals are transmitted continuously. In such systems, a common approach to determining the integer number of the subcarrier offset is to detect the discrepancy between the expected pilot symbols and the actual demodulated symbols. Below we discuss these two initial frequency estimation methods.

Initial Frequency Estimation Using TDM Pilots

TDM pilot symbols or their equivalents are used for initial acquisition in commercially deployed 801.11 Wi-Fi and LTE WAN OFDM communications systems as discussed in Chapter 3. They can also be used for initial frequency offset estimation. Here we briefly review their properties in the context of frequency offset estimation.

1. 802.11a/g

As has been shown in Section 3.6.2, in an 802.11a/g system, 10 short preamble symbols, each of which contains 16 OFDM channel samples, are transmitted consecutively at the beginning of each data packet transmission. As the 10 symbols are all identical, the short preamble sequence is periodic with 10 periods. The received signals are sampled at the OFDM channel sample rate $f_s = 1/T_s$. The receiver detects the periodicity in the received signal samples by the so-called *correlation-and-accumulation* method introduced in [25] as described in Section 3.6.2.2. Once such a periodic pattern is found, the short preamble is claimed detected. The results of accumulations generated every 16 samples given by (3.57) are used for the frequency offset estimation.

When there exists a frequency offset Δf, the carrier phases change from one sample to the next by $2\pi\Delta fT_s$. The phase difference between the two accumulator outputs that are separated by 16 samples is equal to $2\pi \times 16T_s\Delta f = 32\pi\Delta f/f_s$. With the maximum phase ambiguity of π, a frequency offset of $\Delta f < f_s/32$ can be detected. In 802.11a/g systems, an OFDM symbol has 64 OFDM samples with the subcarrier spacing of $f_{sc} = f_s/64$. Thus, the frequency-offset detector using the short preamble can detect a frequency offset up to $2f_{sc}$.

The procedure of this frequency-offset estimation method is summarized below.

Once a valid short preamble is detected, the phase differences between the pairs of accumulator outputs separated by 16 samples are calculated. Since the short preamble is 160 samples long, 10 of such accumulator outputs, one per period, are generated. The first and the last one of them are unreliable and discarded. Thus, there are 7 pairs of such outputs, and 7 phase differences can be computed. Let us denote the phase of the jth integrator output by $\hat{\varphi}(j)$, $j = 0, 1,\ldots, 9$. The estimate of the frequency offset is computed as

$$\hat{\Delta f} = \sum_{j=1}^{7} [\hat{\varphi}(j+1) - \hat{\varphi}(j)]_{\mathrm{mod}\ 2\pi}/(7 \times 16T_s) \qquad (5.101)$$

with $\hat{\varphi}(0)$ and $\hat{\varphi}(9)$ discarded.

It may be desirable to use more than one of the outputs per period for phase difference estimation in order to improve the estimation accuracy. However, no more than two samples per period are necessary because the outputs close to each other are correlated.

The frequency-offset estimate from the short preamble can be used to initialize the value of the frequency register in the receiver's AFC loop, which will be discussed in Sections 5.7.2 and 5.7.3.

The long preamble following the short preamble has two identical 64 samples long OFDM symbols with a 32-samples long CP. The received signal samples generated from the long preamble are periodic and can have two complete periods of 64 samples each. In these two periods, any two samples separated by 63 other samples have the same magnitude but with a phase difference of $128\pi T_s\Delta f'$, where $\Delta f'$ is the residual frequency offset. Thus, the estimate of $\Delta f'$ can be computed as

$$\hat{\Delta f'} = \frac{1}{128\pi T_s}\text{Arg}\left[\sum_{k=0}^{63} y^*(n+k)y(n+k+64)\right]$$

which can be used to refine the value of the AFC's frequency register to further reduce the residual frequency offset. The data mode frequency offset estimation will be described in Section 5.5.2.3.

2. LTE

As described in Chapter 3, for initial acquisition, two OFDM symbols, called primary and secondary synchronization signals, PSS and SSS, are sent every 5 ms. However, no specific OFDM symbols are provided for detection of large frequency offsets. As shown in Section 3.6.1, there are methods that can be used to detect synchronization signals under large frequency offset. Once the synchronization signals are detected, it is possible to partition them into multiple segments for performing frequency estimation. Alternatively, a frequency-binning method, described in Section 3.4.4.2, can be used for the initial frequency offset estimation. Combined, these methods can be used to reduce the large initial frequency offset to less than $0.5f_{sc}$ so that the data-mode frequency detection can take over to further improve the estimation accuracy.

The overall implementation of frequency synchronization in LTE systems will be discussed in the examples given in Section 5.8.4.

Detection of Integer Frequency Offset from FDM Pilots

In broadcasting systems, such as in DVB-T, OFDM signals are transmitted continuously. If no special OFDM symbols are included in the transmission, it is necessary to detect the large frequency offset by using FDM pilots. As proposed in [26], it is possible to determine the integer frequency offset by correlating the FDM pilot symbols with the DFT outputs. This approach is described below for OFDM communication over multipath fading channels.

With a frequency offset that is greater than one subcarrier spacing, it can be expressed as $\Delta f = mf_{sc} + \delta f$ and is shown graphically in Figure 5.19.

Figure 5.19 shows the DFT output with and without frequency offset. As shown in Figure 5.19(a), when there's no frequency offset, the kth symbol $X(n,k)$ weighted by the channel frequency response at the kth subcarrier frequency is at the position of the kth output. Figure 5.19(b) shows the received signal that is shifted in frequency by an amount slightly larger than $2f_{sc}$. In this case, the kth DFT output contains the $(k-2)$th modulation symbol while the $(k+2)$th DFT output contains the kth modulation symbol. For example,

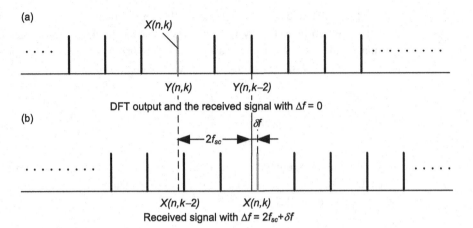

Figure 5.19 DFT output when frequency offset is larger than subcarrier spacing

$$Y(n,k+2) = H(n,k)X(n,k)\text{sinc}(\delta f/f_{sc}) + \sum_{\substack{l \in \text{data/pilot subcarries} \\ l \neq k}} \text{ISI from } H(n,l)X(n,l) + z'(k)$$

If $X(n,k)$ is a pilot symbol, we have

$$X^*(n,k)Y(n,k) = H(n,k-2)X^*(n,k)X(n,k-2)\text{sinc}(\delta f/f_{sc}) + (\text{ISI and noise}) \quad (5.102)$$

and

$$X^*(n,k)Y(n,k+2) = H(n,k)|X(n,k)|^2\text{sinc}(\delta f/f_{sc}) + (\text{ISI and noise}) \quad (5.103)$$

Since $X^*(n,k)$ and $X(n,k-2)$ are uncorrelated, $X^*(n,k)Y(n,k)$ in (5.102) has a zero mean. At the same time, the mean value of $X^*(n,k)Y(n,k+2)$ is equal to $H(n,k)$ scaled by a constant $|X^*(n,k)|^2\text{sinc}(\delta f/f_{sc})$. Below, we consider how to determine the integer frequency offset from (5.102) and (5.103) by using DVB-T as an example.

The DVB-T OFDM signal design defines two types of FDM pilots: the continuous pilots and scattered pilots as shown in Figure 5.20.

As defined in [27], the continuous pilots are in every OFDM symbol on the same subcarriers. There are 45 and 177 continuous pilot symbols in each OFDM symbol for 2k and 8k modes, respectively. The scattered pilots are on subcarriers that are shifted by three positions from one OFDM symbol to the next. In each OFDM symbol, the scattered pilots are modulated on one out of every 12 subcarriers. Because the continuous pilots are on the same subcarrier in every OFDM symbol, the channel frequency responses are the same for OFDM symbols within the channel coherence time, e.g., JT_{Sym} where T_{Sym} is the OFDM symbol duration. For a continuous pilot sequence at subcarrier k_{cp}, we can form a decision metric as

$$D_i = \left| \sum_{j=0}^{J-1} X^*(n+j, k_{cp}) Y(n+j, k_{cp}+i) \right|^2 \quad (5.104)$$

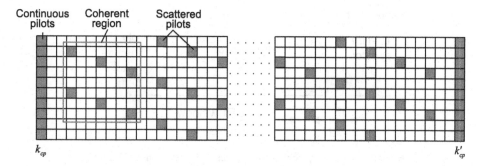

Figure 5.20 DVB-T pilot pattern

where i is in a predetermined range of $[-I_{max}, I_{max}]$ based on the worst-case frequency offset. The decision of the integer frequency shift is equal to the value of i that maximizes D_i.

To obtain a more reliable decision, multiple D_i's can be generated using more than one continuous pilot at $k_{cp,0}, k_{cp,1}, \ldots$, etc. The decision can be made on the sum of these D_i's. Similarly, the scattered pilots can also be used for determining the integer frequency offset if they are within the coherence frequency and time of the channel.

5.5.2.3   Data Mode Frequency Tracking

During the initialization stage, an OFDM receiver learns the approximate offset between the transmitter carrier frequency and the receiver down-conversion frequency. The receiver then corrects its reference frequency based on the approximate estimate. After the correction, the residual frequency offset should be less than half of a subcarrier spacing. In addition, the receiver should also have learned the approximate OFDM symbol timing as will be shown in Chapter 6.

At this point, the receiver is ready to start to demodulate the received OFDM signal. However, the estimated carrier frequency is usually not yet accurate enough at this point to achieve the best possible performance. Therefore, it needs to be refined based on the received signal to ensure satisfactory receiver performance.

In the data mode, the OFDM carrier frequency offset estimation can be done in the time domain or the frequency domain. One of the commonly used time-domain methods, also called *pre-DFT* methods [16], takes advantage of the presence of the CP in OFDM symbols [28, 29]. Carrier frequency offset detection can also be performed in the frequency domain by using FDM pilots in a few different ways. These methods are often called *post-DFT* methods [16]. In practice, both types of methods are used for OFDM carrier frequency synchronization in receivers.

Cyclic Prefix Based Carrier Frequency Offset Detection
In almost all of the OFDM communication systems, there is a CP consisting of N_{CP} OFDM channel samples before the main portion of each OFDM symbol to combat ISI. The samples in the CP are the same as those in the last portion of the OFDM symbol.

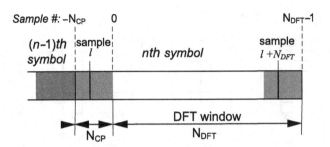

Figure 5.21 OFDM symbols with CP received over a single channel

Recall from Section 1.6 that a DFT window, which contains N_{DFT} received OFDM signal samples that are used as the input to the DFT processing. These samples correspond to the position of the main portion of an OFDM symbol. The segment of N_{CP} samples prior to the DFT Window corresponds to the CP portion of the OFDM symbol. These samples are called CP samples. If the DFT window is defined as starting from sample zero and ending at sample $N_{DFT} - 1$, the indices of CP samples are from $-N_{CP}$ to -1. As shown in Figure 5.21, the sample $y(l)$, $-N_{CP} \leq l \leq -1$, inside the CP sample segment is closely related to the sample $y(l+N_{DFT})$ in the last portion of the DFT window. For a single-path channel, these two samples only differ by a phase rotation caused by frequency offset besides the random variation due to noise.

Each pair of these samples can be expressed as $y(l) = g\tilde{x}_l + z$ and $y(l + N_{DFT}) = e^{j2\pi N_{DFT} T_s \Delta f} g\tilde{x}_l + z'$, where $\tilde{x}_l$ is the transmitter channel sample that appears in both CP and the last portion of the OFDM symbol, g is the channel gain, and z and z' are additive noise terms with a variance of σ_z^2. A length L correlation between $y(l)$ and $y(l + N_{DFT})$, $l =- L,\ldots, -1$, is performed, such that

$$\frac{1}{L}\sum_{l=-L}^{-1} y^*(l)y(l + N_{DFT}) = e^{j2\pi N_{DFT} T_s \Delta f}\left(\frac{1}{L}\sum_{l=-L}^{-1} |g\tilde{x}_l|^2\right) + z'' \qquad (5.105)$$

where z'' is the noise term in the correlation output with a variance of $\left(2E\left[|\tilde{x}_l|^2\right]|g|^2\sigma_z^2 + \sigma_z^4\right)/L$ and $L \leq N_{CP}$. From (5.105), an estimate of the frequency offset can be computed as

$$\hat{\Delta f} = \text{Arg}\left[\sum_{l=-L}^{-1} y^*(l)y(l + N_{DFT})\right]/2\pi N_{DFT} T_s \qquad (5.106)$$

The estimate of the frequency offset given by (5.106) can be used to drive a DFLL, which was described in Section 5.4.7, to perform the AFC function of the receiver. When serving as the FD output to the DFLL, the imaginary, or quadrature, part of (5.105) is often used, i.e.,

$$\text{Im}\left[\frac{1}{L}\sum_{l=-L}^{-1} y^*(l)y(l + N_{DFT})\right] = \left(\frac{1}{L}\sum_{l=-L}^{-1} |g\tilde{x}_l|^2\right)\sin\left(2\pi N_{DFT} T_s \Delta f\right) + z_Q'' \qquad (5.107)$$

The variance of the noise term z''_Q is equal to half of the variance of z''. It can be expressed in terms of the energy and SNR of $y(l)$, i.e., $E_y = |g|^2 E\left[|\tilde{x}_l|^2\right]$ and $\gamma_y = E_y/\sigma_z^2$, as

$$\sigma^2_{z''_Q} = \left(2E\left[|\tilde{x}_l|^2\right]|g|^2\sigma_z^2 + \sigma_z^4\right)/2L = E_y^2\left(2\gamma_y^{-1} + \gamma_y^{-2}\right)/2L$$

This simplified form of the frequency detector is similar to the QCFD described in Section 5.4.7.2 for implementation of the DFLL in single-carrier systems. However, since the FD outputs of consecutive OFDM symbols are generated from different received signal samples, their noise terms z''_Q are uncorrelated. As a result, there are no high-pass filtering effects as those in the QCFD shown previously. The DFLL's output frequency variance is proportional to the variance of $\sigma^2_{z''_Q}$ multiplied by the DFLL's noise bandwidth. Below, we examine two properties of the CP-based frequency-offset detection.

1. Detection Range

Since the range of the estimated phase $\text{Arg}\left[\sum_{l=-L}^{-1} y^*(l)y(l + N_{DFT})\right]$ in (5.106) is from $-\pi$ to π, the frequency detection range without aliasing is

$$\pm(2N_{DFT}T_s)^{-1} = \pm f_{sc}/2 \tag{5.108}$$

Thus, it is necessary to reduce the frequency error to within this range before the data mode tracking starts. Otherwise, due to aliasing, this type of frequency detector cannot distinguish between δf and $\delta f + m f_{sc}$, where m is an integer. Note, however, if the initial frequency offset is larger than $m f_{sc}$, this CP-based frequency offset detector still converges to the closest integer multiple of f_{sc}. After it has converged, the ambiguity of $m f_{sc}$ can be resolved by using the FDM pilot correlation method described in the previous section. Therefore, the CP-based method can be used as the preprocessing stage for FDM pilot-based integer frequency-offset detection, such as in OFDM broadcasting systems.

2. Length of Integration

The accuracy of the frequency detection using the CP depends on the length of integration L. However, the integration length should be selected based on the channel delay spread. If there is no delay spread, i.e., the channel has a single path, L can be set to be equal to the entire CP length N_{CP} to achieve the maximum integration gain. It has been shown in [30] that for a single-path AWGN channel, the CP-based method generates an ML estimate of the frequency offset.

However, if the channel is a multipath channel with a delay spread of $L_c T_s$, the signal energy from the previous OFDM symbols overlaps with the CP of the current symbol. Thus, there is ISI if the integration length L is larger than $N_{CP} - L_c$. As a result, the accuracy of the estimate of frequency offset degrades. To avoid such interference, the integration length L should be no greater than $N_{CP} - L_c$. However, the reduction in integration length reduces the integration gain. In practical implementations, given that the tail of a realistic channel usually has low energy relative to its front portion, choosing L equal to 1/3 to 2/3 of the length of CP would be a good compromise.

FDM Pilot-Based Carrier Frequency Offset Detection

Let us consider the case that the kth subcarriers of the nth and the $(n+N_p)$th OFDM symbols are modulated by pilot symbols. Recall that $Y(n,k)$, the kth DFT output, can be expressed as $Y(n,k) = H(n,k)X(n,k) + z'(n,k)$. If $X(n,k)$ is known, an estimate of $H(n,k)$, can be generated by $\hat{H}(n,k) = X^*(n,k)Y(n,k)$. Similarly, the estimate $\hat{H}(n + N_p, k)$ of $H(n+N_p,k)$ can be generated in the same way.

If the nth and the $(n+N_p)$th OFDM symbols are within the channel coherence time, $\hat{H}(n + N_p, k)$ is approximately equal to $\hat{H}(n, k)$ with a phase rotation. Because the elapsed time between these two OFDM symbols is equal to $T_s N_p(N_{DFT} + N_{CP})$, the estimates of the kth subcarrier coefficients from the nth and $(n+N_p)$th OFDM symbols satisfy

$$\hat{H}(n + N_p, k) \approx e^{j2\pi\Delta f N_p(N_{DFT}+N_{CP})T_s} \hat{H}(n, k)$$

When the estimates $\hat{H}(n, k_j)$ and $\hat{H}(n + N_p, k_j)$, where k_j's are the indices of the pilot subcarriers and $j = 0, \ldots, M-1$, are known from channel estimation, an estimate of the frequency offset can be computed by

$$\hat{\Delta f} = \frac{\mathrm{Arg}\left[\sum_{j=0}^{M-1} \hat{H}^*(n, k_j)\hat{H}(n + N_p, k_j)\right]}{2\pi N_p(N_{DFT} + N_{CP})T_s} \tag{5.109}$$

The estimate of the frequency offset is limited to

$$\pm\frac{1}{2N_p(N_{DFT} + N_{CP})T_s} = \pm\frac{1}{2N_p(1 + N_{CP}/N_{DFT})}f_{sc} \cong \pm\frac{1}{2N_p(1 + \alpha_{CP})}f_{sc} \tag{5.110}$$

where $\alpha_{CP} = N_{CP}/N_{DFT}$ denotes the ratio of the CP length divided by the length of DFT length.

Comparing (5.110) with (5.108), we conclude that the detection range of the FDM pilot-based method is reduced by a factor of $N_p(1 + \alpha_{CP})$ relative to the CP-based method. Even in the best case that $N_p = 1$, i.e., the pilots are available in every OFDM symbol to generate the channel estimates, the detection range would still reduce by a factor of $1 + \alpha_{CP}$. Because the alias frequency is not equal to an integer number of subcarrier spacings, the FDM pilot-based frequency estimator cannot be directly used as the preprocessor for resolving the integer frequency offset.

When the FDM pilots are on the same subcarrier in the closest OFDM symbols that contain pilots, e.g., as shown in Figure 5.16(a), the generation of $\hat{H}(n, k_j)$ and $\hat{H}(n + N_p, k_j)$ for the computation of (5.109) is straightforward. However, if the FDM pilots are staggered, e.g., as in Figure 5.16(b), the channel estimates of the same subcarrier from the FDM pilots will be separated farther from each other. The frequency detection range would be reduced if only the pilot subcarrier coefficient estimates on the same subcarriers are used to compute the channel phase difference between them.

For example, in LTE, N_p is equal to 6 or 7 if only the pilots on the same subcarriers in the two closest OFDM symbols that contain pilots are used. In such a case, the frequency detection range would be significantly smaller than the range when the CP-based pre-DFT method is used. N_p can be reduced to 3 or 4 if this requirement is removed as discussed below.

One approach for improving the detection range is to use the interpolated frequency-domain channel estimates. For example, let us consider the case that subcarrier k is a pilot subcarrier in an OFDM symbol but not in the next closest OFDM symbol that contains FDM pilots. For performing data recovery, the estimates of all of the subcarrier coefficients are usually computed through interpolation. After interpolation, the estimate of the kth subcarrier coefficient will be available in the next OFDM symbol that contains pilots. Such an estimate can be used for generating the frequency-offset estimate in the same way as described above.

Carrier frequency-offset can also be estimated based on the estimated time-domain coefficients, which are generated by the inverse Fourier transform from the FCEs as given by (5.91). If both OFDM symbols are within the coherence time, their corresponding coefficients should be approximately equal to each other except for their phases. The estimate of the phase difference can be computed by using a pair of such estimated coefficients with the same time index. The estimates computed from multiple, say m, coefficient pairs with different time index could be averaged to reduce the variance of the overall estimate. The frequency-offset estimate is computed by dividing the estimate of the phase difference by the time difference between these two OFDM symbols. The noise variance in the phase estimates is inversely proportional to m, i.e., the number of terms combined to generate the estimate of the phase difference. It is also inversely proportional to the magnitude of the time coefficient. Thus, the time coefficients with large magnitude should be used in the estimation.

In all these cases, the correlation between the time-separated frequency domain or time-domain channel estimates, e.g., $\sum_{j=0}^{m-1} \hat{H}^*(n, k_j) \hat{H}(n + N_p, k_j)$ in (5.109), can be used to drive the DFLL to form the AFC block of the receiver. When used as the FD of a DFLL, the imaginary part of the correlation is usually used. The output noise of these FDM pilot-based frequency offset detectors exhibits the high-pass characteristics as analyzed in Section 5.3. As a result, the variance of the DFLL output is proportional to the square of the loop gain k_2.

The properties of the CP-based and FDM pilot-based frequency offset detection methods and the differences between them are summarized below.

- The CP-based method has a larger frequency offset detection range than the range that the FDM pilot-based detection method has.
- The CP-based detection has an alias range equal to a subcarrier spacing. Hence, it can be used as a preprocessor for FDM pilot-based integer carrier offset detection.
- In a multipath channel, there could be inter-OFDM-symbol interference in the CP-based estimate. This adverse effect can be reduced or eliminated by reducing the integration length.
- As long as the multipath delay does not exceed the CP, there is no ISI to affect the FDM pilot-based frequency offset detector.

- The DFLL with FDM pilot-based FD yields lower output frequency error due to the high-pass filtering characteristic of the error than the DFLL with CP-based FD if both types of FDs have the same output variance.
- Due to the noise and interference existing in the received signal, the usable frequency detection range is smaller than the maximal range limited by aliasing. As a rule of thumb, the usable range is about 0.75 of the maximal range for both types of FDs.

Given all of these factors, the FDM pilot-based FD could be a better choice than the CP-based FD in many cases due to its insensitivity to ISI. However, the latter can provide more robust performance than the former when there exists an extremely high-frequency offset due to the inaccuracy of LOs or Doppler effects. Detailed analysis and simulation should be used to evaluate and compare their performance under specific channel conditions.

5.6   Carrier Synchronization for Communications over Wireline/Quasi-Static Channels

Starting with this section, we consider carrier synchronization in practical communication systems. We first consider synchronization for communication over wirelines as an example of utilizing the techniques discussed so far. What will be described is also applicable to other types of communication systems with quasi-static, such as microwave and satellite, channels.

5.6.1   Characteristics of Wireline Channels

Wireline channels, such as telephone local loop or twisted pair of wires, can be viewed as quasi-static for communication signals. The channel characteristics may change over a long time due to environmental changes such as temperature variation. As a result, receivers still need to adapt to such changes, albeit slowly.

Due to relatively benign channel characteristics, communication over wireline channels can potentially achieve high data throughput. Thus, high-order modulations, e.g., 256QAM, are often used. As a result, highly accurate carrier-phase synchronization is required. Yet, although frequency synchronization is still needed, the tracking requirement is relatively easy to meet. Due to these factors, carrier synchronization in receivers for wireline channels is often implemented in the form of the second-order PLL with a long time constant.

Communication links over wirelines can be point-to-point or centralized. In most cases, the devices, e.g., modems, on the two ends operate asynchronously. In other words, the frequencies of their transmitter symbol clocks are different. The receiver clock in a device should be synchronous to the remote transmitter clock. However, it may not be synchronous to its own transmitter clock, which can be free running at the native frequency of the LO. Under these assumptions, the carrier synchronization loop must operate properly when the frequency error is large, e.g., 100 ppm.

Since the wireline communication channel can be viewed as static, except for the training sequences sent at the beginning of transmission, there are usually no reference signals known to the receiver embedded in the transmission in the data mode. The receiver estimates the remote transmitter carrier phase as well as frequency using the training sequences before the data transmission starts. Once the communication link is established, various receiver functions, including carrier and timing synchronization loops and equalizers, operate in the decision-directed mode for adaptation to slow channel changes. If for some reason, e.g., bursty noise, one or more loops are out of synchronization, the receiver will first make an effort on regaining synchronization before deciding to stop the reception and requesting a retrain.

5.6.2   Carrier Synchronization Process and Operations

Wireline modems' carrier synchronization process can be divided into the following steps.

Initial Signal Detection/Handshaking
This is the first step in establishing communication between two devices. In the case of analog modems, such as the modems specified by the V.34 echo-cancellation modem standard [31], the calling modem and the answering modem exchange different sinusoidal tones that are followed by low data-rate coded waveforms. For Ethernet devices, such as those specified in the 10GBASE-T standard [32], bursts of link pulses with embedded data pulses are exchanged between the devices. Besides determining the data speed and operation modes, these handshaking signals at this stage also establish the timing reference between the communicating devices to facilitate receiver training in the subsequent stages.

Training
In this stage, training sequences are transmitted alternatively by the two devices. These training sequences are known to the receiving device and used for training channel-dependent functions of the devices such as equalization, echo cancellation, timing, and carrier synchronization.

At the beginning of this stage, the receiver only has some rough knowledge about the channel, to which it should adapt. To achieve robust adaptation, low-order modulation symbols, such as BPSK, are commonly used as the training sequence symbols. The carrier synchronization block is usually trained together with the equalizer of the receiver.

Figure 5.22 shows a high-level block diagram of the carrier synchronization block in a wireline receiver. Each equalizer output v_n is an estimate of the training symbol a_n, which is known to the receiver at this stage. The imaginary part of $a_n^* v_n$ forms the input to the loop filter, which has a pole, as shown in 5.4.3. The output of the loop filter adjusts the phase of the signal input to the equalizer. The phase adjustment is reflected at the output of the equalizer to form a negative feedback loop.

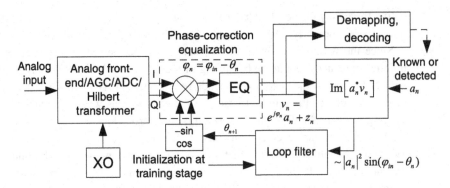

Figure 5.22 Carrier synchronization subsystem of a modem receiver

The description above makes it clear that the carrier synchronization block in the modem is essentially a second-order PLL with a delay in the loop introduced by the equalizer. The analysis of delayed PLL adaptation given in Chapter 4 is directly applicable here.

Data Mode Carrier Tracking

After the training stage, the equalizer generates unbiased estimates of the remote transmitter symbols, and the receiver enters the normal data mode. Decoding metrics are generated from the equalizer output for the channel decoder. The decoder uses the decoding metrics to generate the replica of the transmitted information bits at a low error rate, and carrier synchronization enters the decision-directed tracking mode. Phase errors are computed by using the decisions of a_n, which are generated by re-encoding the decoded bits, instead of known transmitted symbols as in the training mode. Due to the decoding delay, the PLL is operated with delayed update.

5.6.3   Practical Considerations

Below we list a few factors that should be considered in practical implementations.

Phase Tracking during Initial Training

Initially in the training stage, there are relatively large carrier phase and frequency errors. The equalizer and the PLL for carrier synchronization are trained jointly. Their adaptation step sizes during the initial training are usually chosen to be large so that they converge quickly. The equalizer should be able to converge even with the carrier phase and frequency errors. Once it has converged, the equalizer step size can be reduced. The PLL will keep the carrier phase and frequency in lock with those of the remote transmitter so that the equalizer can operate normally.

Decision Reliability in the Data Mode

To reduce the error in phase estimates, the decisions used in the carrier phase detection should be as reliable as possible. Hard symbol decisions of a_n's made directly from v_n's, may not meet this requirement. Reliable decisions are obtained by re-encoding the decision bits from the decoder output to regenerate the transmitted data symbols. Besides the increased complexity and storage requirement, the process of equalization and re-encoding also introduces delays in a PLL update. However, such delays are usually not a problem, because the wireline channel is relatively static and therefore the PLL's time constant can be quite long in the data mode, as discussed in Section 4.5.3.3.

Carrier Phase Recovery in Case of Burst Errors

In the case where there is a burst of noise/interference, the decision from decoder may become unreliable and the PLL may be out of lock. The receiver could request training sequences to be sent again so that it can enter the retraining state. However, this recovery process might take a long time and, as a result, reduce the transmission throughput. A common practice to improve the transmission efficiency in such cases is to temporarily freeze the equalizer. The PLL keeps running but not adapting. It may be desirable to slightly change the estimated frequency to keep the equalizer output rotating at a proper speed. The decoder keeps performing decoding while monitoring the error rate of the decoded bits. Once the phase of the equalizer output approaches the actual channel phase again, the decoding error rate reduces, and the monitor enables the equalizer and PLL adaptations to resume normal receiver operations. With 90- or 180-degree invariant channel codes, the PLL output phase only needs to rotate 90 or 180 degrees to have the opportunity to relock. Retraining will only be requested after it is decided that such a relocking attempt has failed.

Non-Data-Assisted Carrier Synchronization

As discussed in Sections 5.3.1 and 5.4.6.3, non-data-assisted synchronization is possible for both digital and analog communication systems due to the cyclo-stationary property of the communication signals. This method is often used in systems with blind equalization, e.g., systems that use the constant modulus algorithm (CMA), also called the Godard algorithm [33], to train equalizer without sending training sequence.

5.7 Carrier Synchronization in Wireless Communications

Unlike wireline channels, wireless channels are more likely to be disturbed by changes in the environment. In particular, in mobile communications, fast fading is an inherent property of mobile wireless channels. In this section, we will focus our attention on such fast fading channels. Certain types of relatively static wireless channels can be treated in ways similar to wireline channels discussed in the previous section.

It is fair to say that wireless channels are more demanding to communication signal reception than wireline channels. This is especially true from the synchronization point of view. To achieve reliable communications, wireless systems usually include special signals transmitted together with data for achieving robust synchronization. In this section, we look into a few aspects of carrier synchronization in wireless/mobile communications including channel characteristics, signal design for facilitating carrier synchronization, and two possible forms of carrier synchronization implementations.

5.7.1   Wireless Channel Characteristics and System Design Considerations

The characteristics of mobile wireless channels change continuously. Receivers must have reliable estimates of such channel characteristics in order to perform effective data symbol recovery and decoding. One way to facilitate this task is to embed pilot signals, i.e., signals known to the receiver, in the transmitted data. The pilot signal transmitted at the beginning of a new transmission is commonly called a *preamble*. Similar to the training sequences in wireline communications, the purpose of the preamble is to establish a reliable communication link between the transmitter and receiver. The pilot, also called a reference, signal is usually also transmitted during data transmission for receivers to track channel changes after the links are established.

Pilot signals can be transmitted in a burst form or continuously. For example, in GSM transmission, there is a 22-bit sequence, called a *midamble*, at the center of each data slot. On the other hand, WCDMA and cdma2000 base-stations transmit continuous pilot channels in parallel with data channels. The pilot and data channels are scrambled with different codes that are orthogonal or uncorrelated to each other. Thus, in CDMA systems, the interference between the pilot and data channels and between different users are eliminated or significantly reduced. In LTE OFDM systems, FDM pilot symbols are sent periodically in some of the OFDM symbols with the pilot and data symbols modulated on different subcarriers.

The main purpose of the pilot symbols is to generate channel estimates for data symbol recovery. In single-carrier, nonspread spectrum systems, such as Global System for Mobile Communications (GSM), a receiver correlates the known symbol sequences with received signal sample sequences to generate estimates of discrete-time CIRs, as will be described in Section 6.4.3.3. The CIR estimates are used by the receiver to perform data symbol estimation and/or equalization of the received signals. The estimator/equalizer outputs are the estimates of the modulation symbols at the right carrier phase. These estimates are used for recovering the original transmitted information bits.

In CDMA systems, a receiver takes the form of a RAKE receiver with multiple RAKE fingers. Each RAKE finger, which is essentially a single-path demodulator, aligns with one of the channel's multipath components. A RAKE finger correlates a pilot channel spreading sequence with the corresponding chip-rate signal samples from a particular path to generate the estimates of the phase and magnitude of the channel path. At the same time, the RAKE finger uses the data channel spreading sequence to correlate with the signal samples to generate despread data symbols weighted by the complex path gain. The channel path estimate is used to correct the phase of the despread data symbols and to scale them by the magnitude of the path gain for generating decoding metrics.

The channel estimation and data symbol recovery in OFDM systems have been discussed in Section 5.5.1.

In all of these cases, the carrier phases are detected by channel estimation that uses pilot signals transmitted prior to, or along with, data signals for correcting the phase of estimated data symbols. In other words, *channel estimation is used for performing carrier-phase synchronization*. These various channel estimation methods can also compensate for the short-term carrier frequency changes due to Doppler effects.

Besides carrier-phase synchronization, the other tasks of carrier synchronization in most wireless communication systems include acquiring the initial carrier frequency offset between the transmitter and receiver and tracking the average offset during the data mode. In most wireless receivers, carrier frequency synchronization is implemented by employing DFLLs described in Section 5.4.7. This is commonly called automatic frequency control function of the receiver.

During initial acquisition, the receiver's LO may generate large frequency errors up to the worst-case inaccuracy. This is especially true for receivers with LOs implemented using, for cost considerations, XOs without temperature compensation. For temperature-compensated XOs (TCXOs), the errors from the nominal frequencies are usually no greater than 2 ppm (2×10^{-6}). However, for uncompensated XOs, the errors can be $5-10$ ppm or even larger. The frequency offset encountered by the receiver is equal to the relative error of the XOs times the carrier frequency. Therefore, the higher the carrier frequency is, the larger the frequency offset will be. Such large frequency offsets impose serious challenges to system engineers in design and implementation of initial signal and carrier acquisition.

In the data mode, the tasks of frequency synchronization are driving the frequency offset toward zero and maintaining it near zero. These tasks are important for the following reasons.

First, even though channel estimation and/or equalization can handle a certain amount of carrier frequency offset, it would unnecessarily degrade receiver performance. It is much easier and more effective to correct the carrier frequency offset error as a single parameter than to correct the errors of multiple channel or equalizer coefficients.

Second, an accurate carrier frequency estimate provides a good timing frequency reference to reduce the burden of performing timing synchronization. This aspect will be discussed in more detail in the next chapter.

Third, as explained in Section 4.6.2.2, in mobile wireless communications, the accuracy of a mobile transmitter's carrier frequency must meet regulatory requirements. This objective may be difficult to achieve without accurate external references because the mobile LOs are relatively inaccurate. In contrast, carrier frequencies from remote base-stations (BTS/Node-B) are usually much more accurate, e.g., at about 10^{-8}. By synchronizing the mobile transmitter's carrier frequency to the acquired base station's carrier frequency, its accuracy can be significantly improved to meet the necessary requirements.

Given all of these requirements and the possible large initial frequency offset, carrier frequency synchronization is implemented in three steps: initial frequency offset estimation and compensation, DFLL training, and data mode tracking.

In the first step, the carrier frequency offset between the transmitter and the receiver is estimated during initial acquisition. Examples and considerations of initial frequency offset estimation have been discussed in Chapter 3. The resulting frequency offset estimates are used to set the initial values of the frequency registers in the DFLLs.

In the second step, the frequency offsets have been greatly reduced from their initial values but may not be small enough for reliable data symbol estimation and decoding. Using the methods described in 5.4 and 5.5, the remaining frequency offsets are estimated and used to train the DFLLs. In this step, the criterion of selecting the FD and DFLL parameters is for reducing the error of the frequency estimate. More than one set of the parameters may be used consecutively for attaining the balance between fast convergence and low output frequency variance.

At the end of the training, the DFLLs have converged. Their average frequency offset is equal to zero but may still have a higher variance than desired. The receiver enters the data mode and continues DFLL adaptation. The parameters of DFLLs are selected to achieve a low variance of the offset while tracking the changes.

The overall operations of achieving frequency synchronization in wireless receivers will be further illustrated in Section 5.8 by examples of practical wireless receivers. Below we present two architectures for implementation of carrier frequency synchronization in wireless devices.

5.7.2   VC-TCXO-Based Implementation of Carrier Frequency Synchronization

Carrier frequency synchronization in wireless devices, such as cellular phones, is implemented by AFCs, which are essentially DFLLs as described in Section 5.4.7. Among the three basic components of a DFLL, the FDs and the loop filters are always implemented in digital form. However, for a long time, the FCs of the DFLLs in AFCs were implemented by using VC-TCXOs. In other words, the AFC was implemented by mixed-signal DFLLs. A high-level block diagram of the AFC for carrier frequency synchronization based on a VC-TCXO with a digital interface, i.e., a DCXO, is depicted in Figure 5.23.

As shown in the figure, the received RF signal at the carrier frequency f_c from the remote transmitter is down-converted in frequency to baseband in the RF front end of the receiver. A VC-TCXO in the receiver generates a clock with a frequency $\hat{f}_c$, which is nominally equal to the carrier frequency f_c of the remotely transmitted signal. This clock is used as the reference clock for performing frequency down-conversion of the received signal. The difference between f_c and $\hat{f}_c$ results in a frequency offset Δf. The baseband signal is sampled at a rate of f_s, which is also derived from the output of the VC-TCXO.

The frequency offset Δf is detected from baseband received signal samples by an FD with a gain k_d based on the techniques described previously. The estimate of the frequency offset scaled by the loop coefficient α_2, i.e., $\alpha_2 k_d \hat{f}_{offset}$, is accumulated in an integrator and stored in register D_2. The digital value contained in D_2 is converted to an analog voltage by an ADC, whose output controls the output frequency $\hat{f}_c$ of the VC-TCXO, to close the AFC loop.

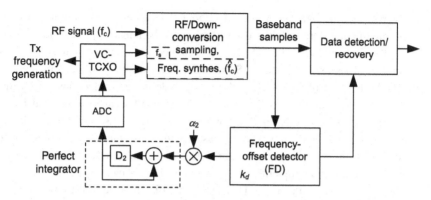

Figure 5.23 VC-TCXO-based carrier frequency synchronization

As an example, the VC-TCXO NT3225SA-19.2M by Nihon Dempa Kogyo Company [34] has a nominal frequency of 19.2 MHz at an input voltage of 1.5 v. Its worst-case frequency error from the nominal value is less than ±2.5 ppm. The adjustment range is between ±9 and ±15 ppm for the input voltage of 1.5 v±1 v. At the carrier frequency of 1.9 GHz, the range of frequency adjustment is between ±17.1 and ±28.5 kHz. The VC-TCXO's gain from its input to the synthesizer output is between 17.1 and 28.5 kHz/volt. Assuming that the output of the ADC is from 0 to 3 v with an input word length of 12 bits from D_2, the nominal digital FC's gain k_d is between 12.5 and 20.9 Hz per every LSB at the input of the ADC. The adjustment resolution is equal to half of the LSB. The resolution can be improved by increasing the word length of the ADC's input. Note that the value of D_2 might have a longer word length than that of the ADC's input for other implementation considerations as discussed in Section 4.5.1.2.

The relationship between the digital values of D_2 and $\hat{f}_c$ may be nonlinear and can be stored as a table. The values stored are predetermined and known to the receiver designer. These values may also be automatically measured and adjusted during the receiver operation in real time. When the receiver turns on, the content of D_2 is set to the value known to generate an $\hat{f}_c$ as close to the true f_c as possible. During the initial acquisition phase, the value of D_2 may be adjusted to reduce the frequency offset Δf.

During training, the frequency offset is detected by the FD and integrated in D_2 to drive $\hat{f}_c$ toward f_c. The loop coefficient α_2 is chosen to be relatively large to result in a short time constant for fast learning. Once the average frequency offset is close to zero, the AFC enters the tracking mode, and the loop coefficient is set to be smaller to reduce the output frequency variation. Such an AFC implementation not only ensures the quality of the reception but also enables the VC-TCXO to generate a local frequency reference with an accuracy close to the remote carrier frequency f_c. The transmitter carrier frequency of the mobile device is generated from the VC-TCXO's output by another frequency synthesizer as described in Section 4.6.2.

The AFC only tracks the average of the observed f_c. It should not respond to short-term frequency variations caused by channel fading due to Doppler effects. In other words, the time constant of the AFC loop in the tracking mode should be longer than the normal fading duration. It is usually chosen to be tens of milliseconds or longer.

The characteristics of the AFC loop described above can be analyzed in the same way the DFLL is as presented in Section 5.4.7. Its design and implementation are very similar to those of the DPLL with the VCO, described in Sections 4.5.1 and 4.5.2.

For many years, the AFCs based on VC-TCXO for carrier synchronization were the de facto implementation in digital wireless receivers. It is conceptually simple because all of the clocks in the receiver, such as system, transmitter, and sampling clocks, can be derived from a single accurate source. Therefore, the receiver architecture is simplified. In a mixed-signal implementation, one VC-TCXO can replace a number of digital signal processing functions. This was cost-efficient when DSPs were expensive and not very powerful during their early years. Moreover, VC-TCXO is usually quite accurate. This simplifies initial acquisition implementation, as sophisticated processing would not be needed.

5.7.3   XO-Based All-Digital Implementation of Carrier Frequency Synchronization

Since the mid-1990s, digital hardware has become much more powerful, and the cost has been constantly going down. As a result, more and more receiver designs have adopted all-digital carrier synchronization implementations. Such all-digital implementations not only reduce cost, but they are also more flexible in complex receiver designs. This is especially true for receivers that need to communicate with multiple systems. The structure, operation, and properties of all-digital carrier synchronization are described below.

Structure, Operation and Implementation Considerations
The block diagram of a typical XO-based implementation of all-digital carrier frequency synchronization, or simply an XO-based AFC, is shown in Figure 5.24. The RF front end and the FD in Figure 5.24 are the same as those in Figure 5.23. The difference is in the implementation of the frequency controller. The VC-TCXO in Figure 5.23 is replaced by an XO in conjunction with signal processing functions implemented in DSP software or ASIC. Below we focus on the blocks that are different from the VC-TCXO-based implementation as shown in the shaded area of the figure.

The XO used in the implementation generates a clock with a fixed-frequency f_{XO}. The clock frequency is converted by a frequency synthesizer to a frequency $f_{c,XO}$ that is nominally equal to the carrier frequency f_c. Low-cost XOs are usually not temperature-compensated[1] and are thus likely to have larger frequency errors relative to their nominal frequencies than those VC-TCXOs have. Typically, their accuracies are in the range of between 5 and 10 ppm or higher.

The received RF signal is down-converted in frequency to baseband by using $f_{c,XO}$ as the reference frequency. The generated baseband signal samples may have a large, but constant, frequency offset that is denoted by $\Delta f_{XO} = f_c - f_{c,XO}$. The offset Δf_{XO} creates

---

[1] To reduce the XO's sensitivity to environment changes, an external temperature sensor and software compensation may be used as a compromise.

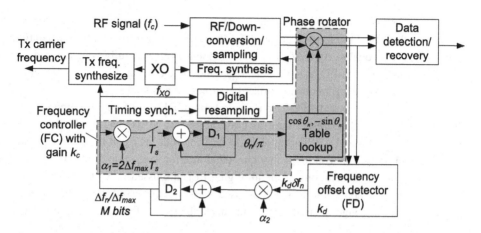

Figure 5.24 XO-based all-digital carrier synchronization

a phase rotation of $2\pi\Delta f_{XO}T_s$ radian per sample. Such sample phase rotations can be compensated for by consecutively rotating the phases in the opposite direction from sample to sample. At the sampling rate of $1/T_s$, the kth complex sample y_k is multiplied by a complex phasor $e^{-j2\pi\Delta f_{XO}T_sk}$ to correct the phase rotation due to Δf_{XO}. The main functions of the XO-based AFC are to estimate the frequency offset Δf_{XO} and to perform phase rotations.

The AFC loop is updated every T_{AFC}, which can be the same as, or longer than, the sampling interval T_s similar to what was discussed in Section 4.5.3.1. As shown in Figure 5.24, at the nth AFC update at nT_{AFC}, the FD generates $k_d\delta f_n$, a scaled estimate of the remaining frequency offset in the phase-rotated samples. After scaled by the loop coefficient α_2, the estimate is integrated and stored in the register D_2. The value of D_2 corresponds to the frequency offset Δf_n, which is an estimate of Δf_{XO} at nT_{AFC}. The functions and operations of the FD and the integrator are the same as those in the VC-TCXO-based AFC.

The content of D_2 usually represents a fractional number between -1 and 1 corresponding to the estimated frequency offset Δf_n, which is between $-\Delta f_{max}$ and Δf_{max} in the unit of Hz. The M MSBs of D_2, which represents the value of $\Delta f_n/\Delta f_{max}$, is scaled by the coefficient $\alpha_1 = 2\Delta f_{max}T_s$ and integrated in the register D_1 every sample interval T_s. The content of D_1 also usually represents a fractional number between -1 and 1, which is equal to the estimated phase θ_n divided by π and used to compensate for the phase rotations due to the frequency offset in the signal samples. Since the phase is a periodic function with a period of 2π, when the value of D_1 underflows or overflows, it simply wraps around to reflect the periodicity of the phase.

The value of D_1, which is equal to θ_n/π, is used to generate $-\sin\theta_n$ and $\cos\theta_n$ by table lookup for performing phase derotation of the signal samples. If the value of D_2 corresponds to the actual frequency offset, i.e., if it is equal to $\Delta f_{XO}/\Delta f_{max}$, the value of D_1 increments by $2\Delta f_{XO}T_s$ every sample to correct the phase increment caused by Δf_{XO}.

If the value of D_2 is smaller than $\Delta f_{XO}/\Delta f_{max}$, the compensation is not sufficient. The FD detects the residual positive frequency offset, and its output δf_{n+1} is positive on average. As a result, the value of D_2 increases to create larger phase corrections. Thus, the AFC loop forms a negative feedback loop. This process continues until the value of D_2 is equal to $\Delta f_{XO}/\Delta f_{max}$ and the frequency offset is fully compensated. At this point, the output of FD is zero on average, and the mean value of D_2 no longer changes. The case for initial frequency offset being negative is similar.

The XO-based AFC loop is also a form of DFLL. Among its components, the FD and the loop filter are the same as those in the VC-TCXO-based AFC loop. The output of the XO-based AFC's digital frequency controller is equal to Δf_n and the input is the M MSBs of D_2. The frequency quantization error is equal to $2^{-M+1}\Delta f_{max}$. The implementation considerations of the all-digital PLL discussed in Sections 4.5.2 and 4.5.3.1 are directly applicable to the XO-based AFC. As discussed there, the word length of D_2 may be longer than M bits for the accumulation of the detected frequency errors.

Utilization of Frequency-Offset Estimates in Other Device Functions

The XO-based AFC generates an accurate estimate of the frequency offset between the XO-based synthesized local down-conversion frequency and the remote transmitter carrier frequency. Thus, it can be used to synthesize the local transmitter frequency with the accuracy of the remote transmitter carrier frequency as shown in Section 4.6.2.2.

Another application is for receiver timing synchronization through digital resampling, which is the subject of Chapter 7. The receiver digital samples are generated based on the XO clock. To achieve timing synchronization, these samples need to be resampled to be synchronous to the remote transmitter symbol rate. In modern receiver implementations, such resampling is usually done by digital interpolation. In most wireless communication systems, a base-station's transmitter symbol clock is synchronous to its carrier frequency. The carrier frequency offset estimated by the local receiver can be used to derive the required sampling frequency.

Advantages and Shortcomings of XO-Based All Digital Design

The original motivation of using the XO-based design was to reduce the cost of devices. This has become especially important in recent years. In the cutthroat competitive environment of commercial wireless devices, even the savings of a few cents per unit could create a competitive advantage. The implementation of the XO-based design has also become feasible due to the increase in capability and decrease in the cost of the digital processing hardware. Besides cost, other advantages of such an implementation have also been recognized.

In particular, modern wireless devices usually need to communicate with multiple communication systems. For example, a mobile device may need to perform voice and data communications with different wireless systems. It may also need to determine its location by accessing GPS signals. All of these systems have their own frequency references, which are not synchronous with each other and may experience different channel variations. There are different requirements for their carrier synchronization. Using a single frequency reference may cause conflicts between the requirements of different systems. A question that modem system designers often encounter is "who is in control of the VC-TCXO?" at this particular time. Using a single clock generated from an XO, each of the synchronization functions for different systems can have a software synchronization block under its own control to make this task more manageable in such a multisystem device.

Despite the advantages, the main issue of the XO-based AFC block is the inaccuracy of the XO's native frequency. Because such an XO generates a large frequency offset initially, the tasks of receiver initial acquisition and frequency synchronization become challenging. This issue was discussed in Chapter 3 and the previous sections in this chapter. More examples will be given when we consider the frequency synchronization of practical systems in Section 5.8.

5.7.4   Implementation of Frequency Binning

During initial acquisition, if the initial frequency offset is too large due to the inaccuracy of XO for a given correlation length to perform initial acquisition, a few remedies can be employed as described in Section 3.4.4. One effective method without degrading the detection performance is frequency binning [2, 35] as described in Section 3.4.2.2. In this section, we describe the implementation of frequency binning using the AFC loop in wireless receivers by an example of the XO-based all-digital AFC.

Let us assume that the frequency offset due to the inaccuracy of the XO is in the range of $(-\Delta f_{max}, \Delta f_{max})$. As an example, we divide the range into five equal-width subranges, each of which is $0.4\Delta f_{max}$ wide. We assume that such a subrange is adequate for the desired correlation length in initial acquisition and for frequency offset estimation. As described in Section 5.7.3, the maximum and minimum values of the frequency register D_2 of the XO-based AFC are equal to -1 and 1, which correspond to $-\Delta f_{max}$ and Δf_{max}, respectively. To perform acquisition using the binning method, searches are performed five times with the value of D_2 set to be equal to -0.8, -0.4, 0, 0.4, and 0.8. Since the actual frequency offset should fall into one of the subranges, there is at least one of the searches encountering a frequency offset with a magnitude no larger than $0.2\Delta f_{max}$.

When performing the searches, the adaptation of the AFC loop is disabled, i.e., the AFC is free running. Moreover, each search should cover all of the possible timing offsets in order to perform all of the necessary hypothesis tests. Once a hypothesis test is successful, the value of D_2 that yields the successful test stays in D_2 as the initial frequency-offset estimate for the subsequent receiver operations.

Such an implementation can be applied to the VC-TCXO-based AFC as well. However, the actual width of each subfrequency-range will depend on the parameters of the VC-TCXO.

One disadvantage of the binning method is that the worst-case acquisition time is proportional to the number of the subranges if the received signal samples need to be continuously fed to the detector during the multipass acquisition. This method may not be desirable when the acquisition time is critical according to system performance requirements.

To reduce the required acquisition time when employing the frequency-binning method, the received signal samples needed for a complete set of hypothesis tests are first buffered. The buffered samples run through the phase rotator to perform detection multiple times with a different value in D_2 each time. The required time for the acquisition depends on the processing speed of the digital hardware. If the multiple passes of detections can be completed within the time T_b needed to collect and buffer the complete set of signal samples, the overall acquisition time only increases by T_b. More samples can be collected and buffered while the multipass detections are being performed. Therefore, no additional processing time is needed.

5.8   Examples of Carrier Synchronization Implementations in Wireless Communication Systems

To illustrate the principles and techniques described in this chapter, we provide a few examples of implementations of carrier synchronization in wireless receivers. As stated above, in most wireless receivers, carrier phase and frequency synchronization are implemented separately. The phase synchronization is done by channel estimation, and the frequency synchronization is realized by AFCs based on DFLLs. Either of the AFC designs described in Sections 5.7.2 and 5.7.3 can be used in the implementations described below. The selection criterion is based on practical considerations such as the receiver hardware platform and cost.

Most components involved in such implementations have been discussed previously. In this section, we will only describe the overall synchronization processes and procedures in the receivers of typical wireless systems.

5.8.1   CDMA2000–1x

Examples of DS-CDMA systems with continuous pilots are cdma2000–1x and WCDMA for cellular communications. We first describe the cdma2000–1x system, simply referred to as *cdma2000* below.

In DS-CDMA systems with RAKE receivers, the estimate of a channel path coefficient in each RAKE finger is used as the carrier phase references for recovery of the data symbols transmitted over the path. Below, we focus our discussion on carrier-frequency synchronization.

The carrier-frequency synchronization in cdma2000 starts from the initial acquisition stage. As described in Section 3.5.1.2, at the beginning of the initial acquisition, also called initial search, a searcher correlates a sequence of L pilot chips to the received signal sample sequences with different time offsets. The samples involved in the correlations are T_c spaced, where T_c is the *chip interval*. The starting time of the sample sequences usually increments every $0.5T_c$, i.e., $kT_c+\hat{\tau}$, $(k+0.5)T_c+\hat{\tau}$, $(k+1)T_c+\hat{\tau}$, $(k+1.5)T_c+\hat{\tau},\ldots$, where $0 \le \hat{\tau} < 0.5T_c$ is an arbitrary, but fixed, sampling time delay. This process continues until a correlation peak is found. Such a peak indicates that the samples involved in the correlation are likely generated by the pilot chip sequence, with which they are correlated.

Assume that a correlation peak is found with the sample sequence starting from $nT_c + i_0T_c/2 + \hat{\tau}$, $i_0 = 0$ or 1, and the hypothesis test is verified. The complex value

$\hat{g}(n, i_0)$ of the correlation peak is an estimate of the coefficient of a channel path at time $nT_c + i_0 T_c/2 + \hat{\tau}$. Both the pilot and sample sequences are then shifted by N positions to perform another correlation, where N may or may not be equal to L. The output value $\hat{g}(n + N, i_0)$ is another estimate of the path coefficient at $(n+N)T_c + i_0 T_c/2 + \hat{\tau}$. For a channel coherence time longer than NT_c, the channel coefficient after NT_c only changes in phase due to the frequency offset Δf. The mean of the phase difference between $\hat{g}(n, i_0)$ and $\hat{g}(n + N, i_0)$ is equal to $NT_c \Delta f$. An estimate of frequency offset can be obtained as

$$\hat{\Delta f} = \frac{(\text{Arg}[\hat{g}(n + (k + 1)N, i_0)] - \text{Arg}[\hat{g}(n + kN, i_0)])_{\text{mod } 2\pi}}{2\pi N T_c} \tag{5.111}$$

In most practical implementations, the FD of an AFC loop uses the channel estimates $\hat{g}(n + kN, i_0)$, $k = 0, 1, \ldots$, to train the receiver AFC to estimate $\hat{\Delta f}$ and achieve frequency synchronization. The FD based on $\hat{g}(n + kN, i_0)$ in the AFC most likely takes the QCFD form, i.e., $\text{Im}[\hat{g}(n + kN + N, i_0)\hat{g}^*(n + kN, i_0)]$, given in Section 5.4.7.2. Note that when the FD is in the QCFD form, we can eliminate $2\pi L T_c$ in the denominator in (5.111) by treating it as a part of the detector gain.

When used for initial training of AFC, the value of L should be selected based on the worst-case initial frequency offset. As discussed in Sections 3.3.1 and 5.4.7.2, if the length of the correlation LT_c is equal to $0.6/\Delta f$, there will be a loss of energy of 6 dB at the integrator output. Such a loss is barely tolerable, for example, if the XO in a receiver could have a frequency error of up to 10 ppm and at a carrier frequency of 1.9 GHz, $\Delta f_{max} = 19$ kHz. In such a case, LT_c should be approximately 0.6/19 kHz = 31.5 µs or less. The chip duration of the cdma2000 system is about 0.814 µs. Thus, $L = 32$ is a possible choice. If the XO's accuracy is even worse or if a 6 dB loss is unacceptable, L should be further reduced. However, an integration length L that is too short reduces the detection probability and may degrade the initial acquisition performance to an unacceptable level. The frequency binning method described in Section 5.7.4 can be used if necessary.

During training, the AFC loop usually has a short time constant in order to speed up its convergence. A short time constant increases the frequency variance at the AFC output and degrades receiver performance. To reduce the impact, multistage training with multiple time-constant values discussed in Section 4.5.3.2 can be used. Moreover, the integration length can also be increased in the later stages of training to reduce the residual frequency error.

Once the initial acquisition is completed, and the system is acquired, the receiver enters the data mode. In the data mode, the AFC loop that has converged with zero-mean residual frequency offset enters the tracking stage. The operations of the AFC block are the same as those in the training stage with a few differences stated below.

First, since the residual frequency offset in the data mode is small, the integration length to generate the channel estimate can be increased. The integration length is mostly limited by the Doppler frequency. For cellular applications, the Doppler frequency is usually no higher than a few hundred Hertz. Thus, the integration length could be quite long. In cdma2000 receivers, the base Walsh code length is 64 chips, which is also usually chosen as the integration length. Multiple integration outputs can be combined for the frequency-offset detection by the FD [13].

Second, a RAKE receiver usually has more than one active RAKE finger. The channel estimates generated by these filters are used to detect the frequency offset. The results from multiple fingers can be combined to improve the estimation accuracy of the frequency offset.

Finally, the time constant of the AFC loop should be long in the data mode. A long time constant not only reduces the variance of the estimated frequency but also tracks the average of the remote transmitter's frequency better by reducing the effects of short-term channel fading. This is especially important for maintaining the device's transmitter carrier-frequency accuracy and stability. An analysis of the DFLL used for AFC in CDMA receivers can be found in [13].

The above procedure is appropriate for DS-CDMA systems with only continuous pilots for initial acquisition, such as cdma2000. For WCDMA, which has special signals transmitted for initial acquisition, it is possible to initialize the AFC loop by taking advantage of such special signals as shown below.

5.8.2   WCDMA

WCDMA, often used as a synonym of UMTS, is another DS-CDMA cellular communication system. Its basic system design and initial acquisition procedure were discussed in Section 3.5.2. Similar to a cdma2000 receiver, a WCDMA receiver uses a continuous pilot for channel estimation by its RAKE fingers once in the data mode. Each estimated channel path coefficient is used for data symbol estimation. However, since every Node-B cell employs a different spreading sequence, a receiver cannot immediately acquire the pilot channel when it is turned on. Hence, WCDMA uses a combination of the PSC and SSCs for cell signal detection in initial acquisition. The detection results can be used to facilitate initial carrier frequency estimation.

As discussed in Sections 3.5.2.2 and 3.5.2.3, the receiver first searches for the 256 chips long sample sequences generated by the PSC. If the initial frequency offset is not too large, the searcher can perform 256 chip correlations for initial detection of the desired signal. If the worst-case frequency offset is excessive, it will be necessary to partition the sample sequence into a few segments. Multiple short correlations are performed, and their outputs are noncoherently combined, as discussed in Section 3.5.2.2.

The chip duration of the WCDMA signal is approximately equal to 0.26 μs, and the length of the PSC is equal to 66.7 μs. As shown in Section 3.3.1, if a coherent integration loss of 6 dB is tolerable and the frequency offset is less than 9 kHz, correlations can be done using the 256 chips of the entire PSC. However, if the frequency offset is larger than 9 kHz or the 6 dB correlation loss is unacceptable, shorter correlations with noncoherent combining should be used.

Once the PSC is acquired, an initial frequency-offset estimate can be generated by correlations of multiple PSC segments with corresponding sample sequences, as described in Section 3.5.2.2. The frequency-offset estimate generated from the detected PSC can be used to initialize the frequency register of the receiver AFC to reduce the residual frequency offset. The PSC detection is followed by detections of SSC and the

P-CPICH. Reduction of the frequency offset based on the detected PSC is essential to ensure the satisfactory detection performance of SSC and P-CPICH. If the correlation lengths for detecting both SSC and P-CPICH are equal to 256 chips, the detection loss can be expressed by $20 \log [\text{sinc}(256 T_c \Delta f)]$ as shown in Section 3.3.1. For the loss to be less than 1 dB, the residual frequency offset must be less than 3.9 kHz.

Once the P-CPICH is acquired, the searcher uses the pilot signal of the sector to search for the channel paths and to assign RAKE fingers. A RAKE finger despreads the P-CPICH to generate the estimate of a channel path coefficient. The estimates from the fingers are used to generate frequency offset estimates, which are combined to drive the receiver AFC.

The AFC in a WCDMA receiver goes through a training stage and then enters the tracking stage in the data mode. These operations are similar to those of the AFC in cdma2000 receivers. In WCDMA receivers, the length of despreading is usually chosen to be 256 chips. As a result, the estimate of the channel path coefficient is generated every 66.67 μs, and the update rate of the AFC is 15 kHz. The residual frequency offset when starting AFC training should be less than 7.5 kHz, e.g., $(0.6 \sim 0.75) \times 7.5$ kHz.

If the initial frequency offset is too large due to the inaccuracy of the XO, the PSC detection performance will significantly degrade. In such a case, the frequency-binning method described in Sections 3.4.4.2 and 5.7.4 can be used at the cost of higher computation complexity and/or longer acquisition time.

5.8.3   802.11 a/g

As was described in Section 3.6.2, the first step of an 802.11a/g Wi-Fi device in establishing a communication link with an AP is to acquire the short and long preambles. During the acquisition process, the device performs coarse and fine initial frequency offset estimations using the preambles as described in Section 5.5.2.2. An accurate channel estimate is also generated at this stage. Once the system is acquired, the receiver can start receiving data from the AP.

As specified in the 802.11 a/g standard [36], each OFDM symbol contains 4 FDM pilots, i.e., four pilot tones, which are not sufficient to perform accurate channel estimation in the data mode. The channel estimate generated during initial acquisition is used as the subcarrier phase references for generating data symbol estimates throughout the entire data frame, during which the Wi-Fi channel can be viewed as static. However, since the estimate of the frequency offset is not exact, the residual frequency error, no matter how small it is, will cause the phase error to accumulate. The receiver performance will eventually degrade if the data frame is not short enough. This problem can be addressed by using the four FDM pilot tones at subcarriers -21, -7, 7, and 21 to correcting the accumulated phase error as described in [16, 36] and is summarized below.

Let us denote the channel coefficient of the kth subcarrier by $H(k)$, its estimate generated during initial acquisition by $\hat{H}(k)$ and the pilot subcarrier indices by $k_{p,i}$, $i = 0, 1, 2, 3$. Assume that when the nth OFDM symbol is received, the accumulated phase error due to the frequency offset is equal to θ_n. With the additive noise term omitted, the DFT output corresponding to the ith pilot can be expressed as

$$Y(n, k_{p,i}) = e^{j\theta_n} X(n, k_{p,i}) H(k_{p,i}) \tag{5.112}$$

Since the pilot symbols $X(n, k_{p,i})$ are known and real valued, by using the estimate of $H(k_{p,i})$ obtained during initial acquisition, we can generate the estimate of θ_n as

$$\hat{\theta}_n = \text{Arg}\left[\sum_{i=0}^{3} \hat{H}^*(k_{pi}) X(n, k_{pi}) Y(n, k_{pi})\right] \tag{5.113}$$

It can be used to compute a finer estimate of the frequency offset.

The estimated fine frequency offset is used to compute and correct the accumulated carrier phase by unrotating the DFT output for data symbol estimation and decoding. This is a type of *feedforward phase correction* described in Section 5.4.4.1. It is also possible to perform feedback phase correction by using the estimated phase errors to drive a PLL.

5.8.4   LTE

Carrier-phase synchronization of the LTE UE is accomplished by the FDM pilot-based channel estimation as commonly done in OFDM systems and has been discussed in Section 5.5.1. Below we mainly discussed its carrier frequency synchronization.

During initial acquisition, the UE receiver first acquires the PSS as discussed in Section 3.6.1. As the length of the PSS is equal to 67 μs, which is the same as the length of the PSC in WCDMA, their tolerances for worst-case frequency offsets are similar at the same carrier frequency.

Once the PSS is acquired, an initial frequency-offset estimate can be generated by either the frequency binning method or the correlations of multiple sample segments of an PSS, as described in Section 3.4.4. This estimate of the frequency offset is used to initialize the AFC's frequency register to facilitate the receiver operations that follow.

The next receiver function after PSS detection is to acquire the SSS. The length of the SSS is the same as that of the PSS. With the initial frequency offset estimated and corrected, the receiver should be able to detect it by using the entire SSS as described in Section 3.6.1.4.

After both PSS and SSS are acquired, the receiver is ready for the data mode operations. In the data mode, the carrier frequency synchronization is achieved by the AFC described in Section 5.7. Either the VC-TCXO- or XO-based AFC can be employed. The frequency error detection can be performed either by using the pre-DFT CP correlation method or by the post-DFT FDM pilot-assisted method discussed in Section 5.5.2.3. As the FDs are most likely to be used for driving the AFC of the receiver, the QCFD implementation described in Section 5.4.7.2 should be appropriate. The characteristics of these two approaches and the trade-offs between them are discussed below.

CP Correlation-Based Frequency-Offset Detection

The normal LTE signal has a 15 KHz subcarrier spacing, f_{sc}. Using the CP-based method, the maximum detection range is equal to $\pm 0.5 f_{sc} = \pm 7.5$KHz. Due to the estimation errors caused by noise and interference, the detection range would be reduced. Empirically, the maximum frequency offset that can be detected reliably should be reduced to about 75 percent or less of the maximum value. The detected frequency offset is generated for every OFDM symbol. The FD output rate, which is usually also the AFC update rate, is equal to the OFDM symbol rate. For $f_{sc} = 15$KHz, the AFC update rates are equal to 14 kHz or 12 kHz depending on the CP length. The CP-based FD should be able to deal with a data mode Doppler frequency up to ± 5.25kHz or ± 4.5kHz.

As considered in Section 5.5.2.3, the issue with the CP-correlation-based FD is the possible high inter-OFDM-symbol interference, which results in high AFC output variances. To reduce the impact of ISI, only the last portion of the CP in the received signal should be used. With a short correlation length, the total energy of the samples involved in the correlation is reduced. Thus, the correlator's output SNR will be low, and the AFC output variance will be high. This issue of ISI is especially serious for signals with normal CP since the average CP energy is only about 8 percent of the energy of the data portion.

FDM Pilot-Based Frequency-Offset Detection

FDM pilots are also being used for frequency-offset detection in LTE receivers. Note that in the LTE's normal data mode there are two FDM pilot patterns. For the LTE signal with the extended CP, FDM pilots exist in one out of every three OFDM symbols. For signals with normal CP, FDM pilots exist in one out of three or four OFDM symbols alternatively. The locations of the FDM pilots are different in the adjacent OFDM symbols that contain them. Figure 5.25(a) and (b) show the pilot patterns of antenna port 0 in OFDM symbols with the normal and extended CPs. These pilots always exist in LTE signals and are suitable for the detection of frequency offset.

If the estimates of the pilot subcarrier coefficients with the same subcarrier index, e.g., b and b′ in Figure 5.25(a), are used for frequency offset detection according to (5.109), the shortest time between them is equal to one slot duration, i.e., 0.5 ms. As a result, the frequency offset that can be detected without aliasing is within ± 1 kHz. Practically, if the 0.75 empirical reduction factor mentioned above is used, the range of frequency offset that such an AFC can handle is about ± 750Hz or less. Such a range may not be sufficient for vehicles that move at high speeds.

To extend the detection range, the frequency offset can be estimated by using the pilots in the nearest OFDM symbols that contain pilots. Since the pilots in the two adjacent OFDM symbols are not at the same subcarrier frequencies, the estimates of the subcarrier coefficients from interpolation should be used. An example is shown in Figure 5.25(a), where the estimate of the subcarrier coefficient at b″ is generated from the other estimates in the same OFDM symbol by interpolation. The estimates of the subcarrier coefficients at b, b″ and b′ can be used for the detection of the carrier phase change. Alternatively, the frequency detection can be done in the time domain using the time-domain channel coefficients described in Section 5.5.2.3.

By using such an implementation, the time between these two OFDM symbols is approximately 0.214 ms or 0.286 ms for the normal CP and 0.25ms for the extended

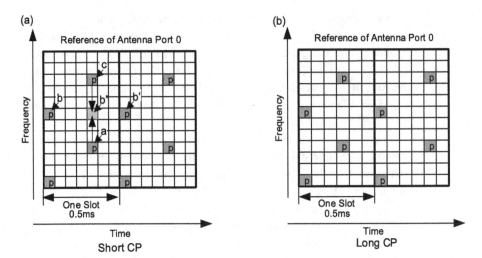

Figure 5.25 FDM pilot patterns in LTE OFDM symbols

CP. The corresponding maximum detection ranges are 2.33, 1.75, and 2 kHz, respectively. These ranges should be sufficient under most mobile channel conditions. However, using the empirical formula that we mentioned previously, the practical detection ranges should be reduced at least by a factor of 0.75.

If the vehicle speed and carrier frequency are even higher, however, using such a design may still encounter problems. For example, a mobile device that operates in the 3.5 GHz band and moving at a speed of 320 kmph on high-speed railroad trains could experience a Doppler frequency higher than 1000 Hz. When the device is passing through a nearby transmitter, the Doppler frequency change will double. This could be beyond the detection ability of the receiver AFC using the FDM pilot-based FDs.

Comparison

To compare these two data-mode FD implementations, the CP-based approach is preferable in high Doppler and frequency offset situations. It can also be used if the signal SNR is low, so the impact of ISI is relatively small. In contrast, the FDM pilots-based method should work well for low carrier frequency offset and/or Doppler frequency in a high SNR environment. Extensive analysis and simulations for various system and channel conditions are needed in order to make proper design trade-offs. If the mobile devices are expected to encounter extreme channel conditions, a combination of the FDM pilot- and CP-based frequency-offset detections may provide the most robust performance under all of these conditions.

5.9   Summary and Remarks

Carrier synchronization is one of the receiver functions that have significant impacts on receiver performance. Carrier synchronization can be done by processing the passband

(RF or IF) signals or baseband signals. In the early days of digital communications, carrier synchronization was mostly achieved by processing passband signals in the analog domain. More recently, in modern communication systems, most of the processing is implemented by digital processing of the digital samples of baseband received signals.

In this chapter, we began with presenting the analog processing of passband signals in the non-data-assisted form, which is more popular than its data-assisted counterpart. These realizations of carrier synchronization have theoretical importance and they are still used today in certain-types of receiver implementations. However, carrier synchronization is now mostly performed by digital means. Thus, this chapter mainly focused on the processing of digital baseband signal samples to achieve carrier synchronization. Both data-assisted and non-data-assisted forms were presented.

Digital communication systems can be classified based on their communication technologies as single-carrier communication and multicarrier, i.e., OFDM, communication. While single-carrier and OFDM communications have many aspects in common from a carrier synchronization viewpoint, there are also major differences. Synchronization for single-carrier and multicarrier communications were treated separately in this chapter.

According to their applications, communication systems can be divided into wireline and wireless systems. As a communication channel, the wireline can be viewed as quasi-static. It has its own special requirement but is generally benign for achieving carrier synchronization. In contrast, wireless communication channels are much more dynamic. Such channels present challenges to the design and implementation of receiver synchronization subsystems. Due to their widespread utilization, a significant portion of this chapter was devoted to the analysis, design, and implementation of carrier synchronization in wireless receivers. In addition, examples of carrier synchronization in practical wireless systems were provided.

This chapter has covered many aspects of the subject of carrier synchronization. The following is noteworthy on this topic.

- Passband signal carrier synchronization is mostly implemented by analog processing means. Digital implementations, if used, are just an alternative way of performing the same tasks and are not discussed in this chapter. Most popular implementations of passband carrier synchronization are non-data-assisted. Due to the cyclo-stationarity of the digital communication signals, the passband carrier synchronization can be implemented by nonlinear processing techniques.
- The baseband carrier phase estimates are maximum likelihood if the samples generated by a receiver filter that matches the received signal. In most cases, the receiver filter is not a true matched filter but its approximation. We call the estimates based on such generated samples approximate ML.
- If the combined frequency response of the MF and channel satisfies the Nyquist criterion, the carrier phase estimate based on MF output samples satisfies the Cramer–Rao bound.
- The ML receiver front end for carrier synchronization can also be used for other receiver functions such as initial acquisition and equalization.
- The objectives of carrier synchronization in baseband processing are to correct channel phase error and frequency-offset in the baseband signal samples.

- Correlating known data symbols and the corresponding received digital samples yields an ML/approximate-ML channel estimate. A phase detector that computes the phase of such a channel estimate as an ML/approximate-ML carrier phase estimate is called an MLPD.

- To simplify computation, the imaginary part of the channel estimate can be used as an approximate channel phase estimate. The phase detector in PLLs that produces such approximate phase estimate for achieving carrier synchronization is called a quadrature phase detector.

- The frequency-offset estimate of the received signal can be computed by dividing the difference between two time-separated signal carrier phase estimates by the time elapsed between them. A frequency detector that employs MLPD for phase detection is called MLFD.

- A simplified frequency-offset detector called a quadri-correlator frequency detector described in Section 5.4.5.2 is commonly used in DFLL implementations.

- One property of MLFDs and QCFDs is that, in most cases, the output noises have a high-pass characteristic. As a result, the output noise of DFLLs is significantly lower than the DFLL output noise if the input noise is white, as often assumed.

- In OFDM systems, carrier-phase detection/correction is performed by channel estimation. Frequency offset detection is performed separately from phase detection.

- There are two types of OFDM frequency offset detector implementations, i.e., pre- and post-DFT methods. Their analysis and implementation were presented, and trade-offs between them were made.

- In most cases, carrier synchronization in wireline communication modems is implemented in the form of second-order phase-locked loops.

- In wireless communications, the detection and correction of carrier (channel) phase and frequency offset are usually implemented separately.

- The channel phase detection is performed by channel estimation using RAKE fingers in CDMA systems and by frequency-domain channel estimation based on FDM pilot in OFDM systems.

- Carrier frequency synchronization is implemented in the forms of frequency locked loops that should be able to operate in the presence of the highest possible frequency offset.

References

[1] H. Nyquist, "Certain Topics in Telegraph Transmission Theory," *AIEE Transactions*, vol. 47, pp. 617–44, 1928.

[2] J. G. Proakis and M. Salehi, *Digital Communications*, 5th edn, New York: McGraw-Hill, 2008.

[3] L. E. Franks, "Carrier and Bit Synchronization in Data Communication – A Tutorial Review," *IEEE Transactions on Communications*, Vols. Comm-28, no. 8, pp. 1107–21, August 1980.

[4] R. D. Gitlin, J. F. Hayes, and S. B. Weinstein, *Data Communications Principles*, New York: Plenum Press, 1992.

[5] W. C. Lindsey and M. K. Simon, *Telecommunication System Engineering*, Englewood Cliffs, NJ: Prentice Hall, 1973.

[6] J. P. Costas, "Synchronous Communications (Classic Paper – Republication)," *Proceedings of the IEEE*, vol. 90, no. 8, pp. 1461–6, August 2002.

[7] H. Osborne, "A Generalized "Polarity-Type Costas Loop for Tracking MPSK Signals," *IEEE Transactions on Communications*, vol. 30, no. 10, pp. 2289–96, October 1982.

[8] G. D. Forney Jr., "Maximum-Likelihood Sequence Estimation of Digital Sequences in the Presence of Intersymbol Interference," *IEEE Transactions on Information Theory*, vol. 18, no. 3, pp. 363–78, May 1972.

[9] G. Ungerboeck, "Channel Coding with Multilevel/Phase Signals," *IEEE Transaction on Information Theory*, vol. 28, no. 1, pp. 55–67, January 1982.

[10] U. Mengali and A. N. D'Andrea, *Synchronization Techniques for Digital Receivers*, New York: Plenum Press, 1997.

[11] A. D'Andrea and U. Mengali, "Design of Quadricorrelator for Automatic Frequency Control Systems," *IEEE Transactions on Communication*, Vols. Comm-41, no. 6, pp. 988–97, June 1993.

[12] F. D. Natali, "AFC Tracking Algorithms," *IEEE Transactions on Communication*, Vols. Comm-32, no. 8, pp. 935–47, August 1984.

[13] F. Ling, "Convergence and Output MSE of Digital Frequency Locked Loop for Wireless Communications," in Proceeding of 1996 Vehicular Technology Conference (VTC '96), Atlanta, GA, 1996.

[14] L. F. Wei, "Rotationally Invariant Convolutional Channel Coding with Expanded Signal Space – Part I: 180 Degrees," *IEEE Journal on Selected Areas in Communications*, vol. 2, no. 5, pp. 659–71, September 1984.

[15] Y. G. Li and G. Stuber, Eds., *Orthogonal Frequency Division Multiplexing for Wireless Communications*, New York: Springer Science+Business Media, 2006.

[16] J. Heiskala and J. Terry, *OFDM Wireless LANs: A Theoretical and Practical Guild*, Indianapolis, IN: SAMS, 2002.

[17] R. Prasad, *OFDM for Wireless Communications Systems*, Norwood, MA: Artech House, 2004.

[18] M. O. Pun, M. Morelli, and C.-C. J. Kuo, *Multi-Carrier Techniques for Broadband Wireless Communications – A Signal Processing Perspective*, London: Imperial College Press, 2007.

[19] Y. Li, "Pilot Symbol Aided Channel Estimation for OFDM in Wireless Systems," *IEEE Transactions on Vehicular Technology*, vol. 49, no. 7, pp. 1207–15, July 2000.

[20] A. A. Hunter, R. Hasholzner, and J. S. Hammerschmidt, "Channel Estimation for Mobile OFDM Systems," in Proceedings of Vehicular Technology Conference, VTC 1999 Fall, Amsterdam, 1999.

[21] R. Crochiere and L. Rabiner, *Multirate Digital Signal Processing*, Englewood Cliffs, NJ: Prentice Hall, 1983.

[22] J. Lin, J. G. Proakis, F. Ling, and H. Lev-Ari, "Optimal Tracking of Time-Varying Channels: A Frequency Domain Approach for Known and New Algorithms," *IEEE Journal on Selected Areas in Communications*, vol. 13, no. 1, pp. 141–54, January 1995.

[23] R. Vijayan, A. Mantravadi, and K. K. Mukkavilli, "Staggered Pilot Transmission for Channel Estimation and Time Tracking". *USA Patent*, vol. 7, no. 907, p. 593, March 15, 2011.

[24] J. H. Stott, *The Effect of Frequency Errors in OFDM*, BBC Research Department, Report No RD 1995/15, London, January 1995.

[25] T. M. Schmidl and D. C. Cox, "Robust Frequency and Timing Synchronization for OFDM," *IEEE Transactions on Communications*, vol. 45, no. 12, pp. 1613–21, December 1997.

[26] M. H. Hsieh and C. H. Wei, "A Low Complexity Frame Synchronization and Frequency Offset Scheme for OFDM System over Fading Channels," *IEEE Transactions on Vehicular Technology*, vol. 48, no. 5, pp. 1596–1609, September 1999.

[27] "Digital Video Broadcasting (DVB); Framing Structure, Channel Coding and Modulation for Digital Terrestrial Television," Sophia Antipolis Cedex, France, 2009.

[28] J. J. Van de Beek, M. Sandell, M. Isaksson, and P. O. Börjesson, "Low-Complex Frame Synchronization in OFDM Systems," in International Conference on Universal Communications ICUPC'95, Tokyo, 1995.

[29] M. Sandell, J. J. van de Beek, and P. O. Börjesson, "Timing and Frequency Synchronization in OFDM Systems Using the Cyclic Prefix," in International Symposium on Synchronization, Essen, Germany, 1995.

[30] J. van de Beek, M. Sandell, and P. O. Börjesson, "ML Estimation of Time and Frequency Offset in OFDM Systems," *IEEE Transactions on Signal Processing*, vol. 45, no. 7, pp. 1800–5, July 1997.

[31] Telecommunication Standardization Sector of ITU, "A Modem Operating at Data Signalling Rates of up to 33 600 Bit/S for Use on the General Switched Telephone Network and on Leased Point-to-Point 2-Wire Telephone-Type Circuits," Geneva, 1998.

[32] IEEE Computer Society, LAN/MAN Standards Committee, "Part 3: Carrier Sense Multiple Access with Collision Detection (CSMA/CD) Access Method and Physical Layer Specifications Amendment 1: Physical Layer and Management Parameters for 10 Gb/s Operation, Type 10GBASE-T," New York, 2006.

[33] D. N. Godard, "Self-Recovering Equalization and Carrier Tracking," *IEEE Transactions on Communications*, vol. 28, no. 11, p. 1867–75, November 1980.

[34] NDK (Nihon Dempa Kogyo), [Online]. Available: www.datasheetlib.com/datasheet/641139/nt3225sa-19.2m-nsa3391e_ndk-nihon-dempa-kogyo.html.

[35] G. R. Lennen, "Receiver Having a Memory Based Search for Fast Acquisition of a Spread Spectrum Signal," USA Patent US Patent 6,091,785, July 18, 2000.

[36] IEEE LAN/MAN Standards Committee, "Part 11: Wireless LAN Medium Access Control (MAC) and Physical Layer (PHY) Specifications," New York, 2012.

[37] J. G. Proakis, *Digital Communications*, 3rd edn, New York: McGraw-Hill, 1995.

[38] S. Speth, S. A. Fechtel, G. Fock, and H. Meyr, "Optimum Receiver Design for Wireless Broadband Systems Using OFDM. I," *IEEE Transactions on Communications*, vol. 47, no. 11, pp. 1668–77, November 1999.

[39] W. C. Lindsey and C. M. Chie, Eds., *Phase-Looked Loops*, New York: IEEE Press, 1986.

6   Timing Synchronization

6.1   Introduction

In a digital communication system, the *channel modulation symbols* to be transmitted are modulated on spectrum/pulse shaping waveforms and sent at fixed time intervals. For single-carrier communication without spectrum spreading, the channel symbol interval is equal to the transmitter data symbol interval T.[1] Its reciprocal is the *transmitter channel modulation symbol rate*, or simply *transmitter symbol rate*, $1/T$. After passing through the communication channel, the baseband received signal should be sampled at a rate synchronous to the transmitter symbol rate. The received signal sampling frequency, or rate, is typically equal to m/T, where m is an integer. Moreover, the sampling time should have a fixed offset, called *timing phase*, relative to the time instant, at which the transmitter symbols are transmitted, to compensate for the delay introduced during transmission.

The baseband received signal samples are processed to generate estimates of the transmitter channel symbols for recovering the transmitted data. In different communication systems, the operations to recover the transmitted data from the received signal samples may be different. However, the common objectives of *timing synchronization* are to synchronize the transmitter and receiver timing frequencies and to determine and maintain the proper timing phase. In a receiver, timing synchronization functions are performed in its timing synchronization block, also called the *timing synchronizer*. In the literature, the receiver operation of achieving timing synchronization is often called *timing recovery*.

The implementations and the characteristics of timing synchronization in different communication systems' receivers, such as various single-carrier receivers and OFDM receivers, have similarities and differences. In particular, there are significant differences between the realizations of timing synchronization in single-carrier and OFDM receivers. There are also differences in the realizations between single-carrier systems with and without spectrum spreading. In this chapter, we present the requirements, properties, and implementations of timing synchronization of these communication systems. This chapter is organized as follows.

---

[1] What is discussed here is also applicable to DS-CDMA and OFDM communications with the understanding that the channel symbol/sample intervals in these systems are different from the data symbol intervals.

After this introductory section, Section 6.2 provides an overview and discusses a number of fundamental aspects of timing synchronization in communication systems. Sections 6.3 and 6.4 present a few typical non-data-assisted and data-assisted timing synchronization implementations for nonspread spectrum single-carrier communication systems. Timing synchronization in DSSS/DS-CDMA and OFDM communication systems is treated in Sections 6.5 and 6.6. Finally, Section 6.7 summarizes what has been discussed and provides remarks on the key aspects treated in this chapter.

6.2   Fundamental Aspects of Timing Synchronization

An overview of timing synchronization was given in Section 1.4.3. In this section, we consider some fundamental aspects of timing synchronization. We focus on the non-spread spectrum single-carrier systems as timing synchronization in such systems has been thoroughly studied in both academia and industry. However, the principles presented here are also applicable to other communication systems.

6.2.1   Baseband Signal Model and ML Timing Phase Estimation

Recall from Sections 2.2 and 5.4.1 that the baseband received signal can be expressed as

$$r(t) = e^{j\theta_t} \sum_{k=-\infty}^{\infty} a_k g_c(t - kT - \tau) + z(t) \cong e^{j\theta_t} x(t, \tau) + z(t) \tag{6.1}$$

where $g_c(t)$ is the *composite channel impulse response* of the transmitter filter and the communication channel, θ_t is the carrier phase at time t, τ is an unknown but fixed time delay with a value between 0 and T, and $z(t)$ is an AWGN random process with the power spectrum density of N_0. Both θ_t and τ are introduced during transmission. The objective of *timing phase* synchronization is to determine the desired timing phase, i.e., sampling time offset $\hat{\tau}$, from the observation $r(t)$.

6.2.1.1   ML Estimate of Timing Phase with Known Transmitted Data Symbols

Assume that the *receiver filter* has an impulse response of $g_c(-t)$. It can be shown that for $r(t)$ given by (6.1), such a receiver filter is the *matched filter* of $r(t)$. As shown in Section 2.2.2, the log-likelihood function of θ_t and τ given the observed signal $r(t)$, which contains the known transmitted data symbols $\{a_k\}$, $n - L < k \leq n$, denoted by vector $\mathbf{a}_{n,L}$, during the timing phase detection is

$$l(\tau, \theta_n) = \frac{2}{N_0} \text{Re} \left[e^{-j\theta_n} \sum_{k=n-L+1}^{n} a_k^* y(kT + \tau) \right] = \frac{2}{N_0} \text{Re} \left[e^{-j\theta_n} \mathbf{a}_{n,L}^H \mathbf{y}_{n,L}(\tau) \right] \tag{6.2}$$

where τ is the timing phase to be estimated, θ_n is the carrier phase at nT, which may be known or unknown but is a constant during the detection time, $y(kT+\tau)$ is the receiver filter output $y(t)$ sampled at $kT+\tau$, and $\mathbf{y}_{n,L}(\tau)$ is the sample vector with the elements $\{y(kT+\tau)\}$, $k = n - L + 1, \ldots, n$. In this chapter, we define $kT + \tau$ as the *sampling time* of sample $y(kT+\tau)$

and τ as the *sampling time offset*, or the *timing phase*. Furthermore, to simplify notation, we define the complex scalar quantity $u_n(\tau) = \sum_{k=n-L+1}^{n} a_k^* y (kT + \tau) = \mathbf{a}_{n,L}^H \mathbf{y}_{n,L}(\tau)$ which is the correlation between $\{a_k\}$ and $\{y(kT+\tau)\}$, for $n - L < k \le n$.

In the case that θ_n is known and the receiver filter is the MF of $r(t)$, the ML estimate of the timing phase is the sampling time offset, or delay, τ of $y(kT+\tau)$ that maximizes $l(\tau, \theta_n)$ given by (6.2).[2] Namely,

$$\hat{\tau}_{ML}(y) = \arg \max_{\tau} \left\{ \mathrm{Re} \left[e^{-j\theta_n} u_n(\tau) \right] \right\} \tag{6.3}$$

If the actual transmission delay is equal to τ_0, the sample $y(kT+\tau_0)$ is generated at the optimal sampling delay of the MF output, such that the average signal energy of symbol a_k in $y(kT+\tau_0)$ is maximized. It can be shown that the mean of $\hat{\tau}_{ML}$ is equal to τ_0.

6.2.1.2   ML Timing Phase Estimation for Signal with Unknown Carrier Phase

In most practical cases, θ_n is unknown but can be assumed to be uniformly distributed between 0 and 2π. An ML estimate of the timing phase τ for such unknown θ_n is defined as the sampling delay that maximizes the likelihood function $\exp(l(\tau, \theta_n))$ averaged with respect to θ_n over the range from 0 to 2π [1]. Following the derivation given in Section 3.2.1, it can be shown that the ML estimate of τ is the sampling delay that maximizes $|u_n(\tau)|$. Hence, $|u_n(\tau)|$ or $|u_n(\tau)|^2$ can be used as the decision metric to determine the ML timing phase estimates for the unknown θ_n.

In this case, to achieve ML timing phase estimation, the transmitted data symbols $\{a_k\}$, $n - L < k \le n$, and the channel response $g(t)$ must be known, even though the carrier phase may be unknown. During the training mode or in the decision-directed mode, these data symbols are known. However, practically, the receiver filter may not match $r(t)$. In such a case, $|u_n(\tau)|$ or $|u_n(\tau)|^2$ generated from the non-ideal receiver filters is used as the metric for approximate ML timing phase estimation.

6.2.1.3   Approximate ML Timing Phase Estimation When neither the Transmitted Symbols nor the Carrier Phase Is Known

From (2.26) the likelihood function of τ, θ_n and $\mathbf{a}_{k,L}$ can be expressed by

$$L(\tau, \theta_n, \mathbf{a}_{k,L}) = \exp \{ l(\tau, \theta_n, \mathbf{a}_{k,L}) \} = \exp \left\{ \frac{2}{N_0} \mathrm{Re} \left[e^{-j\theta_n} \sum_{k=n-L+1}^{n} a_k^* y(kT + \tau) \right] \right\} \tag{6.4}$$

In low SNR environments, the exponent $\mathrm{Re} \left[(2/N_0) e^{-j\theta_n} \sum_{k=n-L+1}^{n} a_k^* y(kT + \tau) \right]$ in (6.4) has a magnitude much smaller than one. For unknown θ_n and $\{a_k\}$, assuming that θ_n is uniformly distributed between 0 and 2π and a_k's have zero means, as shown in

---

[2] The sampling-time offset τ in $y(kT+\tau)$ can be viewed as a negative delay relative to time kT and, thus, τ will be referred to as *sampling delay* as commonly called in the literature.

[2, 3] the expectation of $L(\tau, \theta_n, \mathbf{a}_{k,L})$ with respect to θ_n and $\{a_k\}$ is approximately equal to

$$L(\tau) = \underset{\theta_n, \mathbf{a}_k}{E} [L(\tau, \theta_n, \mathbf{a}_{k,L})] \approx C' \sum_{k=n-L+1}^{n} |y(kT + \tau)|^2 \qquad (6.5)$$

where C' is constant. Thus, an approximate ML timing phase estimate with unknown transmitted data symbols and carrier phase is the sampling delay τ that maximizes the average of $|y(kT+\tau)|^2$.

6.2.2   Optimal Timing Phase Synchronization

Below we consider the optimal timing phase selection for a few types of channels often encountered in single-carrier communications.

6.2.2.1   Single-Carrier Transmission over ISI-Free Channel

For single-carrier systems with channel frequency response that satisfies the Nyquist criterion, the ML timing phase estimate is optimal, and there is no ISI in the generated samples. A channel that satisfies Nyquist criterion, e.g., with a RCOS frequency response, has an overall CIR similar to a sinc function. For such a channel, the mean of ML timing phase at $nT + \hat{\tau}$ is at the maximum of the waveform generated by the nth transmitted symbol as shown in Figure 6.1.

The ML criterion is also applicable to the determination of CDMA RAKE-finger sampling delays as will be discussed in Section 6.5.

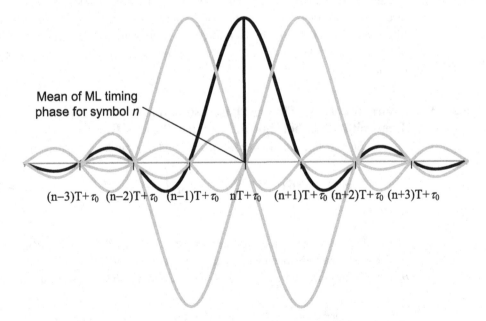

Figure 6.1 Optimal timing phase for T-spaced signal samples over ISI-free channel

6.2.2.2   Single-Carrier Signaling over Severe ISI Channels

For channels that have severe linear distortion, equalizers are usually employed to improve the output SNR to achieve better receiver performance. For equalizers with input samples at the transmitter symbol rate, called symbol rate equalizers (SREs), or T-spaced equalizers [3, 4], the performance of such equalizers is sensitive to the timing phase of the input samples. Unfortunately, for arbitrary ISI channels, there is no known method that can effectively determine the optimal SRE timing phase. Some of practical, albeit suboptimal, timing synchronization schemes applicable to such cases will be discussed in Sections 6.3 and 6.4.

To reduce or eliminate timing phase sensitivity, fractionally spaced equalizers (FSEs) are commonly used in the receivers for communication over such channels [3, 4]. The inputs to FSEs are signal samples generated with a sampling rate that is higher than $1/T$. This rate is often in the form of m/T, where m is an integer greater than 1. Typically, $m = 2$. Since such sampling rates usually satisfy the Nyquist sampling theorem, i.e., there is no aliasing of signal in frequency domain, the equalizer performance is insensitive to the samples' timing phase [5]. Thus, the exact timing phase of the samples is not important as long as the timing phase is stable or varying slowly so that adaptive FSEs can track it. Using FSEs can simplify the design and implementation of timing synchronization. However, since the optimal timing phase of an FSE is not unique, the FSE's coefficients could drift and eventually cause degradation of the equalizer performance [4, 6]. This issue and its remedies will be discussed in Section 6.4.3.

6.2.2.3   DS-CDMA Signal Timing Phase Synchronization

DSSS, or DS-CDMA, belongs to single-carrier communications. The unique feature of DS-CDMA receivers is that they are commonly implemented in the form of RAKE receivers that have been discussed in Section 1.5.3. A RAKE receiver consists of multiple RAKE fingers, each of which demodulates signals passing through a channel path. Because DS-CDMA receivers have high processing gains achieved by despreading, the ratio of signal to ISI and noise will be significantly enhanced. As a result, each path can be viewed as a single-tap AWGN channel and the ML timing phase synchronization can be implemented. The design and implementation of DS-CDMA timing synchronization will be discussed in Section 6.5.

6.2.2.4   OFDM Signal Timing Phase Synchronization

In OFDM systems with guard carriers, the receiver sampling frequency $f_s = 1/T_s$, where T_s is the OFDM channel sample rate, satisfies the Nyquist sampling theorem. Thus, sub-T_s timing phase adjustment is unnecessary. The timing synchronization in OFDM receivers is equivalent to proper DFT window positioning as will be described in Section 6.6.

6.2.3   Timing-Frequency Synchronization

In most cases, timing frequency synchronization is not as demanding as carrier-frequency synchronization. This is because a carrier-frequency offset is equal to the carrier frequency multiplied by the relative frequency error of the local XO. Conversely, the timing-frequency offset is simply determined by XO's inaccuracy. For example, consider a communications system with a symbol rate of 3.84 MHz operated at 1.9GHz. If the XO's inaccuracy is equal to 10 ppm, the timing frequency error is only 38.4 Hz, while the carrier-frequency error can be up to 19 kHz. Compared to the symbol rate, the timing frequency error is 0.001 percent, while the carrier-frequency error is about 0.5 percent. Carrier and timing-frequency errors due to Doppler frequency shift have the same relationship.

Because of the benign characteristic of the timing-frequency error, timing synchronization is usually achieved by a single TLL, also called a TCL or DLL.

A TLL is implemented as a first- or second-order PLL. Moreover, in wireless systems, the transmitter carrier and symbol frequencies are usually derived from the same clock reference. When this is the case, the estimated carrier-frequency offset by the frequency synchronization block in the receiver can be used to correct the timing-frequency offset, and no separate timing frequency detection and correction are needed. If there are additional carrier-frequency up-conversion and down-conversion involved during transmission, the estimated carrier-frequency offset may not be exactly equal to the timing-frequency offset but would still be a good approximation for the latter. However, it may be advantageous to take some additional measures, such as to implement a second-order TLL to compensate for the residual timing-frequency offset.

6.2.4   Implementation Considerations of Timing Synchronization

As mentioned, the receiver timing synchronization block is usually implemented as a first- or second-order TLL, a form of PLL. Similar to any kind of PLL, a TLL consists of a timing phase detector (TPD), a loop filter, and a TPC. When a TLL is implemented as a second-order loop, it achieves both timing phase and frequency synchronization.

A TLL is driven by the timing phase error, i.e., the difference between the desired timing phase and the timing phase of the sample, detected by the TPD. Timing frequency and phase synchronization may be jointly implemented but should be optimized independently. The objective of timing frequency synchronization is to track the long-term average transmitter channel symbol frequency. At the same time, the timing phase could be adjusted based on the short-term channel change or timing clock skipping, e.g., during the entering and exiting receiver sleep states. Such adjustments should not cause a sudden timing frequency change.

A TLL can be implemented with analog components, digital components, or mixed-signal components. In the early days of digital communications, most of the TLLs were implemented in analog or mixed-signal forms. A few types of non-data-assisted classical analog TLLs will be described in Section 6.3. More recently, in new receiver designs, TLLs are mostly implemented in all-digital forms.

A high-level TLL block diagram is shown in Figure 6.2. When it is compared to Figure 4.23 in Section 4.4.2, it is easy to recognize that the TLL shown is essentially a second-order DPLL with some different interpretations of its quantities. The implementation considerations of the TLL and the DPLL are also very similar.

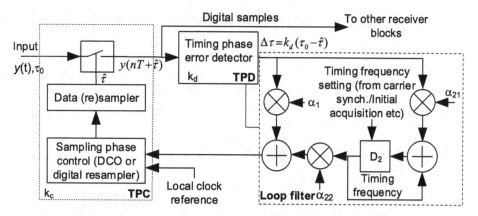

Figure 6.2 The high-level block diagram of a timing locked loop

Below, we describe the structures and operations of the TLLs and emphasize on the TLLs' unique features.

As shown in the figure, the symbol-rate data sampler generates samples $y(nT + \hat{\tau})$, where the sampling time $nT + \hat{\tau}$ is controlled by the TPC. If the input is an analog signal, the sampler is implemented by an ADC. Otherwise, in an all-digital implementation, the input would be digital samples at a rate that may be different from $1/T$. The sampler is implemented as a *digital resampler*, which is the subject of Chapter 7 of this book. If fractional-T samples are needed, multiple samples are generated every T. We use the symbol-rate sampler as an example in this section.

The TPD of the TLL receives samples $y(nT + \hat{\tau})$ from the sampler to generate the timing error $\Delta\tau$, which is an estimate of the timing phase difference between $\hat{\tau}$ and the desired sampling delay τ_0 in digital form. The TPD is characterized by the gain k_d, which is equal to the value of the digital sample $\Delta\tau$, divided by the timing phase difference $\tau_0 - \hat{\tau}$ detected from $y(nT + \hat{\tau})$ that $\Delta\tau$ represents. The main differences between TLLs in different communication systems are the realizations of their TPDs. Discussions of the timing detection algorithms of various systems in the next four sections constitute the main part of this chapter.

The phase errors generated by TPD are fed to the loop filter, whose operations are the same as that of the loop filter in DPLLs described and analyzed in Section 4.4.1.2. Similar to the second-order DPLL described there, the register D_2 in the TLL contains the estimated timing-frequency offset. The coefficient α_{22} should be chosen based on the required maximum frequency offset at which the TLL needs to operate. The value of $\alpha_2 = \alpha_{21} \times \alpha_{22}$ determines the second-order loop gain.

The loop filter output, which is equal to the value of $\alpha_{22}D_2$ plus $\alpha_1\Delta\tau$, is fed to a TPC, which generates digital data samples at the desired timing phase for the TPD and other receiver blocks. The TPC is usually implemented by a DCO or a digital resampler. The design example given in Section 4.5.2 is directly applicable to the implementation of the TLL discussed here with proper interpretation. Examples of all-digital TLL implementations for DS-CDMA receivers will be shown later in this chapter. More examples will be given in Chapter 7 where the TPCs that employ digital resamplers are discussed.

6.3   Timing Synchronization with Unknown Transmitter Data Symbols

In this section, we describe a few classical timing synchronization algorithms that do not require training sequences. These algorithms have played an important role in the development of digital communication technologies. They are still being used in certain types of digital communication receivers today.

6.3.1   Squarer-Based Timing Synchronization

The approximate ML timing phase estimation discussed in Section 6.2.1.3 indicates that, when the transmitter data symbols are unknown, the optimal timing is related to the square of the receiver filter output $y(t)$. Similar to the non-data-assisted carrier phase detection discussed in Section 5.3, the optimal timing phase can be determined by nonlinear processing using the cyclostationarity of the digital communication signals [6]. Below we describe such a timing phase–detection method of a real baseband signal using a squarer for preprocessing.

The expectation of $y^2(t)$ with the real baseband signal $x(t,\tau)$ in (6.1) can be shown to be

$$E\left[y^2(t)\right] = E\left[x^2(t,\tau)\right] + \sigma_z^2 = E\left[a_k^2\right] \sum_{k=-\infty}^{\infty} g^2(t - kT - \tau) + \sigma_z^2 \qquad (6.6)$$

where we have assumed that a_k's are independent and with zero mean.

Note that $E[x^2(t,\tau)]$ is equal to $E[x^2(t + mT, \tau)]$ for any integer m. Therefore, it is periodic with a period of T. It can be shown by using the *Poisson summation formula* [6] that

$$E\left[x^2(t - \tau)\right] = \frac{\overline{a_k^2}}{T} \mathrm{Re}\left[\sum_l D_l e^{\frac{j2\pi l(t-\tau)}{T}}\right] \qquad (6.7)$$

In (6.7), $\overline{a_k^2} \cong E\left[a_k^2\right]$ and

$$D_l = \int_{-\infty}^{\infty} G(l/T - f)G(f)df \qquad (6.8)$$

where $G(f)$ is the Fourier transform of $g(t)$, i.e., the frequency response of the overall channel. For most practical communication signals, the excess bandwidth is less than $1/T$, e.g., $\beta < 1$ for signals with RCOS spectra. For such signals $D_l \neq 0$ only if $l = -1, 0, 1$. In these three terms, D_0 is a constant and $D_1 = D_{-1}$. Thus, (6.7) can be expressed as

$$E\left[y^2(t)\right] = \text{const.} + \left(\overline{a_k^2}/T\right)D_1 \cos\left[2\pi(t - \tau)/T\right] \qquad (6.9)$$

Hence, the component $\cos\left[2\pi(t - \tau)/T\right]$ can be extracted for detecting the timing phase τ. The spectrum of the signal at the squarer input is shown in Figure 6.3. From the figure and (6.8), we observe that the coefficients D_{-1} and D_1 are proportional to the areas of

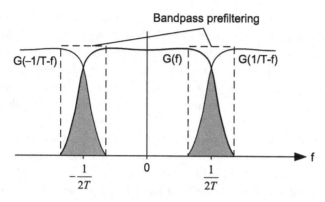

Figure 6.3 Prefiltering for squarer-based timing synchronization

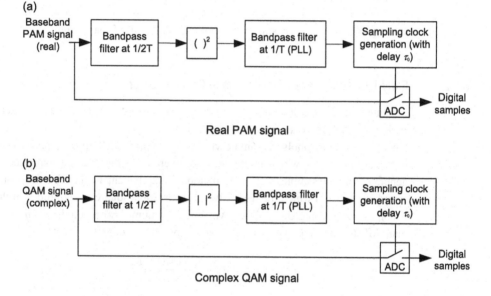

Figure 6.4 Squarer-based timing synchronization

the overlapped band-edge regions between $G(f)$ and $G(\pm1/T-f)$. Thus, the performance of timing synchronization can be improved if the signals and noises outside these transition bands, which constitute interference, are removed.

An example of the implementation of a squarer-based timing synchronization block is shown in Figure 6.4(a).

As shown in the figure, the input $y(t)$ is first prefiltered by a band-pass filter centered at $1/2T$ to remove the DC and noise or interference components. The prefiltered output is processed by a squaring device. The component $\cos\left[2\pi(t-\tau_0)/T\right]$ in (6.9) at the squarer output is extracted by another band-pass filter, which can be realized by a PLL centered at $1/T$. The extracted sinusoid is used to generate the sampling clock.

The squarer-based timing synchronization of PAM signals can be extended to certain types of signals with complex modulation symbols such as QAM. The implementation is shown in Figure 6.4(b), which is very similar to the implementation shown in Figure 6.4(a) for PAM signals. The only difference between these two implementations is that, instead of using the square of the prefiltered output, the square of the complex signal envelope is used. The remaining operations are the same as those in the PAM case.

The squarer-based timing synchronization does not require the knowledge of the transmitter data symbols in the received signal. It can operate independently of demodulation and decoding. As a result, it was widely used in wireline modem implementations. Since only the transitional bands at the band edges of the signal contain timing information, it does not work well for signals with very narrow transitional bands. Moreover, for signals with strong ISI, the squarer-based timing synchronization could lock to a fixed timing phase that may not be optimal for SREs. Therefore, it is most suitable to work with FSEs for such signals. Finally, squarer-based timing synchronization is more suitable for analog processing and is not efficient for DSP-based implementations.

6.3.2   Early-Late Gate-Based Timing Phase Error Detector

Assume that the overall channel frequency response satisfies the Nyquist criterion. For such a channel, the CIR is close to a sinc function. As shown in Section 6.2.1.3, when the transmitted symbols are unknown, an approximate ML timing phase estimate is the sampling delay, which maximizes the average of the squared magnitudes of the generated signal samples [2]. This method is equivalent to that for determining the sampling delay, at which the derivative of the average squared magnitudes of the signal becomes zero. Using the well-known results in numerical methods, such an approximate ML timing phase estimate can be computed recursively by

$$\hat{\tau}_{n+1} = \hat{\tau}_n + \Delta \frac{d\left(y_n^2(\tau)\right)}{d\tau}\bigg|_{\tau = \hat{\tau}_n} \tag{6.10}$$

where $y_n(\tau) \cong y(nT + \tau)$ and Δ is a small positive number that controls the convergence. The square operation can be replaced by other nonlinear operations such as the absolute value or $\ln[\cosh(\cdot)]$ functions. This is another way to implement the squarer-based timing synchronization block described above and leads to the Early-Late Gate, or E-L Gate, approximate ML timing phase estimator described below.

Three samples, $y_n(\hat{\tau}_n)$, $y_n(\hat{\tau}_n + \delta)$, and $y_n(\hat{\tau}_n - \delta)$, are generated every T. For $\delta \leq T/2$, the derivative in (6.10) can be approximated by a difference operation, i.e.,

$$\frac{d\left(y_n^2(\tau)\right)}{d\tau}\bigg|_{\tau = \hat{\tau}_n} \approx \frac{y_n^2(\hat{\tau}_n + \delta) - y_n^2(\hat{\tau}_n - \delta)}{2\delta} \cong \frac{m(\hat{\tau}_n)}{2\delta} \tag{6.11}$$

where, $m(\hat{\tau}_n) = y_n^2(\hat{\tau}_n + \delta) - y_n^2(\hat{\tau}_n - \delta)$ is called the timing *discriminator function* of the E-L Gate algorithm.

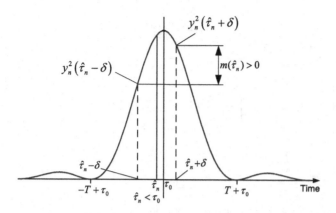

Figure 6.5 Early-late timing phase error detector

When the channel has little ISI, $y_n^2(\tau)$ is close to a squared sinc function of τ and is approximately symmetric about the peak. The difference $m(\hat{\tau}_n)$ between $y_n^2(\hat{\tau}_n + \delta)$ and $y_n^2(\hat{\tau}_n - \delta)$ is a measure of the timing phase error for signals received from such channels. As shown in Figure 6.5, if $\hat{\tau}_n < \tau_0$, we have $y_n^2(\hat{\tau}_n + \delta) > y_n^2(\hat{\tau}_n - \delta)$, i.e., $m(\hat{\tau}_n) > 0$. Conversely, if $\hat{\tau}_n > \tau_0$, $m(\hat{\tau}_n) < 0$.

Thus, the timing phase error can be corrected by forming a negative feedback loop such that

$$\hat{\tau}_{n+1} = \hat{\tau}_n + \Delta m(\hat{\tau}_n) \tag{6.12}$$

and it constitutes a first-order TLL. With the loop-coefficient Δ being a small positive number, the loop will converge to $y_n^2(\hat{\tau}_n + \delta) = y_n^2(\hat{\tau}_n - \delta)$ in steady state and $\hat{\tau}_n$ will be an unbiased estimate of the ideal sampling delay τ_0. Once the loop converges, the middle samples $y_n(\hat{\tau}_n)$ are used for estimation of data symbols.

Figure 6.6 shows an implementation of the E-L Gate-based timing synchronization given by (6.11) and (6.12). If the loop filter is a constant scaling factor Δ as given by (6.12), such a timing synchronizer forms a first-order loop. It becomes a second-order loop if the loop filter is redesigned to contain an integrator.

The E-L Gate-based timing synchronizer can be implemented in either analog or digital form. However, for the loop with the E-L Gate timing phase detector to perform reliably, it is desirable that δ is much smaller than $T/2$. Otherwise, there could be significant interference from the adjacent symbols' signal pulses, and there is no way to distinguish the contributions from a_n and $a_{n\pm 1}$, as shown in Figure 6.1. Hence, the synchronizer based on the E-L Gate timing-error detector is not ideal for digital implementation due to the required high-sampling resolution. Moreover, its tracking range is smaller than other timing synchronization methods.

To reduce computational complexity, it is advantageous to implement timing synchronization with a lower sampling rate as the one described below.

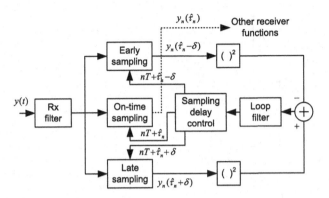

Figure 6.6 Early-Late Gate-based timing synchronization

6.3.3   Gardner's Algorithm

Based on the squarer-timing phase detection, a timing synchronization algorithm was proposed by Floyd Gardner in 1986. It is known as the *Gardner algorithm* [7]. The Gardner algorithm has a number of advantages over the E-L Gate-based algorithm. It has been studied by many researchers and widely used in practice. Below, we derive it from the approximate ML timing estimate with unknown data symbols given by (6.5).

As discussed in the previous section, the sampling delay attains the ML timing phase if the derivative of the average of $L(\tau)$ given by (6.5) is equal to zero. Namely,

$$\frac{d}{d\tau}\left[\sum_{k=n-L+1}^{n}|y_k(\tau)|^2\right] = 2\mathrm{Re}\left[\sum_{k=n-L+1}^{n}y_k(\tau)\frac{dy_k(\tau)}{d\tau}\right] = 0 \qquad (6.13)$$

Using the finite difference to approximate the derivative in (6.13), we have that if

$$\sum_{k=n-L+1}^{n}\mathrm{Re}\left[y_k^*(\hat{\tau})\{y_k(\hat{\tau}+\delta) - y_k(\hat{\tau}-\delta)\}\right] = 0 \qquad (6.14)$$

$\hat{\tau}$ is the ML timing phase. By letting $\delta = T/2$, the timing discriminator function $m_G(\hat{\tau}_n)$ of the Gardner algorithm can be constructed as:

$$m_G(\hat{\tau}_n) = \mathrm{Re}\left[y_n^*(\hat{\tau}_n)\{y_n(\hat{\tau}_n + T/2) - y_n(\hat{\tau}_n - T/2)\}\right] \qquad (6.15)$$

Similar to $m(\hat{\tau}_n)$ in the E-L Gate-based timing synchronization, $m_G(\hat{\tau}_n)$ can be used as the input to a TLL, i.e.,

$$\hat{\tau}_{n+1} = \hat{\tau}_n + \Delta m_G(\hat{\tau}_n) \qquad (6.16)$$

The sample timing converges to the optimum value τ_0 as n goes to infinity.

Because Δ is a small positive number, the loop shown by (6.16) performs long-term averaging. Thus, the summation in (6.14) does not need to be performed explicitly.

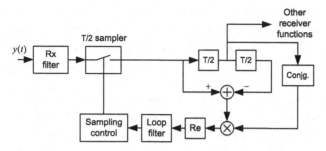

Figure 6.7 Implementation of Gardner algorithm

An example of the implementation of the Gardner algorithm is shown in Figure 6.7.

Similar to the E-L Gate-based timing synchronizer described above, if the loop filter is a constant Δ, the timing synchronization implementation shown in Figure 6.7 is a first-order loop. Timing synchronization based on the Gardner algorithm can also be implemented as a second-order loop with a perfect integrator included in the loop filter.

It is worth noting that the Gardner algorithm has a few interesting properties that are described in [7].

For example, it has been pointed out that just like the squarer-based implementation of timing synchronization, the useful information of the timing phase error is contained in the band-edge regions of the signal frequency response as shown in Figure 6.3. Thus, it is possible to improve the performance of the Gardner algorithm by prefiltering the received signal $y(t)$, as shown in [8]. The downside of this property is that, like the other squarer-based algorithms, it does not work well for a signal with very narrow transitional bands.

In addition, because it uses data samples generated at every $T/2$, only two samples are needed in every symbol interval instead of three as for the E-L Gate algorithm. Such a sampling rate is compatible with that of other receiver functions such as the FSEs. As a result, the receiver implementation can be simplified. Moreover, it has been shown that the performance of the Gardner algorithm is better than the E-L Gate algorithm [8, 9]. Another useful property of the Gardner algorithm is that the operation of the algorithm is not affected by the carrier phase. Thus, it can be used to acquire timing before the carrier phase is acquired. This property provides flexibility to the overall receiver synchronization implementations.

6.4 Data/Decision-Assisted Timing Synchronization

In Section 6.3, three timing synchronization algorithms, which can operate with unknown transmitted data symbols, have been presented. However, it has been shown that data- or decision-assisted timing synchronization algorithms can outperform the non-data-assisted methods [3, 10]. Moreover, for severe ISI channels, known transmitted data symbols can help to overcome issues exhibited in non-data-assisted timing synchronization methods. Thus, it would be advantageous to use the known data symbols or decisions to assist timing synchronization when they are available.

In this section, we present two data-assisted timing synchronization methods: the Mueller–Müller algorithm and the data-assisted Early-Late Gate algorithm. Also discussed are the utilizations of estimated channel and/or equalizer coefficients in timing synchronization.

6.4.1 Mueller–Müller Algorithm

To reduce the complexity of timing synchronization implementation, all-digital processing at a low sampling rate is preferred. Symbol rate, i.e., $1/T$, is the practically lowest possible rate for performing such processing. Hence, a symbol-rate timing synchronization algorithm, known as the *Mueller–Müller algorithm*[11], has been widely used in wireline modems and other single-carrier digital receivers. Below, we will call it the *M-M algorithm*.

Recall that the baseband signal at the receiver filter output can be expressed as

$$y(t) = \sum_{k=-\infty}^{\infty} a_k g(t - kT - \tau) + z(t) \cong x(t; \tau) + z(t) \tag{6.17}$$

where $g(t)$ is the CIR of the overall channel including the transmitter filter, the receiver filter, and the channel, and a_k's are the transmitted data symbols known to the receiver.

Digital samples $\{y_k(\hat{\tau}_k)\}$ are generated by sampling the receiver filter output at $kT + \hat{\tau}_k$. To detect the optimal sampling delay τ_0, a timing discriminator function $m_{MM}(n; \hat{\tau}_n)$ based on the samples $y_{n-1}(\hat{\tau}_{n-1})$ and $y_n(\hat{\tau}_n)$ is constructed by using the known data symbols a_{n-1} and a_n such that,

$$m_{MM}(n; \hat{\tau}_n) = a_{n-1}^* y_n(\hat{\tau}_n) - a_n^* y_{n-1}(\hat{\tau}_{n-1}) \tag{6.18}$$

It can be shown that the expectation of $m_{MM}(n; \hat{\tau}_n)$ is a function of the phase difference $\delta\hat{\tau}_n \cong \tau_0 - \hat{\tau}_n$, i.e.,

$$E[m_{MM}(n, \delta\hat{\tau}_n)] = E\left[|a_n|^2\right][g(-\delta\hat{\tau}_{n+1} + T) - g(-\delta\hat{\tau}_{n-1} - T)] \tag{6.19}$$

We assume that the overall CIR $g(t)$ satisfies the Nyquist criterion, e.g., it has an RCOS frequency response,

$$g(t) = \operatorname{sinc}\left(\frac{t}{T}\right) \frac{\cos(\pi\beta t/T)}{1 - 4\beta^2 t^2/T^2} \tag{6.20}$$

where β is the roll-off factor, $0 < \beta < 1$, and that the offset of the sampling time changes slowly, i.e., $\hat{\tau}_{n-1} \approx \hat{\tau}_n$. By substituting $g(t)$ given by (6.20) into (6.19), the expectation of $m_{MM}(n, \delta\hat{\tau}_n)$ normalized by the energy of a_n is given by

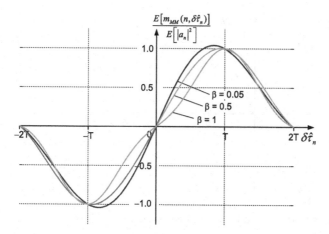

Figure 6.8 Timing error detector characteristics of the Mueller–Müller algorithm

$$\frac{E[m_{MM}(n,\delta\hat{\tau}_n)]}{E\left[|a_n|^2\right]} = \text{sinc}\left(\frac{-\delta\hat{\tau}_n}{T} - 1\right) \frac{\cos\left(\pi\beta(-\delta\hat{\tau}_n/T - 1)\right)}{1 - 4\beta^2(-\delta\hat{\tau}_n/T - 1)^2}$$

$$- \text{sinc}\left(\frac{-\delta\hat{\tau}_n}{T} + 1\right) \frac{\cos\left(\pi\beta(-\delta\hat{\tau}_n/T + 1)\right)}{1 - 4\beta^2(-\delta\hat{\tau}_n/T + 1)^2}$$

which is plotted in Figure 6.8 for β = 0.05, 0.5, and 1.0.

Similar to the non-data-assisted timing phase detectors described above, the output $m_{MM}(n, \delta\hat{\tau}_n)$ can be used to drive a TLL for timing phase and frequency synchronization. Below, we examine a few of the salient features of the M-M algorithm.

Observe from Figure 6.8 that each of the curves presents an S-shape function of $\delta\hat{\tau}_n$. The slope of an S-curve in the monotonic region is a function of β. Since $m_{MM}(n,0) = 0$ in the steady state, the mean of $\delta\hat{\tau}_n$ is equal to 0. Namely, y_n is at the peak of the RCOS CIR. Thus, for signals over channels that satisfy the Nyquist criterion, the sampling delay in steady state is optimal by using the M-M algorithm as the PD in TLLs. For many ISI channels, the timing phase generated by the M-M algorithm may be stable but may not be optimal when SREs are employed.

The M-M algorithm is a data-assisted timing phase estimation algorithm. Similar to any data-assisted algorithms, transmitted symbols and their rough timing should be known to the receiver. Thus, the M-M algorithm can be used during the training stage and when the pilot symbols are present. Tentative decisions may be used instead of true symbols after reliable data reception has been established. One advantage of the M-M algorithm is that the S-curve span goes from $-2T$ to $2T$. Thus, it could converge even if the initial sampling delay is off by up to $2T$ from the optimal phase. Practically, the range is smaller due to the existence of noise and interference.

6.4.2   Data-Assisted Early-Late Gate Timing Phase Detector

When transmitter data symbols a_k's are known to the receiver, such as during training with preamble or pilot sequences presented, or during the decision-directed adaptation,

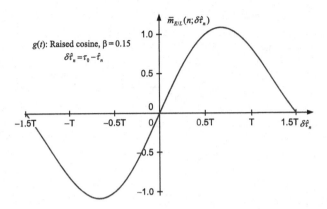

Figure 6.9 S-curve of data-assisted Early-Late Gate algorithm

it is possible to perform data-assisted timing error detection similar to the squarer-based Early-Late Gate algorithm described in Section 6.3.2.

Let us first consider the phase-corrected baseband signal $e^{-j\theta_0}y(t)$ given by (6.1). The signal is sampled at $nT + \hat{\tau}_n$ to yield the sample $e^{-j\theta_0}y_n(\hat{\tau}_n)$, which is an estimate of the phase-rotated symbol a_n. In addition, two more samples, $e^{-j\theta_0}y_n(\hat{\tau}_n - \delta)$ and $e^{-j\theta_0}y_n(\hat{\tau}_n + \delta)$, called early and late samples, are generated. If the channel phase is known or the timing phase detection is for real signals transmitted over real channels, i.e., $\theta_0 = 0$, a timing discriminator function $m(n; \hat{\tau}_n)$ can be constructed as

$$m_{E/L}(n; \hat{\tau}_n) = e^{-j\theta_0}a_n^* y_n(\hat{\tau}_n + \delta) - e^{-j\theta_0}a_n^* y_n(\hat{\tau}_n - \delta) \qquad (6.21)$$

The expectation of (6.21) with respect to a_n is a function of $\delta\hat{\tau}_n = \tau_0 - \hat{\tau}_n$, given by

$$\bar{m}_{E/L}(n; \delta\hat{\tau}_n) = E\left[|a_n|^2\right] [g(-\delta\hat{\tau}_n + \delta) - g(-\delta\hat{\tau}_n - \delta)] \qquad (6.22)$$

As in the case discussed previously, we assume that the channel satisfies the Nyquist criterion, $g(t)$ is a raised cosine function, and $\delta = T/2$. The function $\bar{m}_{E/L}(n; \delta\hat{\tau}_n)$ takes the shape of an S-curve, which is shown in Figure 6.9 for the roll-off factor $\beta = 0.15$.

The output of the data-assisted Early-Late (E-L) Gate timing phase detector can be used directly to drive a timing loop or a multiple of such outputs can be accumulated to generate a timing phase error estimate, similar to the timing phase detectors discussed previously.

The data-assisted E-L Gate timing phase detector with $\delta = T/2$ has a maximum detection range of $-1.5T$ to $1.5T$. Practically, the usable range could be smaller but is still larger than the range of the non-data-assisted E-L detector. The main advantage of the data-assisted E-L algorithm over its non-data-assisted counterpart is that the ISI is reduced after averaging as the data symbols are usually uncorrelated.

Note that the data-assisted E-L Gate algorithm is very similar to Gardner's algorithm. If we compare (6.21) and (6.15), the only difference between them is that $e^{-j\theta_0}a_n^*$ in (6.21) replaces $y_k^*(\hat{\tau})$ in (6.15). If $\delta\hat{\tau}_n$ is not too large, $y_k^*(\hat{\tau})$ can be considered as an estimate of $e^{-j\theta_0}a_n^*$. Therefore, the Gardner algorithm can be viewed as an approximate form of the data-assisted E-L algorithm. Given that the carrier phase in $y_k^*(\hat{\tau})$ cancels the phases in $y_k(\hat{\tau} \pm T/2)$, the Gardner algorithm does not need the phase correction explicitly.

In the case where the channel phase is not known, a discriminator function for complex data symbols can be constructed as

$$
m'_{E/L}(n; \hat{\tau}_n) = \left| \sum_{l=0}^{L-1} a_{n+l}^* y(nT + lT + \hat{\tau}_n + \delta) \right|^2 - \left| \sum_{l=0}^{L-1} a_{n+l}^* y(nT + lT + \hat{\tau}_n - \delta) \right|^2
$$

$$(6.23)$$

where we have assumed that the carrier phase of the samples inside the summation does not change. The summations on the right side of (6.23) are necessary for reducing the ISI in the phase detector output. For example, with $L = 1$, i.e., no accumulation is performed, the data-assisted E-L Gate timing phase detector degenerates to the squarer-based E-L Gate detector for a real signal or a complex signal with unknown phase.

The data-assisted E-L Gate timing phase detector is the basis of timing synchronization in DS-CDMA-based communication systems such as cdma2000 and WCDMA cellular systems. We will provide a more detailed analysis and describe the implementations of timing synchronization in DS-CDMA systems in Section 6.5.

6.4.3   Timing Synchronization Utilizing Estimated CIR/Equalizer Coefficients

The timing synchronization algorithms presented above were originally developed for single-carrier wireline or similar communication systems. They were intended for continuous data transmission over channels that are nearly static and with no or little ISI. The objective of these algorithms is to align the timing phase with the peak of the CIR of the communication channel. They are optimal or near optimal for data transmission over such channels.

However, in challenging communication environments often encountered today, such as rapidly time-varying and/or severe ISI channels, using only the techniques described above may not be sufficient to achieve effective and robust timing synchronization. Additional methods and algorithms have been developed to implement new forms of timing synchronization algorithms or to complement the existing ones.

In this section, we provide high-level descriptions of a few of such techniques based on estimated channel and/or equalizer coefficients for adverse channel conditions. Details of such techniques and algorithms specifically for DSSS/DS-CDMA and OFDM communications will be described in Sections 6.5 and 6.6.

6.4.3.1   Determination of Desired Sample Timing for Symbol Rate Equalizers over ISI Channels

For channels that are nearly static but with severe ISI, such as wireline local loops, wired Ethernet, and microwave channels, the timing synchronization techniques described above are still applicable in most cases. Using these techniques, the timing phase could lock to one of the CIRs' peaks, even if it is not the unique peak. As a result, the timing phase of the samples so generated is stable relative to the actual transmitter symbol timing.

To achieve the best possible performance for communication over such channels, especially at high SNR, receivers usually employ some type of equalizers. As mentioned earlier, the performance of an FSE is insensitive to the sampling delay [5]. The only requirement for FSEs to operate properly is that the sampling delay, i.e., the sample timing phase, is stable. Thus, previously described timing synchronization methods can be used without modification in most cases. However, the estimates of the channel and/or equalizer coefficients can be used to improve the receiver performance and stability.

On the other hand, it is well known that the performance of an SRE is sensitive to the sample timing phase [5, 12]. Using the timing algorithms described above cannot guarantee adequate performances of SREs. Additional steps are needed to determine the timing phase of the samples to achieve the best possible performance. Below we briefly describe two of such methods that are useful for practical implementation.

Heuristic SRE Timing Phase Selection

One of the methods is to generate sample sequences at multiple timing phase offsets. For example, the samples can be generated at a rate of an integer multiple of $1/T$ to form multiple T-spaced sequences. Tests are performed on these T-spaced sequences to determine which one of the sequences is the best for the SRE.

Such tests can be based on different criteria, such as maximal band-edge energy of the equalizer frequency response [13] or maximal CIR energy. The simplest method is to compare the energy of the sample sequences at different timing phases and pick the largest one. The last method is not as accurate as the other two but can be implemented with the lowest complexity. A few possible methods were discussed in [14]. Similar approaches have also been used in wireless receivers that employ equalization [15].

Timing Phase Derived from Equalizer Coefficients

It has been shown in [16] that the optimal timing phase of SREs' input samples can be derived from the converged equalizer coefficients. This method looks promising for the channel model considered there. However, it is not clear how robust it is under diverse channel conditions.

6.4.3.2   Equalizer Coefficients Position Control

In single-carrier receivers, timing synchronization may take multiple steps. The first step is to determine an approximate time phase at a resolution of one symbol interval T through initial hand-shaking sequences. The objective of this step is to determine the position of the received signal samples that contain the data symbol to be recovered. The second step is to determine a stable timing phase within a symbol interval T. For the SRE, the timing phase should also be selected based on the best achievable performance within T when possible. Both of these steps have been discussed above.

After the timing phase has been established, the equalizer coefficients are then generated through initial training or derived from estimated CIR coefficients, or taps. As a third step to ensure that the receiver achieves even better performance, determine

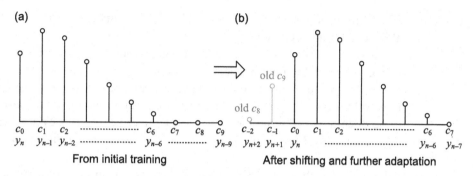

Figure 6.10 Equalizer coefficient position adjustment

the proper alignment of the received signal samples and the estimated equalizer coefficients so that the received signal energy is fully utilized.

Figure 6.10 illustrates the operations of this third step. Figure 6.10(a) shows the magnitudes of the equalizer coefficients after initial training. Let us consider an equalizer with up to 10 coefficients. Due to the inaccuracy of the initially determined alignment between samples and the decisions at the equalizer output, the energies of the equalizer coefficients after training are weighted more on the left side. This indicates that there is unused energy in the samples that arrived later than y_n, which is associated with the first coefficient c_0. To utilize the signal energy in samples of y_k, $k > n$, all of the coefficients are circularly shifted by two positions. Namely, the coefficients c_8 become the first coefficient c_{-2} and is associated with sample y_{n+2}. After further adaptation, the coefficients would be as shown in Figure 6.10(b) and the signal energy is fully utilized.

In summary, the desired alignment of the signal samples and the equalizers coefficients should ensure that the samples containing the energy from a data symbol contributes to the estimate of the symbol. If the magnitudes of the initially generated coefficients were concentrated on one side, it would be necessary to reassign coefficients to the proper samples. The same principle is applicable to the case where the equalizer coefficients are generated from CIR estimates. The implementation is different as will be shown later in this section.

After the initial adjustment, the position of the equalizer coefficients may still slowly change. For example, the position may drift if the receiver sampling clock and the remote transmitter symbol clock frequencies are not exactly synchronized [17]. Such a difference in frequency, no matter how small, could generate a nonnegligible timing phase shift over time. The performance of a receiver that employs an FSE would not be impacted immediately. However, eventually, the weight of the coefficients will shift to become similar to what is shown in Figure 6.10(a), and the receiver performance will degrade.

While the tap-leakage algorithm can be used to avoid the drifting of the FSE coefficients [4, 18], adjustment of the equalizer coefficient positions may still be needed. Hence, the weight distribution of the FSE coefficients should be monitored, and circular adjustments of the positions of the coefficients should be made when necessary. Alternatively, the receiver TLL can control the positions of the equalizer coefficients slowly and gracefully based on the detected shifts of the weight of the coefficients.

6.4.3.3 CIR/Equalizer Coefficients-Based Timing Synchronization in Wireless Mobile Receivers

Mobile wireless communication environments are much more challenging to receiver timing synchronization due to a number of factors.

For example, due to receiver mobility, the distance between a transmitter and a receiver keeps changing, and so is the relative signal propagation delay. Quantitatively, if the distance between the transmitter and the receiver changes by 1 kilometer, the delay changes by 3.33 microseconds. This corresponds to 3.33 T timing change for the symbol rate of 1 MHz.

In addition, with multiple distributed transmitters in mobile communication systems, such as cellular communication systems, a receiver may observe that the signal from a transmitter appears or disappears during reception depending on the relative locations of the transmitter and the receiver, as well as on the landscape. In such cases, receiver timing synchronization involves determining the existence of the useful signals and their delays.

Another possibility is that a mobile device may enter the "sleep" state from time to time to reduce its power consumption for improving the battery life. During the sleep state, the local XO that generates the receiver reference clock may shut down. As a result, the timing phase determined before entering the sleep mode could become invalid when the device wakes up and needs to be reacquired.

Thus, for mobile communications, a receiver needs to constantly acquire and/or track the synchronization signals to ensure reliable operations. The transmitters in such systems invariably transmit signals known to the receiver, i.e., pilot or reference signals, together with data to facilitate receiver operations.

Mobile communication channels are usually multipath channels due to the nature of radio wave propagation. Each of the paths, which are generated by the reflections of the radio wave, can be characterized by its phase, magnitude, and delay. Since multipath channels create ISI, a mobile receiver must be able to achieve satisfactory performance over such ISI channels. To effectively combat ISI, the most popularly adopted system and receiver technologies are OFDM, DS-CDMA, and single-carrier with equalization. Timing synchronization for OFDM and DS-CDMA will be treated in the next two sections. Below, we consider timing synchronization in a single-carrier receiver by using a GSM mobile receiver as an example.

As defined in GSM system specifications [19], GSM data transmission is organized as *hyperframes*, *superframes*, *multiframes*, and *TDM frames*. A TDM-frame, or simply a frame, consists of eight time slots. Each slot is 576.9 µs long, which is equal to 156.25 GSM bit intervals, each of which is equal to 3.69 µs. The physical content of a slot is called a burst, which is the basic physical-layer data transmission unit.

There are four types of bursts defined in GSM standard: normal burst (NB), frequency correction burst (FB), synchronization burst (SB), and access burst (AB). NBs carry data traffic during normal transmission. FBs and SBs are mainly for initial acquisition. FBs are used for frequency synchronization, and SBs for establishing initial timing and other system information. In the data mode, the receiver timing is tracked

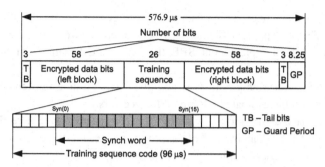

Figure 6.11 The structure of a normal burst

and maintained by using the training sequences, or midambles, in NBs. The structure of an NB is shown in Figure 6.11.

As shown in the figure, there is a training sequence at the center of each NB for performing channel estimation by the receiver. The training sequence, or midamble, is specified by one out of eight *training sequence codes* (TSCs) [20]. A TSC consists of 26 binary symbols. The center 16 symbols form a synch word *syn*. The synch words are specifically designed to have the following autocorrelation properties:

$$R_j = \sum_{i=0}^{15} syn_i syn_{(i+j)|_{\mathrm{mod}\,16}} = \begin{cases} 16 & j = 0 \\ 0 & 0 < |j| < 16 \end{cases} \tag{6.24}$$

A TSC is generated by the cyclic extension of the synch word on both sides of each of five symbols. Thus, the correlation between the TSC and its synch word is an impulse if the shift relative to the peak is less than ± 5 symbol intervals [21].

In the data mode, a GSM receiver correlates the synch word with the received signal samples and searches for the correlator outputs with large magnitudes, or peaks. The process is the same as the initial search procedures described in Chapter 3. If one or more peaks are found, the correlator outputs around the peaks, including those of the peaks, constitute a CIR estimate. Due to the correlation property of the synch word, there is no ISI from the adjacent symbols if the channel delay is less than $5T$, i.e., 18.45 μs. The timing for signal demodulation is determined by the sampling delays of the samples that generate the CIR.

In order to prevent the loss of information in channel estimation, the sampling rate should be higher than $1/T$, typically equal to $2/T$, i.e., two samples per symbol interval. The timing spacing between the coefficients of such a generated CIR estimate is equal to $T/2$. The estimated CIR is used for facilitating the receiver to perform equalization, which can take different forms. For example, the $T/2$ spaced CIR coefficients can be used to form a matched filter as a part of an MLSE equalizer such as the one described in [22].

Alternatively, the estimated CIR can be used to generate a set of linear or decision feedback equalizer coefficients by various well-known methods. The equalizer convolves these coefficients with the data samples to generate estimates of transmitter data symbols. The $T/2$-spaced CIR can also be viewed as two interleaved T-spaced CIRs. If it is desirable to implement an SRE, the T-spaced CIR with the highest energy should be selected.

Below we provide an example to show how the sample timing for demodulation is determined based on the estimated CIR in GSM receivers.

Using the baseband signal model given by (6.1), to generate the CIR estimates we consider the transmitter channel symbols a_{n+j}, $0 \le j < 16$, are the reference symbols and $a_{n+j} = syn_j$. The signal samples can be expressed by

$$y_m \cong y(mT + \hat{\tau}) = e^{j\theta_0} \sum_{k=m-K+1}^{m} a_k g(mT - kT + \hat{\tau} - \tau) \overset{i=m-k}{=} e^{j\theta_0} \sum_{i=0}^{K-1} a_{m-i} g(iT + \delta\tau)$$

(6.25)

where we have assumed that $g(t)$ is not zero only in the region $[-T/2 + \delta\tau, KT + \delta\tau]$, $\delta\tau = \hat{\tau} - \tau$ and $|\delta\tau| < T/2$. We have also ignored the noise $z(t)$ in (6.25), as it only affects the variance of the estimates but not the mean. The CIR coefficient estimates are computed by correlating the synch word with the signal samples that contain the synch word symbols such that

$$\sum_{j=0}^{15} syn_j y_{n+j+l} = e^{j\theta_0} \sum_{i=0}^{K-1} \sum_{j=0}^{15} syn_j a_{n+j+l-i} g(iT + \delta\tau)$$

$$= e^{j\theta_0} \sum_{i=0}^{K-1} \sum_{j=0}^{15} syn_j syn_{j+l-i} g(iT + \delta\tau)$$

(6.26)

since $a_{n+j} = syn_j$. From (6.24), the summations on the right side of (6.26) are zeros if $i \ne l$. Thus, the estimates of the even coefficients c_l^e of the $T/2$-spaced CIR are

$$c_l^e = \sum_{j=0}^{15} syn_j y_{n+j+l} = e^{j\theta_0} 16 g(lT + \delta\tau), 0 \le l \le K$$

(6.27)

The estimates of odd coefficients c_l^o are generated by correlating the samples $y_m' = y(mT + T/2 + \hat{\tau})$ with the synch word, such that

$$c_l^o = \sum_{j=0}^{15} syn_j y_{n+j+l}' = e^{j\theta_0} 16 g(lT + T/2 + \delta\tau) \quad 0 \le l \le K - 1$$

(6.28)

A $T/2$-spaced CIR estimate can be formed by combining the interleaved $\{c_l^e\}$ and $\{c_l^o\}$. However, to meet the ISI-free condition in the CIR estimation, the time span of the coefficient vector should be no greater than $5T$, i.e., $K = 5$. As a result, the CIR coefficient vector estimate would have no more than 11 coefficients, e.g., $\mathbf{c}_{est} = \begin{pmatrix} c_0^e & c_0^o & c_1^e & c_1^o & c_2^e & c_2^o & c_3^e & c_3^o & c_4^e & c_4^o & c_5^e \end{pmatrix}^t$. If necessary, a longer estimate can be generated, but the ISI-free condition will be violated.

The coefficient vector estimate c_{est} can be used to perform matched filtering by correlating c_{est} with $T/2$ spaced signal sample sequence. The output of the matched filter are the estimates of transmitted channel symbols with timing indices determined by the timing indices of the samples that generate c_{est}. For example, if c_{est} is generated according to (6.27) and (6.28) with the samples $\{y_{n+j+l}\}$ and $\{y'_{n+j+l}\}$ and the input sample vector is

$$\mathbf{y}_{n-L} = \left[y_{n-L} \; y'_{n-L} \; y_{n-L+1} \; y'_{n-L+1} \; y_{n-L+2} \; y'_{n-L+2} \; y_{n-L+3} \; y'_{n-L+3} \; y_{n+4} \; y'_{n-L+4} \; y_{n-L+5} \right]^t$$

the matched filter output $c_{est}^H \mathbf{y}_{n-L}$ is a scaled estimate of symbol a_{n-L}.

If a linear or decision feedback equalizer is employed, its coefficients can be derived from the estimated CIR. Alternatively, the coefficients may be directly computed from the received signal samples that contain the reference symbols through training. In the process of generating equalizer coefficient, the relationship between the positions of the received signal samples and the transmitted data symbols to be estimated is also established. In both cases, the objective of timing synchronization is achieved. Moreover, the generated channel/equalizer coefficients, which contain the channel phase information, are used for coherent demodulation of the received signal. Thus, GSM carrier-phase synchronization is also achieved.

6.5 Timing Synchronization in DS-CDMA Systems

As introduced in Section 1.5, RAKE receivers [23] are commonly used for the demodulation of DS-CDMA, or simply CDMA, signals. A RAKE receiver consists of multiple RAKE fingers, each of which demodulates the CDMA signal that passes through a path of a multipath channel. The received signals are usually corrupted by strong additive noise, inter/intra-cell and intersymbol interference, which are suppressed by the PN-code despreading performed at each of the RAKE fingers. After despreading with high processing gain, the output of each RAKE finger can be viewed as the received signal passing through a channel that satisfies the Nyquist criterion, i.e., ISI free.

The performance of a RAKE receiver depends on the following two factors that are related to timing synchronization. First, an initial signal timing phase, or sampling delay, is established for each finger after the correlation peak, generated from the proper sample sequence correlated with the given spreading chip sequence, is determined. Second, the timing phase should be refined and maintained to be at the peak of the CIR of the path. These are the two essential functions of the timing synchronization in CDMA receivers.

6.5.1 Initial Timing Determination

When a device is turned on, it is not yet connected to any network. As a result, the device has no idea about the system timing. The first task of the device is to acquire the signal from a base station that belongs to a network with which the receiver could communicate.

Recall from the initial acquisition process described in Chapter 3 that to acquire the desired transmitter signal, the receiver performs a series of hypothesis tests. Specifically, a sequence of CDMA pilot chips known to the receiver is correlated with received signal sample sequences. If the magnitude of the correlator output is above a predetermined threshold, the receiver assumes that there could be a match between the received signal samples and the pilot chip sequence. Once confirmed by further verification, the receiver declares the hypothesis test successful.

In addition to determining the existence of the desired signal, the successful hypothesis test also provides an initial estimate of the timing phase of the samples. As discussed in Sections 1.5.3, signal samples that a RAKE finger receives are the transmitted chips weighted by the path gain and corrupted by the additive noise and interference. A peak at the correlator output indicates that these samples are generated by the transmitted chips that contain the pilot chips involved in the correlation. Hence, the time position of a sample is an estimate of the timing of the pilot chip that the sample multiplies during the correlation that yields the successful hypothesis test.

The hypothesis tests during initial acquisition are usually performed on the received signal sampled at twice the channel symbol rate, i.e., chip rate for CDMA receiver, i.e. $2/T_c$. Since this rate satisfies the Nyquist sampling theorem, there would be no loss of information. If the effects of the noise and interference were ignored, at the sampling rate of $2/T_c$, the maximum timing error is half of the sampling interval, i.e., $T_c/4$. Such a timing error without correction would cause a performance loss of up to about $10\log[\text{sync}(0.25)] \approx -0.91\text{dB}$.

To better illustrate the initial timing acquisition outlined above, let us look at two practical CDMA wireless communication systems.

cdma2000

As described in Section 3.5.1, in cdma2000 systems, the pilot channel signals transmitted from all of the sectors are based on the same short PN sequences with different shifts, also called offsets or delays, from each other. The pilot signal is periodic with a period of 2^{15} chips. During initial acquisition, a searcher correlates a segment of the common pilot chip sequence with the received signal samples without concerning the transmitting base station. Let us assume that first the L chips of a period of the pilot sequence are used for performing correlation. Once a peak is found, the timing phase of the first sample in the sequence involved in the correlation defines the start of the period of the pilot sequence. This information is used to set the state of a counter that counts the samples generated. The content of the counter can be used as the timing reference of the receiver to keep the receiver and system timings synchronized.

WCDMA

Unlike cdma2000, the pilot signals from different base stations, called *Node-Bs*, in a WCDMA system are scrambled by different scrambling codes. To facilitate the initial acquisition, a Node-B sends a PSC and an SSC. The combination of the PSC and SSC defines a set of the P-CPICHs with eight members in the set. PSC and SSC are the first

256 chips in every 2560 chips long slot, which spans approximately 0.667 ms. The P-CPICH is periodic with a period of 38,400 chips long with a time span of 10 ms.

During initial acquisition, a WCDMA receiver searches the PSC, SSC, and P-CPICH sequentially by correlating these codes with the received signal samples as described in Section 3.5.2. When the PSC is found, the timing phase of the first sample of the sequence involved in the correlation marks the beginning of the slot that contains PSC and SSC. The timing phase learned, even if it is not very accurate, can be used to reduce the range of the SSC search. A successful SSC search provides a more accurate estimate of the timing of the beginning of the same slot. The obtained timing estimate is used as the timing reference for the search of P-CPICH. The three sequential searches provide the estimates of the sample timing phase with improved accuracies from one to the next. The timing phase estimate obtained from the search of P-CPICH should be accurate enough for initial data demodulation. The detected P-CPICH also uniquely defines the Node-B, which transmits the signal involved in the correlation.

In either cdma2000 or WCDMA systems, after initial acquisition, the timing phase error should be within a fraction of T_c. However, the timing estimate may need to be refined to achieve the best possible performance of data demodulation. Moreover, there could be a timing-frequency offset, which would cause the estimated timing phase to drift if there is no correction. Thus, upon entering the data mode, further timing phase estimation is performed to improve the accuracy of the initial estimate, to track the timing phase drift, and to reduce the existing frequency offset.

6.5.2 Timing Phase Tracking

After successful initial acquisition, an acquired channel path is assigned to a RAKE finger with a timing phase accuracy within a fraction of a chip interval T_c. The RAKE finger uses the estimated magnitude and phase of the path to demodulate the data signal received from the assigned path. Concurrently, as discussed in in Section 1.5.3, the synchronization block of the finger continues to estimate the timing phase to reduce the residual error and to track the possible drift. In a RAKE finger, the timing-phase error correction and tracking are performed by using a timing-tracking loop commonly called a DLL [24].

6.5.2.1 Delay Locked Loops in RAKE Fingers

A DLL is a specific implementation of PLLs described in Chapter 4. All of the descriptions and the properties of PLLs apply to DLLs as well. A DLL consists of a TPD, a loop filter, and a TPC. The block diagram of a typical DLL in a RAKE finger is depicted in Figure 6.12.

The input signal is sampled at a rate equal to, or higher than, $2/T_c$. Let us consider that the lth finger in a RAKE receiver processes the samples that contain chips c_{nN+k}, $k = 0,\ldots, N-1$. Regardless of the actual sampling rate, a finger receives samples at $2/T_c$. The finger has a delay line that contains three samples. The center samples, denoted by $y_{nN+k}^{(l)}$, are sampled at $(nN + k)T_c + \hat{\tau}$. Ideally, $\hat{\tau} = \tau_l$, where τ_l is the actual time delay

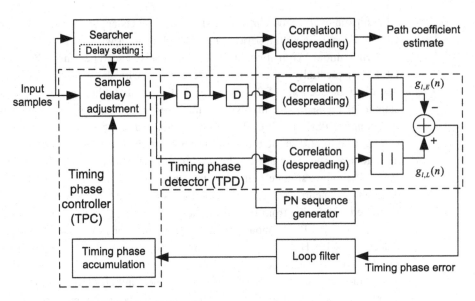

Figure 6.12 A typical DLL in a RAKE finger

of the lth path. The samples received before and after the center sample, called early and late samples, are sampled at $(nN + k)T_c \mp T_c/2 + \hat{\tau}$, denoted as $y_{nN+k,E/L}^{(l)}$. The center samples are used to generate the estimate of the lth path gain, or path coefficient, $\hat{g}_l(n)$, for demodulating the nth data symbols as shown in Section 1.5.3. Namely,

$$\hat{g}_l(n) = \frac{1}{N} \sum_{k=0}^{N-1} p_{nN+k}^* y_{nN+k}^{(l)} \tag{6.29}$$

where p_{nN+k} is the element of the pilot chip sequences in the transmitter chip c_{nN+k}, and $|p_{nN+k}| = 1$. The early and late samples are used to estimate the timing error of the center sample as described below. The interference from other users' signals is eliminated or heavily suppressed as N is commonly chosen to be equal to the length of the longest Walsh code words if variable-length Walsh codes are used. For example, in cdma2000 and WCDMA receivers, N is commonly chosen to be equal to 64 and 256, respectively.

The output of the TPD is filtered by a loop filter, whose characteristics determine the DLL being a first-order or second-order loop. As in PLLs, the output of the loop filter drives a TPC, which adjusts the delay of the samples for timing phase correction. The details of the delay adjustment will be described later in this section and in Chapter 7.

6.5.2.2   CDMA Signal Timing Phase Error detection using Early-Late gate detector

Similar to the path coefficient estimate $\hat{g}_l(n)$ given by (6.29), the average of the path coefficients at $(nN + k)T_c \mp T_c/2 + \hat{\tau}$, $k = 0, \ldots, N-1$, denoted by $\hat{g}_{l,E}(n)$ and $\hat{g}_{l,L}(n)$, are estimated, such that

$$\hat{g}_{l,E/L}(n) = \frac{1}{N}\sum_{k=0}^{N-1} p_{nN+k}^* y_{nN+k,E/L}^{(l)} \tag{6.30}$$

The early/late samples $y_{nN+k,E/L}^{(l)}$ can be expressed as

$$y_{nN+k,E/L}^{(l)} = g_{l,E/L}(n)p_{nN+k}^* + z_{nN+k,E/L}^{(l)} \tag{6.31}$$

where $z_{nN+k,E/L}^{(l)}$ represents the additive noise and interference.

The lth path is modeled as a single impulse with delay τ_l, and the combined frequency response of the transmitter and receiver filters has approximately an RCOS shape. As a result, the CIR of the overall channel is approximately equal to $G_l\text{sinc}((\tau - \tau_l)/T_c)$, where G_l is the lth path gain. If the center samples are generated by sampling the signal at the receiver filter output at $(nN + k)T_c + \hat{\tau}$, the estimates of the path coefficients of the early and late samples are $\hat{g}_{l,E/L} \approx G_l\text{sinc}((\mp T_c/2 + \hat{\tau} - \tau_l)/T_c)$. It is easy to verify that, if the center sample is obtained by sampling the signal at the peak of the receiver filter output, i.e., $\hat{\tau} = \tau_l$, the magnitudes of the early and late path estimates are equal to each other on average. At the same time, if $\hat{\tau} > \tau_l$, we have $|\hat{g}_{l,E}| > |\hat{g}_{l,L}|$ and, conversely, if $\hat{\tau} < \tau_l$, we have $|\hat{g}_{l,E}| < |\hat{g}_{l,L}|$. The quantity $|\hat{g}_{l,L}| - |\hat{g}_{l,E}|$, which is a function of $\delta\hat{\tau} = \tau_l - \hat{\tau}$, is called the *timing discriminator function* of the E-L detector

$$m_l(\delta\hat{\tau}) = |\hat{g}_{l,L}(\delta\hat{\tau})| - |\hat{g}_{l,E}(\delta\hat{\tau})|. \tag{6.32}$$

where $\hat{g}_{l,E/L}(\delta\hat{\tau}) \cong G_l\text{sinc}((\mp T_c/2 - \delta\hat{\tau})/T_c)$. The discriminator function $m_l(\delta\hat{\tau})$ is a metric of timing phase error and is typically used as the output of the TPD in DLLs of CDMA receivers.[3]

The TPD in CDMA DLLs is a type of the data-assisted E-L Gate timing phase detector presented in Section 6.4.2. The distinctive property of the CDMA E-L Gate timing phase detector is that the detector estimates the early and late path coefficients using the samples after they have been despread by the pilot chips with a high processing gain. The despreading process coherently combines the data samples and has a processing gain equal to N, the number of data samples combined. In other words, the pilot to noise/interference ratio increases by a factor of N in power. Such a high processing gain can greatly improve the DLL performance. It is especially important when the channel phase is unknown so that noncoherent detection metrics must be used.

In a typical CDMA system design, the overall frequency response of each path usually takes the RCOS shape. For example, the transmitter filter defined in cdma2000 has approximately a SRCOS frequency response with an excess bandwidth $\beta = 0.125/T_c$. As the receiver filter is usually designed to match the transmitter filter, the overall channel has a RCOS frequency response and its CIR is

---

[3] It is also possible to use the squared magnitudes of the early and late correlator output in (6.32) to compute the discriminator value. However, using the absolute values is preferred because the TPD gain variation would be smaller when the SNR changes in CDMA systems than using the squared values.

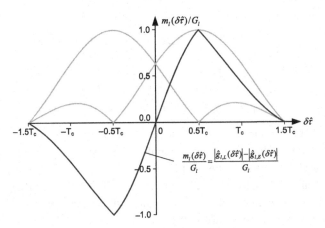

Figure 6.13 Characteristic of Early-Late TPD for DLL

$$g_l(t) = G_l \text{sinc}\left(\frac{t}{T_c}\right) \frac{\cos(\beta \pi t / T_c)}{1 - 4\beta^2 t^2 / T_c^2} = G_l \text{sinc}\left(\frac{t}{T_c}\right) \frac{\cos(0.125 \pi t / T_c)}{1 - 0.0625 t^2 / T_c^2} \cong G_l \tilde{g}(t/T)$$

$$(6.33)$$

which is close to the sinc function. The timing discriminator function of the E-L Gate detector given by (6.32) is plotted in Figure 6.13.

Similar to most phase error detectors, the characteristic of the E-L Gate TPD is in the form of a sideway character S and is called an *S-curve*. Such a TPD can operate properly for $|\delta\hat\tau| < 1.5T_c$, which is the maximum timing phase error that it can detect. However, when the magnitude of the phase error is greater than T_c, the detector is sensitive to the noise in the signal. In addition, the detector gain may become too small to ensure the timely convergence of the DLL. Therefore, the worst-case time phase error should not be beyond $\pm T_c$. The TPD gain normalized by its maximum value at $\delta\hat\tau = 0$, i.e., $k_d(0) \approx 2.255 G_l$, is plotted in Figure 6.14.

A few important properties of the E-L Gate TPD should be noted for being used in DLLs in CDMA receivers. The samples $y_{nN+k,E/L}^{(l)}$ in (6.30) can be expressed as

$$y_{nN+k,E/L}^{(l)} = p_{nN+k} G_l \tilde{g}((\mp T/2 + \hat\tau - \tau_l)/T_c) + z_{l,E/L}$$

$$(6.34)$$

where $z_{l,E/L}$ represents the additive noise and interference in the early and late samples of the lth path. The noise and interference can be modeled as AWGN with a variance of σ_z^2. Thus, the SNR of each sample of the lth path is equal to the G_l^2 / σ_z^2

By substituting $y_{nN+k,E/L}^{(l)}$ in (6.34) into (6.30) and since $|p_{nN+k}| = 1$, we have

$$\hat{g}_{l,E/L}(n) = G_l \tilde{g}((\mp T/2 + \hat\tau - \tau_l)/T_c) + z'_{l,E/L}$$

$$(6.35)$$

It can be shown that the variance of $z'_{l,E/L}$ is equal to σ_z^2 / N. Thus, the accuracy of the TPD output is significantly improved due to the processing gain for large N.

In a CDMA system for voice communications, the signal SNR is usually quite low. Namely, the sample $y_{nN+k,E/L}^{(l)}$ is dominated by the noise $z_{l,E/L}$. Therefore, the variance

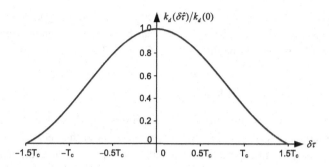

Figure 6.14 Normalized gain of Early-Late TPD

of the samples at the AGC output in (6.34) is determined by the power of the noise and interference $z_{l,E/L}$, and almost independent of the signal power and SNR. Consequently, the variance of the TPD output $m_l(\delta\hat{\tau})$ is also approximately constant.

When the variance of $z_{l,E/L}$ at the AGC output is a constant, the values of $\left|\hat{g}_{l,E}(n)\right|$ and $\left|\hat{g}_{l,L}(n)\right|$ are proportional to the square root of the SNR of the lth finger's signal, i.e., $10^{\gamma_l/20}$, where γ_l is the SNR in dB. From (6.32) it can be shown that the gain k_d of TPD is also proportional to $10^{\gamma_l/20}$. This conclusion is important for the designs of DLLs that employ E-L gate TPDs. For example, if the dynamic range of the signal SNR is equal to 18 dB, the TPD gain can change by a factor of 8. The DLLs employing such TPDs should be designed properly in order for them to perform satisfactorily under such conditions.

Correlations of the early and late samples are computed in parallel. Thus, one value of $m_l(\delta\hat{\tau})$ is generated every N samples. In the early days when digital computation was costly and the computational burden needed to be reduced, early and late correlations were computed alternatively, i.e., one value of $m_l(\delta\tau)$ is generated every $2N$ samples. Such a timing loop is called the *tau-dither loop* (TDL) in the literature [25]. TDL is rarely used in modern CDMA receivers.

6.5.2.3 Loop Filter and Timing Phase Control

The output of the TPD is filtered by a loop filter, which determines the order of the DLL. The loop filter generates the phase adjustment needed to correct the residual timing phase error due to the inaccuracy of the initial timing acquisition or the phase drift caused by a timing-frequency offset. The output of the loop filter is accumulated in the register of a perfect integrator.

The content of the register in the integrator controls the timing phase of the samples through the sample delay adjustment block shown in Figure 6.12. These timing-phase adjusted samples are despread to generate $\hat{g}_{l,E/L}(n+1)$ and quantities for the other receiver functions at $n+1$.

The delay adjustment block changes the delay of its output samples relative to the delay of its input samples. The input samples can have a timing resolution higher than $T_c/2$, e.g., $T_c/8$, the same as the output samples, or simply equal to $T_c/2$. The delay adjustment of the output samples in the first case is discussed in the example of a DLL in a cdma2000 receiver, which will be given in Section 6.5.4. If the input sample timing resolution is equal to $T_c/2$, digital resampling can be used to generate samples with higher timing resolution as will be discussed in Chapter 7.

6.5.3   Searcher Operations in the Data Mode

During initial acquisition, searchers in CDMA receivers are used to detect desired signals for establishing communication links with the transmitters in the network. In the data mode, searchers also play an important role in finding new paths, which carry signals that the device may be able to receive. Below we describe two timing synchronization-related areas, in which searchers play an important role in the data mode.

Finger Management
Soft-handover, or soft-handoff, is an important feature of CDMA systems. To perform soft-handover, more than one base station in a CDMA network transmits to and receives from a device simultaneously when possible and necessary. Such base stations belong to the *active set* of the device as defined in the cdma2000 and WCDMA standards. A CDMA device can receive the signals from the base stations in the active set, demodulate the signals, and combine the outputs of the demodulators to generate estimates of data symbols.

During the data mode, the searcher in a device receiver constantly looks for signals with strength above a threshold. It the searcher finds such a strong signal, the device will decide if this new path should be assigned to a RAKE finger. The function of determining whether to add a signal to and/or to drop a signal from the RAKE fingers is called *finger management*. Finger management plays a critical role in ensuring the satisfactory performance of CDMA receivers. It is probably the most challenging operation that CDMA receivers need to perform. The details of finger management operations are quite complex and depend on the system design. A brief description is provided below.

When a new path is found by the searcher, the receiver decides if it is strong enough to be of interest. If it is and the signal from the path is from a base station that belongs to the active set, the receiver would make a decision if a finger should be assigned to this path and deassigned from the existing path if necessary. To ensure the stability of the device operation, a *hysteresis threshold* is often used in the decision making. Namely, the signal from the new path must be stronger by a certain amount than the signal from the path that is already assigned to warrant the path swapping and to avoid constant flip-flops between the paths.

If the path that carries strong signals from a base station that does not belong to the active set, the device informs the base station, with which it is communicating, that it has a good candidate to connect. The base station in turn informs the cellular network, which decides if this new base station should be included in the active set of the device. The device is informed once a decision is made.

In order to reduce the burden of the searcher in finding new paths, the network maintains a list of the identities of the pilot signals from the base stations near the location of the device. These nearby base stations belong to the *neighbor set* of the device as defined in the cdma2000 standard. It is called the *monitored set* in the WCDMA standard. The device is only required to search the signals from the base-stations belonging to the neighbor or monitoring set. The rest of the entire pilot set needs to be searched only if extra search capability of the device is available.

Additional details of finger management operations can be found in the literature, such as Chapter 11 of [24] for cdma2000 and [26] for WCDMA.

Fat Finger

Fat finger is another challenging channel condition that CDMA receivers face. So far, we have assumed that when a RAKE finger is assigned to a channel path, the DLL of the finger is used to estimate the path delay and to track it as needed. For the DLL to operate properly, the paths must be resolvable, i.e., the delay difference between paths should be no less than one chip time-interval T_c. When the delay difference between two paths is less than T_c, these two paths may not be distinguishable by the DLL. Even if two fingers are assigned to two such paths initially by the searcher, the two fingers will eventually merge into one. This channel condition is usually called *fat finger* in the CDMA technical community.

The DLL that is commonly implemented in CDMA receivers cannot handle the fat fingers well. One approach to dealing with such a situation is for the searcher to assign two fingers to the peaks of the two paths with their DLLs disabled. The searcher monitors the path positions and makes adjustments when necessary. This method can ensure that the signal energies from these paths are collected and utilized. However, such a fat finger channel is not equivalent to a two single-path channel, because the processing gain from despreading cannot completely suppress the ISI introduced between the two close correlated paths.

To better address this problem, some form of equalization should be used. To reduce the computational complexity, simplified processing algorithms have been developed as compromised solutions. The *generalized-RAKE* (G-RAKE) algorithm introduced in [27] is one such example. Essentially, it is a hybrid form of a RAKE receiver and an equalizer. The G-RAKE can reduce the effect of ISI between the paths. Other ad hoc and/or semi-theoretic algorithms have been developed as well. These algorithms address the fat finger issue in a non-optimal but relatively efficient way.

6.5.4   An Example of DLL Implementation in cdma2000 Receivers

So far, we have presented a number of aspects of the timing synchronization in CDMA receivers including the basics of DLLs. Below, we describe the DLL implementation in the RAKE finger of a cdma2000 receiver. Practical designs of receivers vary depending on the system and implementation considerations. What is shown below is only an example of many possible designs. However, the principles described here are applic-able to other CDMA receivers.

Since the principles and components of DLLs have been discussed in previous sections, we only describe the basic operations and some specifics regarding the TPD

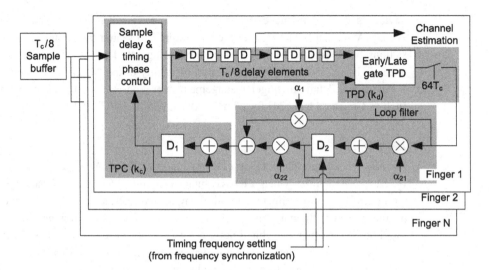

Figure 6.15 A reference cdma2000 DLL design in RAKE receiver

and TPC. The block diagram of an all-digital implementation of a DLL in a RAKE finger is shown in Figure 6.15.

As shown in Section 6.5.2, the RAKE finger receives signal samples at a rate equal to or higher than $2/T_c$. In this example, we assume that the sampling rate is equal to $8/T_c$, i.e. the timing resolution is equal to $T_c/8$. When the CDMA receiver operates under low SNR channel conditions, the timing errors caused by such a resolution will not cause noticeable degradation in performance.

The $T_c/8$ samples are stored in a delay line for introducing the necessary delay to match the path delay before the samples are sent to the TPD. This delay line is also used as a buffer for adjusting the sample timing phase required for proper RAKE finger operation. Given that three $T_c/2$ spaced samples are needed by a RAKE finger in each chip time interval, the finger contains a tapped delay line with 8 registers to store 8 $T_c/8$ samples.

The center sample in the delay line is used by the channel estimator to generate the estimate of the path coefficient for data symbol estimation. The first and the last samples in the delay line are multiplied by the complex conjugate of the corresponding element of the pilot sequence. Normally, after one chip interval, eight new samples are shifted into the delay line to generate another pair of such products. The number of new samples shifted into the delay line could be seven or nine if a timing correction is needed. Since the length of the longest Walsh code word in a cdma2000 system is equal to 64, 64 pairs of these products are added together to yield a pair of early and late path coefficient estimates.

Every 64 T_c, a timing error metric $m_l(n)$ is computed by subtracting the absolute value of the early estimate from that of the late estimate given by (6.30) and (6.32) with $N = 64$. The generated timing error metric is processed by the DLL's loop filter.

If a VC-TCXO is used in frequency synchronization implementation as described in Section 5.7.2, the frequency of the local sampling clock derived from the output of the

VC-TCXO is locked to the remote carrier frequency in the steady state. Consequently, the timing frequency is either equal, or very close, to the remote transmitter symbol clock frequency. In such a case, a first-order DLL should be adequate. However, if an XO based all-digital implementation is employed for frequency synchronization as described in Section 5.7.3, the sampling clock generated from the XO would have a nonnegligible frequency offset from the remote transmitter symbol clock. This offset needs to be compensated digitally. In this case, a second-order DLL is appropriate because it can detect and compensate for the timing-frequency offset as a part of its functions.

In Figure 6.15, a second-order loop is shown for the all-digital DLL implementation. The DLL has the same loop filter as does a second-order DPLL that is shown in Section 4.4.1.2. Below, we focus on the aspects that are specific to DLL operation and implementation.

There are two perfect integrators in the second-order DLL. The register D_1 of the integrator in the TPC contains the sampling phase, which is in the range from 0 to T_c and is represented by an unsigned fractional fixed-point number in the range of $[0, 1)$. As a result, the controller gain k_c is equal to 1 with a unit of T_c. The register D_2 of the integrator in the loop filter contains the estimated timing-frequency offset. The timing-phase change due to the frequency offset is equal to α_{22} multiplied by the value of D_2 and it is added to register D_1, which controls sample timing adjustments. Since the resolution is equal to $T_c/8$, three MSBs of the value in D_1 determine the timing adjustment.

As stated above, the TPC normally shifts eight new samples into the delay line every T_c, unless the value of the three MSBs of D_1 changes. Specifically, if the value of the three MSBs decreases or increases by one (modulo 8), seven or nine samples are shifted into the delay line. Such an implementation is often used in CDMA receiver ASICs in early years. More recently, implementations using digital resampling of the $T_c/2$ samples have become more popular. We will discuss such implementations in Section 7.7.2.

The value of the register D_2 corresponds to the estimated timing-frequency offset, which is in the range from $-\Delta \tilde{f}_{t,\max}$ to $\Delta \tilde{f}_{t,\max}$, where $\Delta \tilde{f}_{t,\max} = \Delta f_{t,\max} T_c$ with $\Delta f_{t,\max}$ being the maximum timing-frequency offset. The maximum value of D_2 corresponds to the normalized maximum frequency offset $\Delta \tilde{f}_{t,\max}$. The content of D_2 usually represents a signed fractional fixed-point number in the range of $[-1, 1)$. Its word length is selected according to the receiver requirement.

The value of α_{22} is determined as follows. To compensate for the maximum normalized frequency offset $\Delta \tilde{f}_{t,\max}$, the timing phase must advance by $\Delta \tilde{f}_{t,\max} \times 64 T_c$ every DLL update. For a value 1 in D_1 corresponding to a timing phase change of T_c, the adjustment is equal to $64\Delta \tilde{f}_{t,\max}$. Since the maximum value of D_2 is equal to 1 at $\Delta \tilde{f}_{t,\max}$, the adjustment at the maximum frequency offset is equal to α_{22} times the maximum value of D_2, i.e., α_{22}. Thus, the coefficient α_{22} should be equal to $64\Delta \tilde{f}_{t,\max}$.

Although only three MSBs are used to control the timing phase, the bit width of register D_1 should be chosen so that it accumulates the phase increment received from the loop filter without the loss of precision. When an overflow or underflow occurs, the register value simply wraps around.

The maximum TPD gain k_d is equal to $2.255G_l$, where G_l is the magnitude of the center path coefficient, as discussed in the example in Section 6.5.2.2. The gain k_d has a unit of $1/T_c$. As also shown there, in a CDMA receiver, the signal samples at the ADC output are normalized in power by the AGC. When operating under low SNR, the total signal power is approximately equal to the total noise power. Thus, the estimates of the early, late, and center path coefficients at the correlator outputs have a constant noise variance. The magnitudes of the early and late estimates are proportional to the square root of the signal SNR.

Moreover, as discussed in Sections 4.3.2 and 4.3.3, the convergence time constant of a first-order loop, as well as of a second-order loop with critical damping and over-damping, is mainly determined by the first-order loop gain $k_1 = k_d\alpha_1 k_c$. For a DLL that updates once every 64 T_c the time constant is approximately equal to $64T_c/k_1$.

The variance at the TPD output is a constant and roughly independent of SNR. As a result, the variance of the timing jitter at the DLL output is a constant and only depends on the first-order coefficient α_1. Yet, the first-order loop gain k_1 is proportional to the TPD gain k_d, which is proportional to the square root of SNR γ_l. Therefore, the loop time constant is inversely proportional to $10^{\gamma_l/20}$. If the dynamic range of γ_l is equal to 18 dB, the time constant is eight times longer at the lowest SNR than that is at the highest SNR. The value of α_1 should be determined according to the requirements of both the phase jitter and time constant over the entire dynamic range.

The DLL's second-order loop gain k_2 is equal to $k_d\alpha_{21}\alpha_{22}k_c$. The gain k_2 is also proportional to the square root of the SNR and affects the DLL's damping factor and the convergence time of the frequency error. The value of α_{21} can be chosen such that the loop is critically damped when the SNR is at the minimum. When the received signal is at a higher SNR, by keeping the same α_{21}, the DLL will be over-damped. However, this is acceptable because the convergence time is actually shorter due to a larger k_1.

6.6   Timing Synchronization in OFDM Systems

OFDM is a form of multicarrier communication. In recent years, OFDM has become the technology of choice in the new designs of wireless data communication systems. While there are some commonalities in timing synchronization between OFDM and single-carrier communications systems, OFDM timing synchronization has a number of distinctive properties, which will be studied in this section.

6.6.1   Fundamentals of OFDM Timing Phase Synchronization

Similar to single-carrier communication systems, the timing synchronization in OFDM systems needs to achieve the timing phase as well as frequency synchronization. To begin with, we look at the general principles of the OFDM timing phase synchronization.

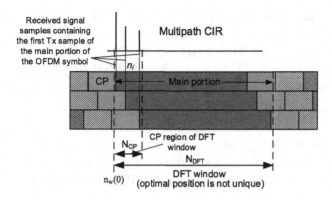

Figure 6.16 Optimal DFT window position

6.6.1.1 OFDM Timing Phase Synchronization – The DFT Window Positioning

With guard carriers included in the OFDM signal design, the sampling of the received signal at the OFDM transmitter channel sample rate f_s satisfies the Nyquist sampling theorem, so there is no information loss. Thus, the timing resolution T_s, which is the OFDM channel sample interval equal to $1/f_s$, is sufficient for optimal timing phase synchronization. In other words, there is no need to optimize the timing phase between the T_s-spaced samples.

Let us first consider the optimal timing phase selection for the CIR that is shorter than the OFDM signal's cyclic prefix (CP). The impact on the receiver performance when this condition is not met will be discussed later. To facilitate the discussion, we first define the following terms:

- DFT window – the position of the segment of received signal samples that are used as the input to the DFT block in an OFDM receiver to recover a transmitted OFDM symbol (Section 1.6);
- CP region of the DFT window – the first N_{CP} samples inside the DFT window;
- Path position (n_i) – the index of the receiver sample generated by the first sample of the main portion of an OFDM symbol through the ith channel path.

The selection of the optimal OFDM timing phase is equivalent to determining the optimal DFT window position. If the CIR is no longer than the signal's CP, with proper DFT window placement, there is no degradation due to ISI. An example of such a DFT window position is shown in Figure 6.16.

As shown in the figure, the optimal DFT window contains a complete period of the main portion of the OFDM symbol received from every path. For this to be true, the path positions must be located inside the CP region of the DFT window as defined above. When the CIR is no longer than the CP, this condition can always be satisfied with the proper placement of the DFT window, and there is no degradation caused by ISI.

Note that the optimal window position is not unique if the CIR is shorter than the CP. If possible, it is desirable to leave some margin before the first and after the last paths inside the CP region of the DFT window. Allocating the margin before the first path, called *first arriving path backoff*, is especially important for the robustness of receiver operations. However, if the CIR is longer than the CP, there will be degradation in receiver performance due to ISI. Such a condition is called *excess delay spread* [28]. Below, we analyze the impact of the excess delay spread on the receiver performance and the proper DFT window placement under such conditions.

6.6.1.2   Effect of Excess Delay Spread

From the discussion above, we conclude that there is no ISI generated by the signal through a path if the path position n_i is inside the CP region of the DFT window, i.e.,

$$n_w(0) \le n_i < n_w(0) + N_{CP} \tag{6.36}$$

When there are paths that do not satisfy (6.36) due to excess delay spread, ISI will occur. As shown in [28], the variance of the interference caused by all of the paths is equal to

$$\sigma_\varepsilon^2 = \sum_i |h_i(n_i)|^2 \left(2\frac{\Delta\varepsilon_i}{N_{FFT}} - \left(\frac{\Delta\varepsilon_i}{N_{FFT}}\right)^2 \right) \tag{6.37}$$

where $|h_i(n_i)|$ is the magnitude of the ith path and

$$\Delta\varepsilon_i = \begin{cases} n_w(0) - n_i & n_i < n_w(0) \\ n_i - n_w(0) - N_{CP} + 1 & n_i \ge n_w(0) + N_{CP} \\ 0 & \text{elsewhere} \end{cases} \tag{6.38}$$

Besides the ISI caused by the excess delay spread, the receiver performance is also degraded due to the loss of energy contributed to the OFDM symbol estimate at the DFT output from the paths that generates ISI. However, the impact of the loss is usually less than that from ISI.

Both types of the impairments are shown graphically in Figure 6.17. Three cases of multipath positions relative to the DFT window position are shown.

The case of $n_i < n_w(0)$ in (6.38) is illustrated by the left-most path a, whose position is ahead of the DFT window. As a result, the front part of the main portion of OFDM symbol n through path a is not included in the DFT window, and its energy does not contribute to the DFT output. In addition, a part of the CP of OFDM symbol $n + 1$ through path a is inside the DFT window and creates ISI. The right-most path c in Figure 6.17 shows the condition of $n_i \ge n_w(0) + N_{CP}$ in (6.38). In this case, the last part of the main portion of OFDM symbol n through path c is outside the DFT window. Thus, its energy does not contribute to the DFT output either. Moreover, the last portion of the OFDM symbol $n - 1$ through path c is inside the DFT window and creates ISI. The two middle paths b_1 and b_2 satisfy (6.36), i.e., they are inside the CP region of the DFT window. Hence, they do not create ISI and cause no energy loss.

Based on (6.37), (6.38), and Figure 6.17, we make the following remarks:

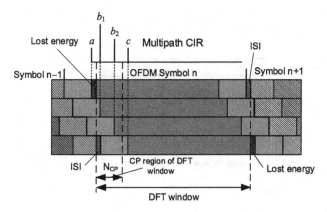

Figure 6.17 Impairments due to excess delay spread

(1) For a path that generates ISI, the interference is proportional to the strength of the path. It is also proportional to the ratio of the length of the portion of the adjacent OFDM symbols inside the DFT window divided by the length of the main portion of the OFDM symbol. Thus, unless the length of the signal portion that causes ISI is long, the interference is relatively limited.

(2) The expression given by (6.37) assumes that the receiver has knowledge of the paths that create ISI, and these paths are included in the channel estimate for performing OFDM signal demodulation. If a path is not included in the estimate, all of the signal energy is lost and becomes interference. Thus, in a sense, the *accuracy of channel estimation is more important than the length of the CP for avoiding ISI.*

(3) For paths *a* and *c*, even though they generate ISI in a similar manner, their estimates are generated differently. Normally, the samples inside the DFT window are used for CIR estimation. Therefore, only the paths inside the DFT window are estimated correctly. Because the DFT/iDFT operations assume that the data are periodic, any path prior to the DFT window is circularly shifted and appears at the end of the channel estimate.

The above remarks are important in deciding the DFT window placement and evaluating the impact on receiver performance. In particular, based on remark (3), it is desirable to leave some margin in front of the CIR estimate that is used for data symbol recovery, i.e., to provide *FAP backoff*. This is because, in real-world operations, it is difficult to tell if a path that appeared at the end of the estimated channel is due to the long channel delay or due to the cyclic shift of the coefficient estimate. Providing the FAP backoff facilitates the timing algorithm to correctly find all of the paths for CIR estimate generation.

6.6.1.3 CIR Estimation for Timing Synchronization

In the previous two subsections, we have presented the general considerations of the DFT window position selection and the effect of the excess delay spread on receiver

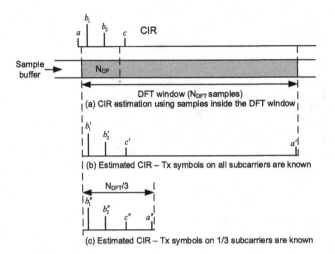

Figure 6.18 Estimated CIRs from known and partially known OFDM symbols

performance. Below, we described the methods of determining the DFT window position based on the CIR estimates. Examples of the practical implementation of these methods can be found in [29].

For setting up the desired DFT window position during initialization or in the data mode, we assume that a rough estimate of the CIR has been obtained, and an initial DFT window position has been decided. To determine a more accurate DFT window position, an FCE is generated from the output of the DFT that is performed using the samples inside the initial DFT window as described in Section 5.5.1. An estimate of the CIR can be obtained by taking the iDFT of the FCE.

There are two scenarios that FCEs and CIRs are generated. In the first case, the data symbols modulated on all of the subcarriers of an entire OFDM symbol are known. The length of the FCE is equal to N_{DFT}. Such a scenario usually occurs during initial acquisition, e.g., in the cases of both the 802.11a/g and LTE. In the second case, only some of the modulation symbols are known to the receiver. For example, assume $1/L_p$ of the modulation symbols in an OFDM symbol are FDM pilots. The length of the FCE is equal to N_{DFT}/L_p. Using the LS estimation given by (5.91) in Section 5.5.1.2, we generate the CIR estimate by taking an iDFT of the FCE. The length of such a generated CIR estimate is equal to the length of the FCE, i.e., N_{DFT}/L_p.

In the case that all of the modulation symbols in an OFDM symbol are known, the CIR estimate generated by taking the iDFT of FCE is shown in Figure 6.18(b). The channel path a before the DFT window shows up at the end of the CIR estimate as a' Since the length of the CIR estimate N_{DFT} is much longer than the possible true CIR length, it is not difficult to tell that a' is most likely a path before the DFT window after the circular shift, rather than an actual path with a long delay. The DFT window should move forward such that it starts before path a so that the CIR can be estimated correctly to properly demodulate the data symbols.

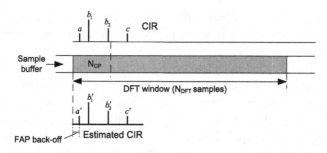

Figure 6.19 An example of desired DFT window placement

Figure 6.18(c) shows the estimated CIR from an OFDM symbol with 1/3 pilot subcarriers. Similar to the previous case in the CIR estimate, path a is circularly shifted to the end and becomes a''. However, as the CIR estimate is not much longer than the actual CIR, it would be difficult to determine if the last path is due to the circular shift of an early path, or it is indeed a path at the end of the CIR estimates. To avoid this problem, one possible solution is to provide the FAP backoff of the CIR estimate. The problem of the misperceived path estimate would not happen as long as any possible newly appeared path is always inside the DFT window due to the FAP backoff.

Based on the above discussion, the criteria for the desired DFT window selection can be summarized as follows.

- All of the significant CIR coefficients are located within the CP region of the DFT window if possible;
- A proper amount of backoff is allocated between the beginning of the DFT window and the estimated FAP;
- If (1) and (2) cannot be satisfied at the same time, the later arriving paths may be located after the CP region of the DFT window. However, it is still desirable to leave some FAP backoff at the beginning of the DFT window.

An example of the channel estimate with a desired DFT window placement based on the above criteria is shown in Figure 6.19.

Adjusting the estimated CIR position can be achieved by moving the DFT window. For example, the desired CIR estimate shown in Figure 6.19 can be obtained by moving the DFT window shown in Figure 6.18(a) backward in time to beyond path a. In other words, the starting point of the DFT window will include additional earlier received signal samples.

6.6.2   Impact of Timing Frequency Offset on OFDM Receiver Performance

Similar to the carrier-frequency offset analyzed in Section 5.5.2, a nonzero-timing frequency offset would shift the frequencies of the OFDM subcarriers, albeit in a different manner. Consequently, perfect orthogonality among subcarriers is not preserved resulting in inter-carrier interference. To determine the accuracy requirement of timing frequency synchronization, we first analyze the impact of the frequency offset on the OFDM receiver performance.

Recall from Section 1.6 that an OFDM symbol of the transmitted signal is the sum of N data symbols modulated on subcarriers at frequencies kf_{sc} with k in the range from 0 to $N_{DFT} - 1$, or, equivalently, from $N_{DFT}/2$ to $N_{DFT}/2 - 1$. Moreover, due to the rectangular windowing effect of the iDFT operation that generates these subcarriers, each of the subcarriers has a spectrum shape of a sinc function centered at kf_{sc}. In real time, the length of the window is equal to $N_{DFT}T_s = 1/f_{sc}$. Therefore, the kth subcarrier has a spectrum in the form of $\mathrm{sinc}(f/f_{sc} - k)$. The spectrum of the baseband transmitter signal is the sum of these sinc functions. The baseband transmitter signal is modulated onto the carrier frequency and transmitted.

At the receiver, the transmitted signal is received and down-converted in frequency to baseband. If both carrier and timing-frequency offsets are equal to zero, the DFT output elements of the received signal are orthogonal to each other. In this section, we consider the degradation caused by timing-frequency offset assuming perfect carrier synchronization.

When the timing-frequency offset is not zero, the receiver sampling frequency is different from the OFDM transmitter channel sampling frequency f_s. Let us denote the actual receiver sample clock frequency by $f'_s \cong 1/T'_s = (1 - \xi)f_s$, where $\xi = \delta f_s/f_s \cong (f_s - f'_s)/f_s$ is the relative timing frequency error. Taking DFT of the input signal sample vector, the kth output is equal to the spectrum of the received signal sampled at $kf'_{sc} \cong kf'_s/N_{DFT} = (1 - \xi)kf_{sc}$. To facilitate the analysis, we will consider the DFT output elements as symmetric around 0. Moreover, only the N_D+1 active subcarriers with frequencies kf'_{sc}, $-N_D/2 \le k \le N_D/2$, not the guard subcarriers, are considered.

Let us first assume that the timing-frequency offset is equal to zero and the noise in the received signal is ignored. Without noise, the kth element of the DFT output $Y(k)$ can be expressed as $H(k)X(k)$, where $H(k)$ is the channel frequency response at the subcarrier kf_{sc}, and $X(k)$ is the kth modulation symbol of the OFDM symbol to be demodulated. Such a case is shown in Figure 6.20(a). However, when a timing-frequency offset exists, the kth DFT output element $Y'(k)$ is the received signal spectrum sampled at kf'_{sc}, as shown in Figure 6.20(b) and (c).

The kth DFT output element can be expressed as

$$Y'(k) = H(kf_{sc} - k\xi f_{sc})X(k)\mathrm{sinc}(k\xi) + \mathrm{ICI}(k) + z'(k) \qquad (6.39)$$

Namely, the magnitude of $X(k)$ is reduced by a factor of $\mathrm{sinc}(k\xi)$ and, in addition to the noise term $z'(k)$, there is an ICI term caused by all of the other subcarriers.

Consider a simple case where the channel has a flat frequency response, i.e., $H(f)$ is constant, and all of the data symbols are uncorrelated with equal energy. Similar to the analysis that leads to (5.96) and (5.97) in Section 5.5.2.1, it can be shown that the signal to ICI ratio (SIciR) of the subcarrier k, denoted by SIciR_k, with the relative frequency offset ξ is equal to

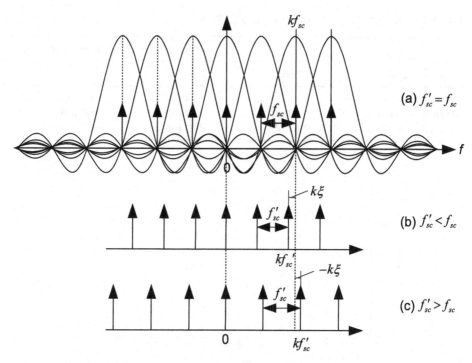

Figure 6.20 Received signal spectrum without and with timing-frequency offset

$$SIciR_k = \frac{S}{ICI}\bigg|_{k,\xi} = \frac{\text{sinc}^2(k\xi)}{\displaystyle\sum_{l=-N_D/2,\, l\neq k}^{N_D/2} \text{sinc}^2((1-\xi)l - k)} \tag{6.40}$$

which is a function of the subcarrier index k, the normalized timing-frequency offset ξ, and the number of total active subcarriers, i.e., N_D+1. As an example, we consider the LTE OFDM signal of 5 MHz channel bandwidth with $\xi = 10^{-4}$. Since it has 300 data subcarriers, $N_D/2 = 150$. The $SIciR_k$ given by (6.40) is plotted in Figure 6.21.

As can be seen from the figure, the $SIciR_k$'s are lower for the subcarriers near the band edges than for the ones at the center, i.e., near DC. This is because the difference between the frequency of the kth DFT output element and the desired subcarrier frequencies becomes larger when the absolute value of the subcarrier index k increases. However, the DFT output elements that have the lowest SIciR are the ones slightly inside the band edges because these elements have more close neighbors contributing to the ICI than the ones at the edges.

OFDM signals with a larger number of active subcarriers are more sensitive to a timing-frequency offset. In Table 6.1, we summarize the worst-case subcarrier SIciR for a signal with 300, 600, 1200 data subcarriers at normalized timing-frequency offsets of 10^{-4}, 10^{-5}, 10^{-6}, and 10^{-7}, i.e., at 100, 10, 1, and 0.1 ppm.

Table 6.1 Worst-Case Subcarrier Signal to ICI Ratio (SIciR)

Number of data subcarriers	Normalized timing-frequency offset			
	$\xi=10^{-4}$	$\xi=10^{-5}$	$\xi=10^{-6}$	$\xi=10^{-7}$
300 (LTE 5 MHz)	31.9 (dB)	51.9 (dB)	71.9 (dB)	91.9 (dB)
600 (LTE 10 MHz)	25.7 (dB)	45.7 (dB)	65.7 (dB)	85.7 (dB)
1200 (LTE 20 MHz)	19.5 (dB)	39.6 (dB)	59.6 (dB)	79.6 (dB)

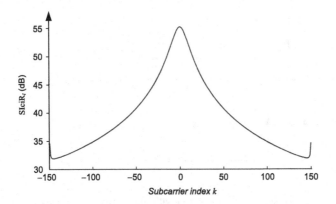

Figure 6.21 Signal to ICI ratio distribution

From the table, we observe that when the number of data carriers increases by a factor of two, the worst-case subcarrier SIciR decreases by approximately 6 dB. However, when the normalized frequency offset decreases by a factor of 10, the worst-case SIciR increases by 20 Db. Overall, OFDM receiver should not be very sensitive to the timing-frequency offset. For example, as shown in Table 6.1, a frequency offset that is less than 10 ppm, which is easy to achieve in practice, is adequate for an LTE 20 MHz channel receiver without significant performance degradation. Moreover, since what we have been considering is the worst-case subcarrier SIciR, the average SIciR would be higher at the same frequency offset and has even less impact on the receiver performance.

Before concluding this section, we would like to point out that the results given in this section were originally presented in [30] with somewhat different but equivalent proof. Readers may find other interesting materials and examples there.

6.6.3 Timing Initialization

Similar to other digital communication systems, OFDM timing synchronization can be divided into two stages: the timing initialization stage and the timing tracking stage. In this section, we discuss the methods of performing timing initialization.

In the initialization stage, the receiver determines the OFDM signal timing, i.e., the proper DFT window position, to an accuracy of within a few channel sample intervals. With such a timing accuracy, most of the OFDM receiver operations can begin. However, the accuracy after initialization may not suffice for the receiver to achieve the best possible performance. Moreover, during the data mode operation, the timing phase may change due to channel condition changes and/or due to a residual timing-frequency offset. Timing phase tracking and refinement are necessary and will be treated in Section 6.6.4.

Timing initialization of OFDM receivers can be performed either based on known OFDM symbols, e.g., TDM pilots, or with unknown data. We first consider timing initialization based on known OFDM symbols.

6.6.3.1 Known OFDM Symbol-Based Timing Initialization

In many packet-data communication systems, special symbols known to the receiver are sent at the beginning of the packet transmission to facilitate the initial system acquisition by devices. These symbols can also be used for initial timing estimation.

During system acquisition, known data symbol sequences, which are the time-domain representation of the known OFDM symbols, are correlated with received signal samples. Detected correlation peaks not only are an indication that the samples are generated by the known data symbols but they also constitute the estimates of the CIR. As shown in Section 6.6.1.1, an initial DFT window position can be determined based on the estimated CIR. Below, as examples, we describe the timing initialization processes for 802.11a/g Wi-Fi system and LTE cellular system. Since many aspects of determining initial timing have been discussed in Chapter 3, we will only briefly review the timing initialization process and elaborate when necessary.

Timing Initialization of 802.11a/g
As has been discussed in Sections 3.6.2 and 5.5.2, 10 short preamble symbols, each of which contains 16 OFDM channel samples, are transmitted at the beginning of each 802.11a/g data packet. The detection of the short preamble symbols provides a rough timing estimate. Following the short preambles are two long preamble symbols, each of which contains 64 OFDM channel samples, together with a 32-sample-long CP. A 64-sample-long CIR estimate can be generated from the samples in the likely location of the long preamble symbols based on the timing determined from the short preamble. Since the actual channel span is usually much shorter than 64 sample intervals, the CIR estimate can accurately determine the DFT window. The details of the operations have been presented in Section 3.6.2.

Timing Initialization of LTE
Recall from Section 3.6.1 that during initial system acquisition, a UE receiver first detects PSS and then SSS. Once the SSS is detected, its data symbols modulated on the subcarriers become known to the receiver. An FCE is obtained by performing element-by-element multiplication of the complex conjugate of these symbols and

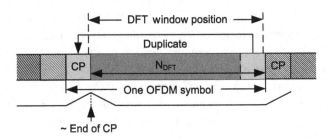

Figure 6.22 CP correlation-based timing estimation

corresponding DFT output. As shown in Section 3.6.1.4, a 64-point iDFT of the FCE results in a CIR estimate, which is used to generate an initial timing estimate.

The initial estimate may not be precise enough for a UE to perform data demodulation. This is because the PSS and SSS have a bandwidth of merely 1.4 MHz, which is often much narrower than the OFDM signals for data transmission. Such a rough initial timing estimate is used by the UE to extract the wideband FDM pilots to generate more accurate CIR estimates and more precise timing.

6.6.3.2   CP-Based Delay-and-Correlation for Initial Timing Estimation

In certain reception scenarios, it may be desirable or necessary to perform initial timing estimation without the knowledge of transmitted data symbols. An effective method in such cases is to perform the CP-based delay and correlation. This approach is similar to the CP-based carrier-frequency offset estimation described in Section 5.5.2.3.

Recall that a CP is a duplication of the last portion of the OFDM symbol. The timing phase detection can be performed by delaying a sequence of N_{cp} samples and correlating with a sample sequence that arrives $N_{DFT}T_s$ later, such that

$$D(n) = \left| \sum_{l=0}^{N_{CP}-1} y^*(n+l) \times y(n+l+N_{DFT}) \right| \tag{6.41}$$

The OFDM symbol timing estimate is determined by detecting $\hat{n} = \arg \max_n |D(n)|$. If the channel has only a single path, the correlator output would have a triangle shape as shown in Figure 6.22. The peak of the output indicates that all of the samples $y(n+l)$ and $y(n+l+N)$ in the correlation are the same up to a phase rotation. In other words, the samples $y(n+l)$'s belong to the CP of an OFDM symbol and $y(n+l+N)$'s belong to the last portion of the same OFDM symbol, where $l = 0, \ldots, N_{DFT}-1$, in the two sequences of samples. The DFT window should start from the first sample after the CP portion of the samples yields the correlation peak.

For a multipath channel, the shape of the magnitude distribution of the DFT output elements depends on the shape of the CIR. As a result, the estimated timing determined by the correlation peak may not be accurate enough for performing OFDM signal demodulation, and further refinement will be needed. However, a rough initial timing estimate is still useful for reducing the processing requirement. Below we provide two examples of the possible usages of such a rough estimate.

DVB-T

In some OFDM communication systems such as DVB-T for terrestrial TV broadcasting, signals are transmitted continuously. There are no special OFDM symbols known to the receiver that are embedded in the transmitted signal. However, a rough DFT window position can be established by using the timing phase information obtained from the CP-correlation-based timing estimation. The FDM pilot subcarriers in the DVB-T signals can be identified from the DFT output as shown in Section 5.5.2.3. Once the positions of the pilot subcarriers are determined, the elements of the DFT output corresponding to the pilot subcarriers can be used to generate CIR estimates. More accurate timing phase information can be derived from the CIR estimates.

LTE

As discussed in Sections 3.6.1, 5.8.4, and the previous section, LTE UE initial acquisition needs to first detect PSS and SSS. Since the PSS and SSS are located in the last two OFDM symbols of every 5-ms half-slot, without knowing the OFDM symbol boundary, the UE needs to search every T_s offset of the entire 5 ms time duration. For a sampling rate of 1.92×10^6 samples/second, $1920 \times 10^6 \times 5 \times 10^{-3} = 9600$ hypothesis tests need to be performed every 5 ms. If a rough timing with an accuracy of plus or minus a few samples is determined beforehand by using the CP-correlation-based method, the number of search operations can be reduced by at least a factor of 10.

6.6.4   Data Mode Timing Tracking

After entering the data mode, the device receiver uses the timing estimated during initial system or data packet acquisition to perform data symbol recovery. However, unless the data packets are very short, it would be necessary to change the DFT window position if time-domain CIRs change during the data mode. We call such adjustments *data mode timing tracking*.

The desired receiver timing is determined based on time-domain CIR estimates. In most OFDM systems, FDM pilots are inserted for generating FCEs in order to perform data symbol recovery. For timing tracking, the estimates of CIRs are generated by taking the iDFT of the FCEs as discussed in Section 5.5.1.2. To reduce the overhead, one FDM pilot subcarrier is inserted into every few, say L_p, data subcarriers. The length of the CIR estimate so generated is roughly equal to $1/L_p$ of the OFDM symbol length.

For example, in the LTE downlink signal, 1/6 of the subcarriers are FDM pilots in an OFDM symbol with pilots. These pilot subcarriers are interleaved between two adjacent OFDM symbols with FDM pilots as shown in Figure 5.25. The length of the CIR estimate generated from the pilots in one OFDM symbol is equal to 1/6 of the length of an OFDM symbol. For slow time-varying channels, the channel estimates from adjacent OFDM symbols that contain FDM pilots can be combined. The length of the CIR estimates after combining is equal to 1/3 of the OFDM symbol length, as shown in Section 5.5.1.2.

During the normal data mode operations, for the DFT window position that is adequate for performing data symbol recovery, the estimated CIR should satisfy the conditions described in Section 6.6.1 and shown in Figure 6.16. Namely, the FAP estimate should be inside the CP region of the DFT window with some backoff. In addition, every CIR estimate generated is compared with the desired position. If the FAP position changes, the DFT window position may need to be adjusted to maintain the proper FAP backoff to ensure robust receiver operation.

The data mode timing tracking is a relatively easy task if the channel is stable such as in regular Wi-Fi and cellular systems. The most challenging environment for the data mode tracking is OFDM communication over single-frequency networks (SFNs), as in multicasting/broadcasting systems/modes. In such cases, a new path from another transmitter may show up before the FAP of the current CIR estimate if the FAP backoff is too small. If this new path appears at the end of the channel estimate, it is difficult to handle as has been discussed earlier. However, maintaining a large FAP backoff is not a good solution either, because it reduces the receiver's ability to fully utilize the CP to combat ISI. It is advisable to devise an optimal strategy for handling such cases based on real-world channel measurements.

6.6.5   A Baseline Design Example

In this section, we provide an example of the implementation of an OFDM timing synchronization block to explain the design principles. An XO-based all-digital realization is shown for illustration of practical considerations.

Let us consider a device for receiving the LTE downlink signal with a 20 MHz channel bandwidth. According to the LTE specification, the OFDM signal consists of 1200 subcarriers. The distribution of the pilot subcarrier in OFDM symbols is as shown in Figure 5.25. The receiver frequency and timing synchronization is implemented based on an all-digital architecture. We also assume that the clock reference of the receiver is generated from an XO-based local oscillator with a 5 ppm frequency accuracy.

In this design, the sampling clock is directly derived from the XO output. From (6.40) it can be shown that for the given XO accuracy and the number of subcarriers, the worst-case subcarrier SIciR is equal to 45.6 dB. Thus, from the UE demodulation performance point of view, no timing-frequency offset compensation is needed. The resolution of the timing phase adjustment is one channel-sample time interval T_s. The block diagram of such a basic OFDM timing synchronization implementation is shown in Figure 6.23.

The OFDM receiver timing synchronization block shown operates as follows. After receiver filtering, the signal is sampled at the nominal timing clock rate derived from the local XO, and N_{DFT} of these samples are stored into a data buffer. A DFT operation is performed on these samples to yield estimates of frequency-domain modulation symbols of the OFDM symbol.

If an OFDM symbol contains FDM pilot subcarriers, the DFT output elements that correspond to these subcarriers are extracted and used to generate an FCE. The estimate

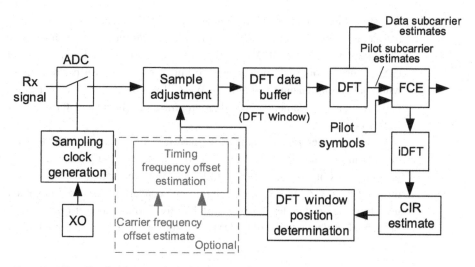

Figure 6.23 Baseline implementation of OFDM timing synchronization

of the timing domain CIR is obtained by performing the iDFT of the FCE. The details of these operations have been discussed in Sections 1.5.3 and 5.5.1.1.

As discussed in Section 6.6.1.1, the desired DFT window position can be determined from the CIR estimate. Normally, at every OFDM symbol interval, the data buffer receives $N_{DFT}+N_{CP}$ new samples and stores the last N_{DFT} samples for processing. However, if the DFT window needs to be adjusted due to the position change of the CIR estimate, the data buffer could receive more or less new samples of the next symbol, whenever such an adjustment occurs. For example, if the DFT window moves backward or forward by K samples relative to the previous window, the buffer receives $N_{DFT}+N_{CP}-K$ or $N_{DFT}+N_{CP}+K$ new samples. Hysteresis thresholds should be used for the DFT window adjustment to avoid random changes due to noise and interference.

If a frequency offset between receiver and transmitter sampling clocks exists, timing adjustments occur regularly even if the position of the actual CIR does not change. For example, if there is a frequency offset of 5 ppm, for every 10^6 transmitted signal samples, the receiver generates 10^6-5 receiver samples. For an LTE signal with a 20 MHz channel bandwidth and normal CP, $N_{CP}+N_{DFT} \approx 2194$ on average. As a result, there is one sample short and the timing needs to be adjusted once for every 91 received OFDM symbols. However, under such conditions, the receiver performance would not be noticeably affected according to (6.40) if the DFT window position is properly adjusted.

The sample adjustments are determined based on the detected CIR position changes as shown above. Alternatively, the timing-frequency offset can be computed based on the estimated carrier-frequency offset and/or from the regular shifts of the estimated CIR positions. The regular adjustments in DFT window positions in small steps, e.g., one sample at a time, can be performed based on the estimated timing-frequency offset. Such an approach can reduce the chance of large adjustments and makes the receiver timing change smoothly.

It should be noted that the implementation described above is only for illustrating the basic operations of timing synchronization in OFDM receivers. Such an implementation should work well for OFDM reception under normal conditions. However, various enhancements could be implemented to improve the receiver performance and to deal with challenging channel conditions. We will not discuss such enhancements here, as they are highly implementation and environment dependent and, thus, beyond the scope of this book.

In the aforementioned example, we have assumed that the ADC that generates the digital samples have a clock at the OFDM nominal sampling frequency $1/T_s$. However, in some cases, it may be desirable to let the ADCs operate at a clock frequency that is different from $1/T_s$. In such cases, digital rate converters discussed in Chapter 7 can be used to generate samples at the rate of $1/T_s$ for normal receiver operation. An example of such an implementation of OFDM synchronization will be presented in Section 7.6.3.

6.7   Summary and Remarks

Timing synchronization is another key receiver function that has significant impact on system performance. Timing synchronization is usually implemented by processing baseband signals and was mostly implemented by analog domain processing in the early days. However, in modern communication systems, the processing is almost exclusively done through the processing of digital baseband signal samples.

In this chapter, ML and approximate ML timing phase estimation for data-assisted and non-data-assisted processing was first introduced to form the foundation of optimal timing synchronization. General considerations of design and implementation of timing synchronization in communication systems using various single-carrier and multi-carrier technologies were then discussed.

After introducing the basics, typical algorithms that implement timing synchronization in single-carrier communication systems were described. Three classic non-data-assisted timing synchronization algorithms: the squarer detector, Early-Late Gate timing detector, and Gardner's algorithm, were presented. These algorithms are all essentially based on the squarer operations and, thus, are approximate forms of non-data-assisted ML timing estimation. Timing synchronization algorithms using known data symbols, including the Mueller–Müller algorithm, the data-assisted Early-Late Gate timing phase detector, and algorithms based on estimated channel/equalizer coefficients were presented.

A significant portion of this chapter was devoted to timing synchronization for DS-CDMA- and OFDM-based digital communications. Various aspects of designs and implementations of timing synchronization for systems employing these technologies were described.

In digital communications, timing synchronization is the receiver function that is most directly affected by channel conditions. Hence, for communications over time-varying channels, the receiver performance, especially its robustness, highly depends on the performance of timing synchronization.

A few noteworthy remarks are given below as a summary of this chapter.

- If the receiver filter is matched to the received signal, the optimal timing phase estimate in the maximum likelihood sense is the sampling delay that maximizes the matched filter output. Such an optimal receiver front end is the same as the likelihood receiver front ends for initial signal detection, carrier phase detection, and equalization.

- To obtain an exact ML timing estimate, the receiver must know both the transmitter data symbols and the parameters of the communication channel. In practice, if only the transmitted data symbols are known, an approximate ML estimate can be obtained by a receiver with a front end that matches the transmitter filter or to the average of the channels.

- If the transmitted data symbols are unknown, it was shown that an approximate ML timing estimate could be generated by using a squarer-based detector. Namely, the desired timing is attained if the squared magnitude of the received signal samples reaches the maximum.

- The non-data-assisted timing-phase detection algorithms presented in Section 6.3 are variations derived from the squarer timing phase detector.

- Data-assisted timing algorithms utilizing the ML or approximate ML receiver filter outputs can achieve better performance than the non-data-assisted algorithms. All of these algorithms are originally designed to obtain the optimal, or nearly optimal, timing phase for ISI-free, or almost ISI-free, channels.

- For severe ISI channels, these classic timing algorithms are often used in conjunction with equalizers. In most cases, these algorithms would lock to a stable timing phase, which may not be optimal for symbol rate equalizers.

- The performance of SREs highly depends on the input signals' sampling-time offset. No known methods that can effectively determine the optimal timing phase of SREs are available for arbitrary ISI channels.

- The performance of fraction spaced equalizers is insensitive to the sampling delay. For receivers employing FSEs, the task of timing synchronization is to maintain a stable timing phase.

- RAKE receivers with multiple RAKE fingers are commonly employed in DS-CDMA receivers. The signal received by a RAKE finger can be viewed as passing through a single path channel. The objective of the DS-CDMA timing synchronization is to determine and track the RAKE fingers' sampling delays. When using the embedded pilot signal in the data, the timing phase detection is approximately ML.

- Another crucial CDMA receiver task related to timing synchronization is *finger management* described in Section 6.5.3. Under dynamic channel conditions, finger management is probably the most important and challenging function of CDMA receivers.

- To perform timing synchronization in OFDM receivers is equivalent to determining the desirable DFT window position. Since the sampling rate in OFDM receivers satisfies the Nyquist sampling rate, the resolution of timing adjustment is equal to the sample time interval T_s. No inter-sample interpolation is needed in most practical cases.

- If the channel delay spread is less than the CP length of the OFDM signal, there is no ISI with optimal DFT window position.

- Channels with delays larger than CP length, called *excess delay spread*, introduce ISI and cause signal energy loss, which degrades receiver performance. However, the amount of interference is proportional only to the amount of excess delay spread relative to the OFDM symbol length, as long as the CIR of the entire channel can be estimated correctly.

- It is desirable to leave some margin before the first arriving path inside the DFT windows, i.e., FAP backoff, to improve the robustness of the receiver operation.

- The ICI energy of OFDM signals due to the timing-frequency offset can be expressed as a function of the offset. The receiver performance would not be significantly impacted if the sampling-frequency offset is not excessive.

- In any digital communication system, receiver timing synchronization is usually implemented in the form of a first- or second-order PLL.

References

[1] A. J. Viterbi, *CDMA Principle of Spread Spectrum Communications*, Boston: Addison-Wesley, 1995.

[2] U. Mengali and A. N. D'Andrea, *Synchronization Techniques for Digital Receivers*, New York: Plenum Press, 1997.

[3] J. G. Proakis and M. Salehi, *Digital Communications*, 5th edn, New York: McGraw-Hill, 2008.

[4] R. D. Gitlin, J. F. Hayes, and S. B. Weinstein, *Data Communications Principles*, New York: Plenum Press, 1992.

[5] R. D. Gitlin and S. B. Weinstein, "Fractionally-Spaced Equalization: An Improved Digital Tranversal Equalizer," *Bell System Technical Journal*, vol. 60, no. 2, pp. 275–96, 1981.

[6] L. E. Franks, "Carrier and Bit Synchronization in Data Communication – A Tutorial Review," *IEEE Transactions on Communications*, Vols. COMM-28, no. 8, pp. 1107–21, August 1980.

[7] F. M. Gardner, "A BPSK/QPSK Timing-Error Detector for Sampled Receivers," *IEEE Transactions on Communications*, vol. 34, no. 5, pp. 423–9, 1986.

[8] A. N. D'Andrea and M. Luise, "Optimization of Symbol Timing Recovery of QAM Data Demodulators," *IEEE Transactions on Communications*, vols. COMM-44, no. 3, pp. 399–406, 1996.

[9] W. G. Cowley, "The Performance of Two Symbol Timing Recovery Algorithm SPSK Demodulators," *IEEE Transactions on Communications*, Vols. COMM-42, no. 6, pp. 2345–55, 1994.

[10] L. Frank, "Synchronization Subsystems: Analysis and Design," in *Digital Communications: Satellite/Earth Station Engineering*, K. Feher (ed.), Englewood Cliffs, NJ: Prentice Hall, 1983.

[11] K. H. Mueller and M. Müller, "Timing Recovery in Digital Synchronous Data Receivers," *IEEE Transactions on Communications*, vols. COMM-24, no. 5, pp. 516–31, 1976.

[12] S. U. H. Qureshi and G. D. Forney, Jr., "Performance and Properties of a T/2 Equalizer," in *National Telecommunications Conference Record*, 1977.

[13] J. E. Mazo, "Optimum Timing Phase for an Infinite Equalizer," *Bell System Technical Journal*, vol. 54, no. 1, pp. 189–201, 1975.

[14] Y. Kakura and T. Ohsawa, "Automatic equalizer capable of surely selecting a suitable sample timing a method for generating sampling clock used for the sample timing and a recording medium usable in control of the automatic equalizer," US Patent 6,314,133 B1, November 6, 2001.

[15] G. Mergen, P. Subrahmanya, N. Kgsturis, and R. Sundaresan, "Equalizer in a wireless communication system," US Patent 8,160,128 B2, April 17, 2012.

[16] S. Haar, D. Daecke, R. Zukunft, and T. Magesacher, "Equalizer-Based Symbol-Rate Timing Recovery for Digital Subscriber Line Systems," in Proceedings of Globecom'02, Taipei, Taiwan, 2002.

[17] G. T. Davis and B. D. Mandalia, "Tap Rotation in Fractionally Spaced Equalizer to Compensate for Drift due to Fixed Sample Rate". US Patent US 4899366 A, February 6, 1990.

[18] R. G. Gitling, H. C. Meadors and S. B. Weinstein, "The Tap-Leakage Algorithm: An Algorithm for the Stable Operation of a Digitally Implemented, Fractionally Spaced Adaptive Equalizer," *Bell System Technical Journal*, vol. 61, no. 8, pp. 1817–39, 1982.

[19] European Telecommunications Standards Institute SMG, "Digital cellular telecommunications system (Phase 2+); Physical layer on the radio path; General description (GSM 05.01 version 5.4.0)," European Telecommunications Standards Institute, Sophia Antipolis Cedex, France, 1998.

[20] ETSI TC-SMG, "Digital cellular telecommunications system (Phase 2+); Multiplexing and multiple access on the radio path (GSM 05.02)," European Telecommunications Standards Institute, Sophia Antipolis Cedex, France, 1996.

[21] M. Drutarovsky, "GSM Channel Equalization Algorithms - Modern DSP Coprocessor Approach," *Radioengineering*, vol. 8, no. 4, pp. 26–31, 1999.

[22] G. Ungerboeck, "Adaptive Maximum-Likelihood Receiver for Carrier-Modulated Data-Transmission Systems," *IEEE Transactions on Communications*, vol. 22, no. 5, pp. 624–36, 1974.

[23] R. Price and P. E. Green, Jr., "A Communication Technique for Multipath Channels," *Proceedings of the IRE*, vol. 46, pp. 555–70, March 1958.

[24] J. S. Lee and L. E. Miller, *CDMA System Engineering Handbook*, Norwood, MA: Artech House, 1998.

[25] H. P. Hartmann, "Analysis of a Dither Loop for PN Code Tracking," *IEEE Transactions on Aerospace and Electronic Systems*, vols. AES-10, no. 1, pp. 2–9, 1974.

[26] Z. Ghadialy, "A Look at Working of Soft Handover (SHO)," March 25, 2005. [Online]. Available: www.3g4g.co.uk/Tutorial/ZG/zg_sho.html. [Accessed June 16, 2015].

[27] G. E. Bottomley, T. Ottosson, and Y. P.-E. Wang, "A Generalized RAKE Receiver for Interference Suppression," *IEEE Journal on Selected Area in Communications*, vol. 18, no. 8, pp. 1536–45, 2000.

[28] S. Speth, S. A. Fechtel, G. Fock, and H. Meyr, "Optimum Receiver Design for Wireless Broadband Systems Using OFDM. I," *IEEE Transactions on Communications*, vol. 47, no. 11, pp. 1668–77, November 1999.

[29] B. Vrcelj, F. Ling, and M. M. Wang, "Methods and apparatus for determining timing in a wireless communication system," US Patent US 7,623,607, November 24, 2009.

[30] J. H. Stott, *The Effect of Frequency Errors in OFDM*, BBC Research Department, Report No RD 1995/15, London, January 1995.

7 Timing Control with Digital Resampling

7.1 Introduction and History of Digital Timing Control

As discussed in Section 6.2, the receiver filter output $y(t)$ is sampled at $kT + \hat{\tau}$ to generate digital samples that are used by other receiver functions for recovering the transmitted data. The desired sample timing phase $\hat{\tau}$ is determined by a timing locked loop in the receiver timing synchronization block. When the TPC in a TLL is implemented in the analog domain, the sampling time is controlled by a sampling clock generator. If the desired timing phase $\hat{\tau}$ needs to change, the phase of the sampling clock is adjusted by the VCO of the TLL.

Since the early 1970s, the pioneering work on digital interpolation by Schafer, Rabiner, Crochiere, Bellanger, and others [1, 2, 3, 4] established the foundation for applying digital signal processing techniques to all-digital implementations of timing synchronization. In particular, the textbook by Crochiere and Rabiner [5] educated many engineers for such implementations in practical communication systems. In the mid-1980s, due to the advance in digital signal processing technologies, implementations of digital timing synchronization finally became feasible, and the migration from analog processing to digital processing started.

In early 1985 *digital resampling* technology was proposed at Codex Corporation, a subsidiary of Motorola, for receiver timing synchronization in V.32 wireline echo-cancellation modems. Based on a project memo authored by Dr. Shahid Qureshi, Senior Director of the R&AD Department at Codex at that time, an all-digital timing synchronization using digital resampling was successfully developed and implemented in the Codex V.32 wireline modem. To the author's knowledge, this modem was one of the earliest commercial modem products, if not the earliest one, that employed timing synchronization using the digital resampling technology. Later on, all-digital timing synchronization was adopted in Codex V.33, V.34, and V.56 modems. A paper that describes the digital timing synchronization technique based on this implementation was published in 1993 [6].

Figure 7.1(a) and (b) show the extracts from the original memo. A copy of the complete set of related memos, which contain design and implementation details, can be found online [7].

Another paper about a technique of digital resampling for timing synchronization in echo-cancellation modems based on the work at AT&T Bell Labs was published in 1988 [8].

(a)

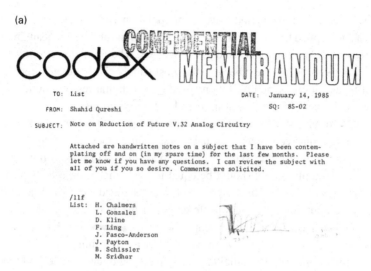

Figure 7.1 (a) Title page of Qureshi's Codex memo on digital timing control

(b)

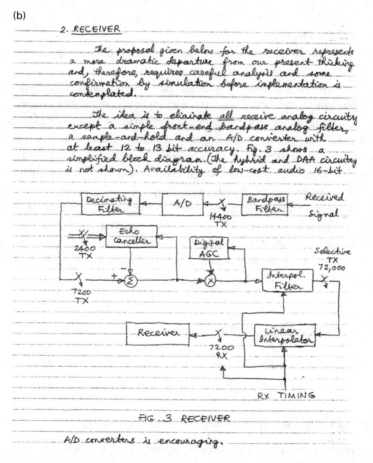

Figure 7.1 (b) Description of digital timing control in Qureshi's Codex memo

During the same time period, digital interpolation techniques were independently proposed and applied to receiver timing synchronization for the European Space Agency projects by Dr. Floyd Gardener and his colleagues. The journal papers on this subject published in 1993 [9, 10] have become the most widely cited references on this topic.

As of today, timing synchronization based on the digital resampling technology has become the de facto form of its designs and implementations.

All-digital timing synchronization is implemented in two steps. In the first step, a digital sample sequence is generated using a sampling clock that is usually asynchronous to the remote transmitter symbol clock. To avoid loss of information, the sampling frequency, or sampling rate, should be equal to, or higher than, the Nyquist sampling rate, of the signal [11, 12]. The digital samples so generated would have arbitrary timing phases and a sampling frequency different from the remote transmitter symbol frequency. In the second step, this sample sequence is converted to another sample sequence by digital resampling and used by other receiver functions.

The samples generated from digital resampling should have the desired timing phase and frequency, which is synchronous to the remote transmitter symbol clock. As the sample rates before and after the resampling are different, the resampling process performs *digital rate conversion* [5], which is the basis of digital timing synchronization discussed in this chapter.

Examples of the digital resampling applications for timing synchronization include receiver sample generation in echo-cancellation modems and in receivers with local frequency reference generated by XOs that have fixed/unadjustable clock frequencies. This chapter covers the theory, analysis, and implementation of digital resampling/rate conversion and its applications to timing synchronization. It is organized as follows.

Section 7.2 presents the fundamentals of digital resampling. The original and a popular form of their implementations utilizing the polyphase FIR filter bank structure is presented in Section 7.3. The design and implementation of digital interpolation filters, a key component of digital resampling, is covered in Section 7.4. Additional digital resampling techniques are presented in Section 7.5. Section 7.6 discusses the details of the implementations of timing synchronization using digital resampling. Such implementations are further illustrated by examples in practical digital communication systems in Section 7.7. Finally, summary and key remarks are given in Section 7.8 to conclude this chapter.

7.2   Basics of Digital Resampling

In this section, we present the basics of digital resampling and discuss some of its applications. A summary of the terms related to digital resampling will also be given.

7.2.1   Nyquist Sampling Theorem and Analog Signal Reconstruction

Let us consider an infinitely long sample sequence $\{x_k\}$, k $= -\infty\ldots, n, \ldots, \infty$, which is generated by uniformly sampling an analog signal $x(t)$ with spectrum $X(f)$ at the sampling rate $f_s = 1/T_s^1$, i.e., $x_k = x(kT_s)$. The objective of digital resampling is to generate one or more digital samples $y_n = x(nT_s+\tau_n)$ from the sequence $\{x_k\}$, where τ_n can be greater or less than zero and its value may be a function of n.

According to the *Nyquist sampling theorem* [12], the original analog signal $x(t)$ can be reconstructed from the sample sequence $\{x_k\}$ without distortion as long as the sampling rate f_s is at least twice the bandwidth of $x(t)^2$ [11]. Such a sampling rate is called the *Nyquist sampling rate* of the analog signal $x(t)$. In other words, if this condition is satisfied, there is no information loss in the sampling process. The original signal can be reconstructed from the sample sequence $\{x_k\}$ such that

$$x(t) = \sum_{k=-\infty}^{\infty} x_k \frac{\sin\left[(\pi/T_s)(t-kT_s)\right]}{(\pi/T_s)(t-kT_s)} = \sum_{k=-\infty}^{\infty} x_k \mathrm{sinc}(t/T_s - k) \qquad (7.1)$$

The equation given by (7.1) is called the *interpolation formula* for the reconstruction of the original analog signal from the signal's samples [11]. Conceptually, it can be understood as follows.

The summation on the right side of (7.1) is the convolution of the sample sequence $\{x_k\}$, which is in the form of an infinitely long impulse train, i.e., $\sum_{k=-\infty}^{\infty} x(kT_s)\delta(t-kT_s)$, and the continuous time function $\mathrm{sinc}(t/T_s)$. The spectrum of the signal generated by the convolution is the product of the spectrum of $\{x_k\}$ and the frequency response of $\mathrm{sinc}(t/T_s)$. Since f_s meets the requirement of the Nyquist sampling rate of $x(t)$, $X(f)$ is equal to zero if $|f| \geq f_s/2$. As known from discrete-time linear system theory, the spectrum of the discrete sequence $\{x_k\}$ is the periodic replications of $X(f)$ with the period equal to f_s. If f_s satisfies the Nyquist sampling rate of $x(t)$, there is no overlap between any two repeated $X(f)$'s. Moreover, the Fourier transform of $\mathrm{sinc}(t/T_s)$ is a rectangular function with the magnitude of 1 from $-0.5f_s$ to $0.5f_s$ and zero elsewhere. The spectrum of the summation in (7.1) is the product of the rectangular function and the spectrum of the sample sequence $\{x_k\}$ and is equal to $X(f)$. As a result, the inverse Fourier transform of this spectrum is identical to $x(t)$. The process of sampling and signal reconstruction is graphically shown in Figure 7.2.

7.2.2 Digital Resampling and Rate Conversion

From (7.1), it can be seen that the sample y_n, which is $x(t)$ sampled with an arbitrary delay τ_n relative to x_n, can be expressed as

$$y_n = x(nT_s - \tau_n) = \sum_{k=-\infty}^{\infty} x_k \mathrm{sinc}(n - k - \tau_n/T_s) \qquad (7.2)$$

---

[1] The Terms T_s and f_s used in this chapter denote the sampling time interval and frequency in general. They should not be confused with the OFDM channel sample interval and frequency.

[2] The analog signal $x(t)$ referred here is a real signal. Only its bandwidth in the positive frequency region is considered. For complex signals, which consist of real and imaginary components, the sampling rate should be at least equal to the signal's total bandwidth in both positive and negative frequency regions.

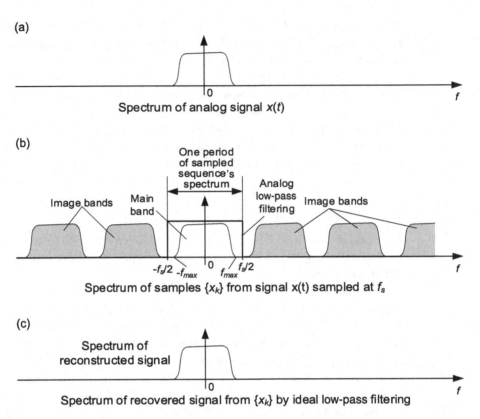

(a)

Spectrum of analog signal x(t)

(b)

One period
of sampled
sequence's
spectrum

Image bands   Main
band

Analog
low-pass
filtering   Image bands

$-f_s/2$ $-f_{max}$ 0 f_{max} $f_s/2$ f

Spectrum of samples $\{x_k\}$ from signal x(t) sampled at f_s

(c)

Spectrum of
reconstructed signal

0   f

Spectrum of recovered signal from $\{x_k\}$ by ideal low-pass filtering

Figure 7.2 Signal reconstruction from digital samples

Equation (7.2) shows that the sample y_n is a linear combination of x_k, $k = -\infty \ldots$, $n, \ldots, \infty$, with coefficients $\text{sinc}(n-k-\tau_n/T_s)$. In the case that τ_n is a constant τ independent of n, the samples y_n can be viewed as the output of the sample sequence $\{x_k\}$ filtered by a discrete-time digital filter with coefficients of $\text{sinc}(m-\tau/T_s) \cong \text{sinc}$ $(m-D)$, where $D = -\tau/T_s$ is the normalized delay. Such a discrete-time filter is called an *ideal delayer*, which has the frequency response of $e^{-j\omega D}$. The operation specified by (7.2) for performing resampling is commonly called *digital interpolation* in the literature [5].

Digital interpolation can be used in a few different ways. If y_n's are generated every T_s according to (7.2), and with τ_n/T_s being a constant, the output is a sequence with a fixed timing-phase shift of the input sample sequence without the change of the sampling rate. Alternatively, the output samples may also be generated with sampling phase change $\Delta\tau$ from one sample to the next, i.e., $\tau_{n+1} = \tau_n + \Delta\tau$, in addition to nominal phase increment T_s. The resulting sample sequence has a sampling rate equal to $f_s' = f_s/(1 + \Delta\tau/T_s) \cong f_s/(1 + \delta\tilde{\tau})$, where $\delta\tilde{\tau} = \Delta\tau/T_s$ is the normalized delay increment per sample. In this case, performing digital interpolation implements a *digital rate converter* [5]. Note that f_s' can be larger or smaller than f_s.

The digital rate converter changes the sampling time by $T_s + \Delta\tau$ from sample to sample. Thus, its function is equivalent to that of a VCO controlled analog sampler. Such a rate converter can be used for implementing timing synchronization, instead of a voltage or digital/numerical controlled oscillator, to achieve the timing phase and frequency control. It is one of the key components in all-digital implementations of timing synchronization.

Even though the principles of digital resampling and rate conversion are straightforward, their implementation can take different forms and could be rather involved. In the ensuing sections, we will consider the details of a few types of digital resamplers that are widely used in practical implementations of timing synchronization.

The interpolation filter used in (7.1) for performing signal reconstruction is not unique if f_{max}, the maximum frequency components in $x(t)$, is less than $f_s/2$. Any filter with a frequency response that is equal to a constant, e.g., 1, for $|f| \leq f_{max}$ and zero for $|f| > f_s/2 - f_{max}$ can perform ideal signal reconstruction. In other words, the interpolation filter can be any low-pass filter that satisfies these conditions and is not necessary in the form of the sampled sinc function. Below we will simply assume that the interpolation filter is such a proper low-pass filter.

7.2.3 A Summary of Digital Resampling–Related Terminology

As (7.2) shows, digital resampling means generating new sample(s) of a continuous signal $x(t)$ with arbitrary sampling time delay(s) directly from an existing sequence of samples $x_k = x(kT_s)$. A resampler is simply a device or an algorithm that generates such samples. During the development of digital resampling technology, many technical terms were used in the literature to describe the operations of the resampler. The meanings of these terms could be confusing because some of the terms may actually mean the same thing or have the concepts that are very close to each other. In this chapter when we describe various techniques and applications of digital resampling, we will use the terms close to those originally appeared in the literature. For clarification, we provide a summary of these commonly used terms below.

Digital Interpolation/Interpolator

As it was originally defined in the digital rate conversion references [3, 5], digital interpolation means to generate a resampled output sequence *with more than one output samples per input sample*, as oppose to digital *decimation*. In other words, it performs sampling rate *up-conversion* as oppose to sampling rate *down-conversion*. The FIR filter for performing such sampling rate up-conversion is often called an *interpolation filter* [3].

However, interpolation is also often used in the broad sense. According to the definition in numerical analysis, *interpolation is a method of constructing new data points within the range of a discrete set of known data points* [13]. Based on this definition, the terms digital interpolation/interpolators are commonly used as synonyms for digital resampling/resamplers.

Digital Rate Conversion/Converter

When a digital resampler generates an output sequence of samples $\{y_n\}$ from the input sequence $\{x_k\}$, the input and output sampling rates are likely to be different. Thus, such a digital resampler performs sampling rate conversion and is called a *digital rate converter*. In the application to timing synchronization, digital resampling/resampler and digital rate conversion/converter are often used as synonyms of each other and may be used interchangeably.

Digital Delayer

This is another way to describe a digital interpolator for performing digital resampling. Since an output sample from the digital interpolator is a linear combination of the input samples, its sampling time is likely to be later than that of the earliest input sample involved in the combining. In other words, the digital interpolator performs the delay operation of the earliest sample. Thus, it is called a *delayer*. At the same time, the sampling time of the interpolator output is ahead of that of the latest sample involved in the interpolation. Therefore, the interpolator can be viewed as performing the sampling time *advance* operation. Whether it performs a sampling time delay or advance operation depends on the sample that is used as the time reference.

7.3   Polyphase FIR Filter Bank for Digital Resampling/Rate Conversion

Digital resampling can be realized through digital interpolation by using either infinite impulse response (IIR) or finite impulse response (FIR) filters. In practice, *polyphase filter banks*, or simply *filter banks*, with multiple subfilters are commonly used. In addition to many attractive properties, FIR filter banks can be computationally more efficient than the IIR filter bank implementations when used for digital interpolation [1]. The structure and characteristics of FIR filter banks and their applications to digital resampling, particularly for performing rate conversion, are described below.

7.3.1   Integer Sampling Rate Up-Conversion with Digital Interpolation

Let us first consider the implementation of digital interpolation that uses an FIR filter to achieve the rate conversion from f_s to $L_I f_s$, where L_I is an integer called the *interpolation factor*. In this section, the term *interpolation* is used in its *narrow sense* as defined in Section 7.2.3 and [3].

Basic Operations of Digital Interpolation

When performing digital interpolation, the objective is to convert the digital sample sequence $\{x_k\}$ discussed in Section 7.2 to a new sample sequence $\left\{x_j'\right\}$, which is identical to the sequence generated by directly sampling $x(t)$ at the rate of $L_I f_s$. As shown in Figure 7.3, to achieve such a rate up-conversion, $L_I - 1$ zeros are first inserted between every two adjacent $x_k's$ to increase the sampling rate to $L_I f_s$. The zero-inserted

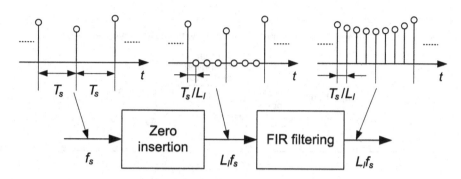

Figure 7.3 Time-domain operations of digital interpolation

sample sequence has a spectrum that is the original signal spectrum repeated L_I-1 times resulting in L_I-1 image bands. These image bands are then rejected by the low-pass filter. Assume that the spectrum of $x(t)$ is zero for frequency $|f| > f_{max}$, and $f_{max} < f_s/2$. To perform ideal interpolation, the frequency response of the low-pass filter should be a constant between $-f_{max}$ and f_{max}, and zero for $|f| > f_s - f_{max}$.

The process of operations described above is called *digital interpolation* (in the narrow sense) [3, 5], because it generates $L_I - 1$ new samples between any two adjacent input samples. Therefore, digital interpolation performs digital resampling of the input sample sequence with a time resolution of T_s/L_I, which is L_I times finer than the timing resolution of the original sequence as shown in Figure 7.3. In other words, it performs the rate up-conversion of the original sequence by increasing its sampling rate, i.e., up-sampling, by a factor of L_I. The frequency-domain characteristics in different processing stages are shown in Figure 7.4.

A Frequency-Domain View of Digital Interpolation

After L_I-1 zeros are inserted between every two adjacent samples, the sampling rate is increased to $L_I f_s$, and so is the period of the spectrum of the zero-inserted sequence. However, its spectrum shape is still in the same form as the original sequence as shown in Figure 7.4(a). The frequency response of a possible low-pass interpolation filter is shown in the same figure. Figure 7.4(b) shows the shape of the spectrum of the sequence after low-pass filtering, which is identical to the spectrum of the sequence generated by directly sampling the analog signal at the rate of $L_I f_s$. Thus, the reconstructed sample sequence at the output of the low-pass filter and the sequence generated by direct sampling of the analog signal are identical.

Implementation of Digital Interpolation Using a Polyphase FIR Filter Bank

During interpolation, the input sequence has L_I-1 zeros in every L_I samples. For an FIR low-pass filter with ML_I coefficients, only the M non-zero samples are involved in the convolution to compute an output sample. Thus, only M multiplications are needed per output sample rather than the $L_I M$ multiplications required for the normal FIR filter that uses all of the filter coefficients to generate an output sample. Taking advantage of this

(a)

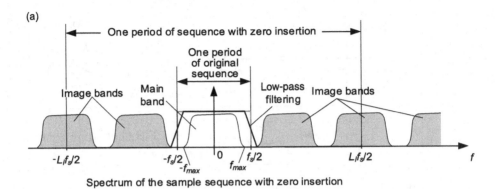

Spectrum of the sample sequence with zero insertion

(b)

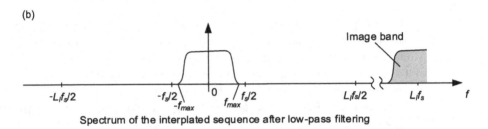

Spectrum of the interplated sequence after low-pass filtering

Figure 7.4 Frequency-domain characteristics of digital interpolation

property greatly reduces the complexity of the implementation of the interpolation process. Such an implementation leads to the *polyphase filter-bank* structure.

When the interpolation process starts, we assume that the 0th, L_I^{th}, . . . , $[(M-1)L_I]$th elements in the zero-inserted input sample vector are not zero. These non-zero elements are the consecutive samples of the sequence $\{x_k\}$ to be interpolated. Denoting the coefficient vector of the FIR low-pass filter by $(c_0,\ c_1,\ c_2,\ \ldots,\ c_{L_IM-1})^t$, these M non-zero elements convolve with the coefficients $c_0,\ c_{L_I},\ c_{2L_I},\ \ldots,\ c_{(M-1)L_I}$ to generate the first interpolated output sample.

In the next step, the same non-zero samples convolve with $c_1,\ c_{L_I+1},\ c_{2L_I+1}, \ldots,$ $c_{(M-1)L_I+1}$ to generate the second output. The process continues until the last non-zero element in the zero-inserted input sample vector is shifted out and a new sample from the sequence $\{x_k\}$ is shifted in. At this point, the coefficients $c_0,\ c_{L_I},\ c_{2L_I}, \ldots,\ c_{(M-1)L_I}$ are used again to generate the (L_I+1)th interpolated sample. Note that the samples generated earlier have a sampling time before that of the samples generated later. In other words, the earlier samples can be viewed as the delayed version of the later samples.

From the above description, it can be seen that during digital interpolation, each sample vector consisting of the M samples from the sequence $\{x_k\}$ is used L_I times. The sample vector convolves with L_I different short coefficient vectors, whose elements are subsets of the FIR filter coefficients, to generate L_I output samples. After the L_I outputs are generated, the last element of the sample vector is shifted out, a new sample is shifted in, and the interpolation process repeats again. Such an interpolation process can

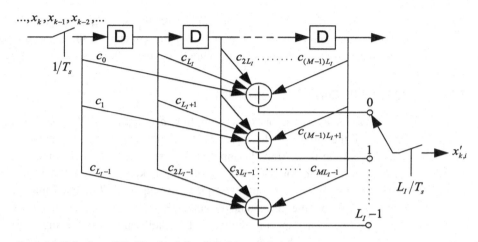

Figure 7.5 Polyphase FIR filter bank for digital interpolation

be implemented by using the polyphase filter bank architecture shown in Figure 7.5. The FIR low-pass filter of the polyphase FIR filter bank that performs digital interpolation is called the *interpolation filter* of the filter bank. In Sections 7.3 and 7.4, we will simply call the digital interpolator implemented by the polyphase FIR filter bank the *digital interpolator*, unless stated otherwise.

The FIR interpolation filter shown in Figure 7.5 can be viewed as consisting of a bank of L_I subfilters. The ith subfilter has coefficients, $c_i, c_{L+i}, \ldots, c_{L(M-1)+i}$. There are a few salient features of such an implementation of the interpolation filter. First, the number of operations required to generate each interpolated sample is equal to M. A total of ML_I operations are needed to perform interpolation in each input sample interval T_s to generate L_I interpolated samples. Second, ideally the subfilters should have the frequency responses that differ by constant group delays between each other but otherwise identical. The difference between the group delays of the ith and $(i+j)$th subfilters should be equal to jT_s/L_I in the passband of the interpolation filter. These features will be examined when we consider the practical design and properties of such an interpolation filter in Section 7.4.

The digital interpolator shown above generates samples that have delays with a resolution of T_s/L_I. It performs up-sampling of a sample sequence by a factor of L_I, which is defined as the *up-sampling factor* also called the *interpolation factor*.

To facilitate the discussion below, we first describe the sample timing of the output samples of the digital interpolator in continuous analog time.

With the input sample vector $\mathbf{x}_k = (x_k, x_{k-1}, \ldots, x_{k-M+1})^t$, the output of the zeroth subfilter, denoted by $x'_{k,0}$, is the sample $x(kT_s - T_0)$, where T_0 is a constant determined by the group delay of the zeroth subfilter. With the same input vector, the ith subfilter output $x'_{k,i}$ is the sample $x(kT_s + iT_s/L_I - T_0)$, which has less delay than $x'_{k,0}$. After the input sample vector $\mathbf{x}_k$ is shifted to drop the last sample x_{k-M+1} and the new sample x_{k+1} shifted in to become its first sample, it becomes $\mathbf{x}_{k+1} = (x_{k+1}, x_k, \ldots, x_{k-M+2})^t$, and the output of the zeroth subfilter corresponds to the sample $x(kT_s + T_s - T_0)$. Thus, when the input vector is $\mathbf{x}_k$, the L_I subfilters' outputs span a T_s-long-time interval

$[x(kT_s - T_0),\ x(kT_s + T_s - T_s/L_I - T_0)]$ with the starting point marked by zeroth subfilter's output. With the entire input sample sequence as the input, the samples at the outputs of the subfilters span the entire timeline with a resolution of T_s/L_I.

7.3.2   Rational Up/Down Rate Conversion

In contrast to interpolation, or up-sampling, *decimation*, or *down-sampling*, reduces the rate of the input sample sequence by keeping only one out of every few samples and discard the rest. After the input sample sequence is down-sampled by a factor of L_D, i.e., only one out of every L_D samples is kept, the time spacing between two adjacent samples of the output sequence is equal to L_D times that of the input sequence. When the interpolated sequence discussed in Section 7.3.1 is down-sampled, the time spacing between two adjacent output samples is equal to $(L_D/L_I)T_s$. The combination of these two steps, i.e., digital interpolation and decimation, generates a sequence with a sampling rate of $(L_I/L_D)f_s$. Thus, it performs digital rate conversion by a factor of L_I/L_D. In general, digital rate conversion can be performed by a conversion factor of any rational number using properly selected values of L_I and L_D.

The FIR filter bank discussed above is commonly used for the rational rate conversion. In such an implementation, one of L_I subfilters convolves with the input sample segment to generate an output sample. With a conversion factor of L_I/L_D, if the nth output sample is the output of the subfilter with the index i_n, the subfilter with the index $i_{n+1} = (i_n + L_D)_{\mathrm{mod}\ L_I}$, where the subscript "mod L_I" denotes the modulo L_I operation, is selected to generate the $(n+1)$th output. Moreover, to generate the $(n+1)$th output, zero, one, or more new samples are shifted into the input buffer, and the same number of oldest ones are dropped. The samples in the input buffer are convolved with the selected subfilter coefficients to generate the next output sample. This provides the means to generate a new sample with a sampling time advance of $(L_D/L_I)T_s$ relative to the previous samples. In other words, the filter bank generates a sample sequence with the sampling rate of $(L_I/L_D)f_s$.

The rational rate conversion could yield a conversion rate that is greater or less than one depending on the relative values of L_I and L_D. Namely, it can either increase or reduce the sampling rate of the output relative to the input. Finally, for timing synchronization, such a filter bank can be used for the sampling phase control with a time resolution of T_s/L_I. The resolution of the sampling phase control can be improved by increasing the value of L_I.

7.3.3   Rate Conversion by an Arbitrary Factor

When the rate conversion is applied to timing synchronization, the locally generated digital samples at the rate f_s are resampled to generate samples at a different rate that is synchronous to the remote transmitter symbol rate. Because the transmitter symbol rate is not known to the receiver, the sampling time of the output samples is determined in real time. Therefore, the time spacing between output samples would be an arbitrary number and may change with time. This is equivalent to performing the rate conversion by a factor equal to an arbitrary real number. Before describing the practical implementation of such arbitrary-rate conversion by digital resampling, we first consider how it can be achieved ideally.

Denoting the conversion rate by r, which is an arbitrary real number, the sample rate after conversion is equal to rf_s. The time spacing between two adjacent output samples is equal to $1/rf_s = T_s/r$. Assume that the nth resampler output y_n is equal to $x(t_n)$, i.e., the input signal $x(t)$ sampled at the time t_n. The next output y_{n+1} should be equal to $x(t_n+T_s/r)$, i.e., $x(t)$ sampled at t_n+T_s/r. These samples can be directly generated from the input sample sequence by using the proper interpolation filter coefficients. However, in practice, it would be too complex and unnecessary to compute the exact interpolation filter coefficients in real time. Thus, some approximation in such rate conversion is necessary and acceptable.

Rate conversion by an arbitrary factor can be implemented approximately by using the polyphase filter bank structure described above. Depending on the requirement of the application and implementation considerations, zero-order interpolation, i.e., sample-and-hold, and first-order, i.e., linear, interpolation are the most popular. Below, we consider the details of their implementations. The worst-case distortion due to the approximations will be analyzed. The application of higher order interpolation will also be considered.

7.3.3.1 Zero- and First-Order Interpolation of the Outputs of FIR Filter Bank

As discussed in Section 7.3.1, when the entire sample sequence $\{x_k\}$, i.e., ..., x_{k-1}, x_k, x_{k+1}, ..., is used as the input of the FIR filter bank with L_I subfilters, its output is the sample sequence $\{x(kT_s + iT_s/L_I - T_0)\}$, for ..., $k-1$, k, $k+1$, ..., and $i = 0, 1, \ldots, L_I - 1$. The samples of the output sequence cover the entire time line spanned by $\{x_k\}$ with delay T_0 and the time spacing between two adjacent samples equal to T_s/L_I. With proper selections of k, i, and introducing a timing offset α as defined below, any arbitrary sampling time instant can be uniquely specified.

Specifying Sampling Time with the Polyphase Filter Bank Outputs
The desired sampling time t_n of y_n can be uniquely expressed as

$$t_n = k_n T_s + i_n T_s/L_I - T_0 + \alpha_n T_s/L_I \tag{7.3}$$

where $0 \le \alpha_n < 1$ is the fractional part of T_s/L_I of t_n. In other words, the sampling time of y_n is between the sampling times of the outputs of the i_nth and the (i_n+1)th subfilters of the filter bank with $\mathbf{x}_{k_n}$ as the input vector. An approximate value of y_n can be computed from the outputs of the i_nth and (i_n+1)th subfilters, denoted by x'_{k_n, i_n} and x'_{k_n, i_n+1}, respectively. To facilitate the descriptions below, we define the normalized sampling time $\tilde{t}_n$ of y_n by $\tilde{t}_n = t_n/T_s$ and omit the constant time delay T_0 to simplify the notation. Namely,

$$\tilde{t}_n = k_n + i_n/L_I + \alpha_n/L_I \cong k_n + \tilde{t}_n^{(r)} \tag{7.4}$$

Note that in (7.4), k_n is an integer and $\tilde{\imath}_n^{(r)}$ is the fractional part of $\tilde{\imath}_n$ with $0 \leq \tilde{\imath}_n^{(r)} < 1$, defined by

$$\tilde{\imath}_n^{(r)} \cong i_n/L_I + a_n/L_I \tag{7.5}$$

As will be shown later, $\tilde{\imath}_n^{(r)}$ is commonly stored and used in computing the sampling phase from sample to sample in the digital timing loop implementations.

One special case is that the subfilter index i_n is equal to $L_I - 1$ in (7.3). In this case, the subfilter with index i_n+1 is actually the zeroth subfilter with the input sample vector $\mathbf{x}_{k_n+1}$, which is the vector $\mathbf{x}_{k_n}$ shifted by one position and prepending a new sample. Thus, the sample y_n is between the subfilter outputs x'_{k_n,L_I-1} and $x'_{k_n+1,0}$. To simplify presentation, we will still use the notations of x'_{k_n,i_n} and x'_{k_n,i_n+1} as in other cases with the understanding that $x'_{k_n,L_I} \cong x'_{k_n+1,0}$.

The sampling time of the next interpolated sample y_{n+1} is equal to $t_n + T_s/r$. The normalized sampling time $\tilde{\imath}_{n+1}$ can be computed by first adding $1/r$ to $\tilde{\imath}_n$, i.e.,

$$\tilde{\imath}_n + 1/r = \tilde{\imath}_{n+1} = k_{n+1} + i_{n+1}/L_I + a_{n+1}/L_I \cong k_{n+1} + \tilde{\imath}_{n+1}^{(r)} \tag{7.6}$$

To generate y_{n+1}, the parameters k_{n+1}, i_{n+1}, and a_{n+1} are determined as follows. First, from (7.6), we observe that k_{n+1} is the integer part of $\tilde{\imath}_{n+1} = \tilde{\imath}_n + 1/r$. It can be computed as

$$k_{n+1} = \lfloor \tilde{\imath}_{n+1} \rfloor = \lfloor \tilde{\imath}_n + 1/r \rfloor \tag{7.7}$$

where $\lfloor \bullet \rfloor$ denotes the floor function, which extracts the integer part of the value inside the floor bracket, and the fractional part $\tilde{\imath}_{n+1}^{(r)} = \tilde{\imath}_{n+1} - k_{n+1}$.

Second, also from (7.6), multiplying the fractional part of $\tilde{\imath}_{n+1}$ by L_I, we have

$$L_I \tilde{\imath}_{n+1}^{(r)} = L_I(\tilde{\imath}_{n+1} - \lfloor \tilde{\imath}_{n+1} \rfloor) = i_{n+1} + a_{n+1} \tag{7.8}$$

Because a_{n+1} is in the range of $[0, 1)$, i_{n+1} is the integer portion of $L_I \tilde{\imath}_{n+1}^{(r)}$. Therefore,

$$i_{n+1} = \left\lfloor L_I \tilde{\imath}_{n+1}^{(r)} \right\rfloor \tag{7.9}$$

Finally, a_{n+1} can be obtained by subtracting i_{n+1} from $L_I \tilde{\imath}_{n+1}^{(r)}$ such that

$$a_{n+1} = L_I \tilde{\imath}_{n+1}^{(r)} - i_{n+1} \tag{7.10}$$

In short, the sampling time of y_{n+1} is between the sampling time instants of the (i_{n+1})th and $(i_{n+1}+1)$th subfilter outputs with input vector $\mathbf{x}_{k_{n+1}}$. The time differences between y_{n+1} and the outputs of these two subfilters are equal to $a_{n+1}T_s/L_I$ and $(1 - a_{n+1})T_s/L_I$.

The input vector $\mathbf{x}_{k_{n+1}}$ is generated by shifting the vector $\mathbf{x}_{k_n}$ by $m_{n+1} = k_{n+1} - k_n$ positions with m_{n+1} new input samples shifted into the input delay line and the same number of old samples dropped. The amount of the relative shift m_{n+1} can also be directly computed from the fractional part of the normalized sampling time $\tilde{\imath}_n^{(r)}$ by adding the increment $1/r$, such that

$$m_{n+1} = \left\lfloor \tilde{t}_n^{(r)} + 1/r \right\rfloor \tag{7.11}$$

The new fractional part of normalized sampling time $\tilde{t}_{n+1}$ is equal to

$$\tilde{t}_{n+1}^{(r)} = \tilde{t}_n^{(r)} + 1/r - m_{n+1} \tag{7.12}$$

As a result, it is only necessary to store the value of $\tilde{t}_n^{(r)}$ instead of t_n or $\tilde{t}_n$ every sample interval.

Moreover, the normalized advance of the sampling time from one sample to the next is a constant $1/r$ when the rate converter has a constant conversion rate of r. In general, the procedure described above can implement sampling time advances that are determined in real time and may change from sample to sample. Such cases will be discussed when considering the implementation of timing synchronization using digital resampling. Examples will be given in Sections 7.6 and 7.7 to illustrate practical implementations.

Below we look into the details of the generation of the samples y_n and y_{n+1} by using the polyphase filter bank–based resampler through the zero- and the first-order, or linear, interpolation.

Zero-Order Interpolation (Sample-and-Hold)
Let us assume that the sample y_n to be generated is at a specific normalized time $\tilde{t}_n$. As (7.4) shows, it is located between the output samples of the two subfilters i_n and i_n+1 of the filter bank of the resampler with input sample vector $\mathbf{x}_{k_n}$. The timing differences between the desired rate converter output and the two closest subfilter output samples, denoted by x'_{k_n, i_n} and x'_{k_n, i_n+1}, are also determined. As long as these two digital filter output samples are relatively close in time, the value of the rate converter output can be approximately computed based on these two samples.

It is most computationally efficient to simply let the output sample of the rate converter take the value of the closest subfilter output. As described above, the time differences between y_n and the earlier and later subfilter outputs are equal to $\alpha_n T_s/L_I$ and $(1 - \alpha_n)T_s/L_I$, respectively. If $\alpha_n < 0.5$, then y_n takes the value of the earlier subfilter output x'_{k_n, i_n}. Otherwise, y_n takes the value of the later subfilter output x'_{k_n, i_n+1}. We call such an approximation method *zero-order interpolation*, i.e., a sample-and-hold operation. Its operations are illustrated in Figure 7.6.

In the figure, the desired sampling time of y_n is between the output samples of the first and second subfilters with the input vector $\mathbf{x}_{k_n}$, denoted by x'_{k_n, i_n} and x'_{k_n, i_n+1}, with $i_n = 1$. Because $\alpha_n < 1 - \alpha_n$, i.e., y_n, is closer to $x'_{k_n, 1}$ than to $x'_{k_n, 2}$, the estimate of y_n, $\hat{y}_n$, is equal to $x'_{k_n, 1}$.

To generate y_{n+1}, the sampling time is advanced by T_s/r. The parameters i_{n+1}, α_{n+1}, and m_{n+1}, which are used to generate the estimate of y_{n+1}, are computed according to (7.6) through (7.11). The input vector to the filter bank is shifted and modified according to the value of m_{n+1}. Based on the value of α_{n+1}, one of the subfilters' output, either $x'_{k_{n+1}, i_{n+1}}$ or $x'_{k_{n+1}, i_{n+1}+1}$, is computed and assigned as the value of y_{n+1}. In Figure 7.6, y_{n+1} is between $x'_{k_{n+1}, 1}$ and $x'_{k_{n+1}, 2}$. Since $1 - \alpha_{n+1} < \alpha_{n+1}$, $\hat{y}_{n+1}$, the

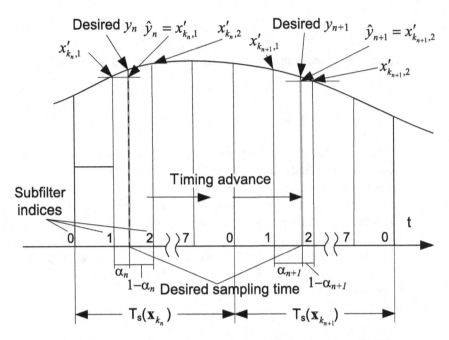

Figure 7.6 Digital rate conversion with zero-order interpolation (sample-and-hold)

estimated value of y_{n+1}, is equal to $x'_{k_{n+1},2}$. This process is repeated for every output of the rate converter.

To reduce the approximation error, it is desirable to have a large L_I, i.e., a large number of subfilters. As a result, the time difference between the outputs of the adjacent subfilters becomes small, and the distortion in the rate converter output due to approximation is reduced.

Zero-order interpolation is computationally efficient. To generate one rate converter output, only one subfilter output needs to be computed. The shortcoming is that a larger number of subfilter coefficients needs to be stored since the number of the coefficients is proportional to L_I. To reduce the coefficient storage requirement, first-order, or linear, interpolation can be used.

Linear (First-Order) Interpolation
The implementation of linear interpolation is outlined below, and it is similar to the implementation of the zero-order interpolation described above.

The two subfilters closest to the desired output sample y_n in sampling time are determined in the same way as are those in zero-order interpolation, and their output samples are computed. However, instead of using the closest subfilter output as the estimate of y_n, the two outputs are linearly interpolated. Specifically, with the two closest subfilter output samples x'_{k_n,i_n} and x'_{k_n,i_n+1}, the estimate of y_n is computed as

$$\hat{y}_n = (1 - \alpha_n)x'_{k_n,i_n} + \alpha_n x'_{k_n,i_n+1} \tag{7.13}$$

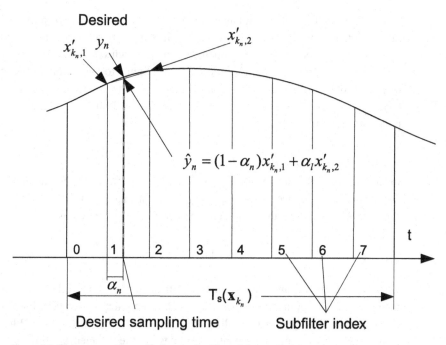

Figure 7.7 Rate conversion with first-order (linear) interpolation

where α_n is the timing differences between y_n and x'_{k_n, i_n}. The value of $\hat{y}_n$ given by (7.13) is the cross point of the sampling time of y_n and the line that connects x'_{k_n, i_n} and x'_{k_n, i_n+1}, as depicted in Figure 7.7.

The parameters for the generation of the next sample y_{n+1}, i.e., i_{n+1}, α_{n+1} and k_{n+1}, are determined in the same way as are those in the zero-order interpolation. The approximate output sample is computed according to (7.13) with the sample index $n + 1$ instead of n.

To generate an output sample, e.g., y_n, by using linear interpolation, two subfilter output samples, x'_{k_n, i_n} and x'_{k_n, i_n+1}, need to be computed. Thus, the computational complexity of the sampling rate converter based on linear interpolation is more than doubled that of the converter with zero-order interpolation. However, the requirement of memory to store the interpolation filter coefficients is significantly reduced. In the next section, we analyze the distortion introduced by the rate converters based on zero and linear interpolations due to approximation.

7.3.3.2 Analysis of the Distortion Due to Approximation in Rate Conversion

So far in this section, we have considered the implementation of the sampling rate conversion based on digital interpolation that employs polyphase FIR filter banks. If the conversion ratio has an arbitrary and/or time-varying value, the zero-order or linear interpolation can be used to generate the approximations of converted samples. Such implementations introduce distortion in the converted signal samples and result in reduced SNRs of the samples at the converter output relative to its input.

The SNR of the sample sequence at the rate converter output has a direct impact on the performance of the targeted application. It is particularly important for the applications in digital communications, such as timing synchronization. In this section, we derive the expressions of the bounds on the signal to distortion ratios at the rate converter outputs as functions of the number of the filter-bank subfilters and the characteristics of the signal samples being converted. Both zero-order and linear interpolations are considered.

Zero-Order Interpolation

In the signal model defined in Section 7.3.1, the spectrum of the signal $x(t)$ is in the frequency range between $-f_{max}$ and f_{max}. According to the discrete-time form of *Parseval theorem* [11], the total signal power can be expressed as

$$P_s = \frac{1}{2\pi} \int_{-\omega_x}^{\omega_x} |X(\omega)|^2 d\omega = \frac{A^2 \omega_x}{\pi} \tag{7.14}$$

where $\omega_x = 2\pi f_{max}/f_s$ and we have assumed that the signal has a flat spectrum with the magnitude $|X(f)| = A$ in the range of $[-\omega_x, \omega_x]$. If the interpolation filter has L_I subfilters, the maximum timing error normalized by T_s is equal to $|\alpha_n/L_I| \leq 0.5/L_I$. When $|\alpha_n/L_I|$ is small, the error of the interpolated samples in the frequency domain due to the approximation is equal to

$$X(\omega)\left(e^{j\omega D} - e^{j\omega(D-\alpha_l/L_I)}\right) \approx -jX(\omega)e^{j\omega D}\omega\alpha_l/L_I \tag{7.15}$$

where D is normalized desired delay.

The worst-case total error power due to the distortion introduced by the zero-order approximation is

$$P_{e,0} = \frac{1}{2\pi} \int_{-\omega_x}^{\omega_x} |X(\omega)e^{j\omega D}\omega\alpha_l/L_I|^2 d\omega \leq \frac{1}{2\pi} \int_{-\omega_x}^{\omega_x} A^2 \left(\frac{0.5}{L_I}\right)^2 \omega^2 d\omega = \frac{A^2 \omega_x^3}{12\pi L_I^2} \tag{7.16}$$

The worst-case signal to the distortion ratio due to the approximation is equal to

$$\gamma_{sd,0} = P_s/P_{e,0} \geq 12L_I^2/\omega_x^2 \tag{7.17}$$

Example 7.1 Assuming that the received signal is sampled at 6.5 MHz and the signal has a raised cosine waveform with a 22 percent excess bandwidth and a symbol rate of 3.84MHz, i.e., $f_{max} = 2.3424$ MHz, we have $\omega_x = 0.72\pi$. If the signal to distortion ratio is greater than 45 dB, i.e., $\gamma_{sd,0} = 31622$, we obtain

$$L_I = \sqrt{\gamma_{sd,0} \times (\omega_x)^2/12} = 0.72\pi \times \sqrt{31622/12} = 116 \tag{7.18}$$

Namely, 116 or more subfilters are needed.

First-Order/Linear Interpolation

When using first-order, or linear, interpolation, the approximation of the rate converter output y_n is given by (7.13). The frequency response of the linear interpolator is equal to

$$X(\omega)\left((1 - \alpha_l)e^{j\omega(D-\alpha/L_l)} + \alpha_l e^{j\omega(D+(1-\alpha)/L_l)}\right) \tag{7.19}$$

The total signal power is the same as that given by (7.14). The error of the interpolated samples in the frequency domain due to approximation can be expressed as

$$E(\omega) = X(\omega)\left(e^{j\omega D} - (1 - \alpha_n)e^{j\omega(D-\alpha_n/L_l)} - \alpha_n e^{j\omega(D+(1-\alpha_n)/L_l)}\right)$$

$$\approx \frac{1}{2}X(\omega)e^{j\omega D}\alpha_n(1 - \alpha_n)\omega^2/L_l^2 \tag{7.20}$$

Since $\alpha_n(1 - \alpha_n) \leq 0.25$, we obtain the upper bound of the total error energy as

$$P_{e,1} \approx \frac{1}{2\pi}\int_{-\omega_x}^{\omega_x}\left|\frac{X(\omega)e^{j\omega D}\alpha_n(1 - \alpha_n)\omega^2}{2L_l^2}\right|^2 d\omega < \frac{1}{2\pi}\int_{-\omega_x}^{\omega_x}\frac{A^2[0.25]^2}{4L_l^4}\omega^4 d\omega = \frac{A^2\omega_x^5}{320\pi L_l^4} \tag{7.21}$$

The signal to distortion ratio is greater than

$$\gamma_{sd,1} \geq \frac{A^2\omega_x}{\pi} \bigg/ \frac{A^2\omega_x^5}{320\pi L_l^4} = \frac{320L_l^4}{\omega_x^4} \tag{7.22}$$

To achieve a given worst-case signal to distortion ratio $\gamma_{sd,1}$, the required number of subfilters is approximately equal to

$$L_l \approx \omega_x \sqrt[4]{\gamma_{sd,1}/320} \tag{7.23}$$

Note that the value calculated according to (7.22) is the worst-case signal-to-distortion ratio. When first-order, i.e., linear, interpolation is used to perform timing synchronization, α goes through all of the possible values from 0 to 1. The average $\gamma_{sd,1}$ would be higher than the worst-case value according to (7.22) for the same L_l. Thus, the actual degradation of the receiver performance caused by the approximation of linear interpolation should be less than what is estimated according to (7.22).

Example 7.2 With the same design requirements given in the previous example, i.e., $\omega_x = 0.72\pi$ and the signal-to-distortion ratio is greater than 45 dB, we have $L_l \approx 0.72\pi \times \sqrt[4]{31622/320} \approx 7$. Thus, it is sufficient for the FIR filter bank to have eight subfilters in order to achieve the worst-case signal-to-distortion ratio of 45 dB. If $L_l = 4$, the worst-case signal-to-distortion ratio is approximately equal to 35 dB. The average would be around 40 dB.

7.3.3.3   High Order Interpolations

Above we have considered using zero-order or first-order/linear interpolation of the outputs of the subfilters of the polyphase FIR filter bank for performing digital resampling to achieve the rate conversion with an arbitrary ratio. Such a resampler also provides a means for generating the approximate samples with arbitrary timing phase, or delay. The SNRs' degradation at the output of the resampler/rate converter due to the approximations can be evaluated analytically as shown above and in [14].

As will be shown in Section 7.5.2, the linear interpolation described above is equivalent to the first-order Lagrange interpolation, which is well known in numerical analysis. Thus, it is possible to use higher order interpolation to interpolate the subfilter outputs of a polyphase filter bank to achieve more accurate results. However, based on the analyses given in Section 7.3.3.2, the performance of the filter bank with linear interpolation should be adequate for most applications. Hence, higher order interpolation is usually not necessary. A better resampling performance can be achieved by increasing the number of the subfilters of the filter bank. In most cases, using a larger number of subfilters together with linear interpolation should be more efficient than using a smaller number of subfilters together with higher order interpolation.

7.4   Design and Properties of Interpolation Filters with Polyphase FIR Filter Bank Structure

The central component of a digital resampler based on the polyphase FIR filter bank is the interpolation filter. As discussed in Section 7.2, the optimal interpolation filter is an ideal low-pass filter. It has a constant-magnitude passband covering the spectrum of desired signal. Its stopband gain should be zero, i.e., with infinite attenuation. To facilitate its implementation, a transition band is defined between the passband and stopbands. Since the optimal interpolation filter has a strictly finite span of its frequency response, it has an infinite time response and is practically unrealizable.

A practically implementable interpolation filter must have a finite time-domain impulse response. Therefore, it cannot have a strictly finite support in the frequency domain, and approximation will be needed. As a result, the SNR of the resampled signal samples would be reduced, and other distortions would be introduced. Such a reduced SNR and introduced distortions are permissible in practice if they do not noticeably degrade the performance of the targeted applications.

To reduce the implementation complexity, the interpolation filter should have a limited length, i.e., not too many coefficients, as long as the degradation due to the distortion and the reduction in SNR of the interpolated samples is acceptable.

It is well known that the FIR filter length depends on the ratio of the sampling frequency and the width of its transition band. Thus, we should take advantage of the characteristics of the communication signal by including a transition band that is as wide as possible when designing the interpolation filter for timing synchronization.

7.4.1 Requirements and Properties of Interpolation Filters for Timing Synchronization

The requirements for the design of the interpolation filter used in digital resampling depend on the signal characteristics of the applications. For timing synchronization, the signal to be resampled is the baseband signal for digital communications as described below.

Signal Model

What is discussed below applies to the digital resampling/rate conversion of the signal in single-carrier digital communication systems. The principles are also applicable to OFDM communication systems to be discussed in Section 7.7.3.

The baseband signal of a single-carrier communication system has a nominal bandwidth of $1/T$, which is the transmitter channel symbol rate. It has an excess bandwidth β/T, where β is usually a number much smaller than 1, e.g., $\beta = 0.22$ for WCDMA [15]. Thus, the received signal has a total bandwidth of $(1+\beta)/T$, which is also the signal's Nyquist sampling rate. To convert the signal to digital samples without loss of information, the sampling rate f_s should be higher than $(1+\beta)/T$. With digital rate conversion, the nominal sampling frequency does not need to be synchronous to the remote transmitter symbol rate.

For the communication signal with the sampling rate described above, the signal spectrum and the low-pass interpolation filter design prototype are shown in Figure 7.8. To reduce the complexity, the interpolation filter is usually implemented as an FIR filter with real coefficients. Therefore, its frequency response is complex-conjugate-symmetric about zero frequency. Only the positive frequency parts of the zero-inserted samples' spectrum and the filter response need to be considered.

As described above, the signal component is in the frequency range between $-(1+\beta)/2T$ and $(1+\beta)/2T$. As an example, we consider the design of an interpolation filter that has a polyphase filter bank structure with four subfilters. To perform

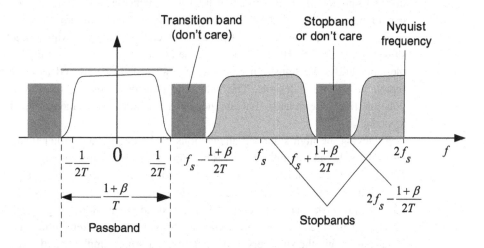

Figure 7.8 Baseband signal spectrum and interpolation filter design prototype

digital interpolation, three zeros are inserted between every two adjacent input samples. Hence, the sampling rate is increased by a factor of four, and both the sampling frequency and the period of the spectrum of the zero-inserted sequence become $4f_s$. The spectrum's positive frequency portion is from 0 to $2f_s$, called the *Nyquist frequency*, which is equal to half of the sampling rate. In addition to the main portion of the signal spectrum that is centered at zero frequency, there are three image bands in each period. The samples at the interpolation filter output are used to generate digital signal samples that are synchronous to the remote transmitter symbols.

Considerations on Interpolation Filter Design

An interpolation filter is simply a low-pass filter. The main portion of the spectrum of the zero-inserted sample sequence is inside the passband of the interpolation filter. For the generic applications of digital resampling, it is important that the passband frequency response of the interpolation filter is constant so as not to introduce distortions into the output signal. This requirement becomes less stringent and different for the timing synchronization of received signals [6, 9]. In most cases, the resampled signals are further processed by the receiver filters and/or the equalizers in receivers. The filters and equalizers can correct the minor linear distortion introduced by the interpolation filter so that the receiver performance would not be affected.

Therefore, for such applications it is more important for all of the subfilters to have the same passband frequency response and only to differ by the constant group delays between them. This is because the interpolated samples will be sequentially generated by all of the subfilters if the sampling clock and the remote transmitter symbol clock are asynchronous. These samples can only be processed equally effectively by the same receiver filter or equalizer if the characteristics of different subfilters do not change except their group delays. The uniformity of the subfilters is determined by the stopband attenuations of the interpolation filter.

Another function of the interpolation filter is to reject the image bands of the zero-inserted samples. If the interpolation filter is implemented as a generic low-pass filter, the stopband that rejects the image bands should start from the lower edge of the first image band, i.e., from $f_s - (1 + \beta)/2T$ and end at $2f_s$, the Nyquist frequency, as shown in Figure 7.8. The band between the end of the passband and the beginning of the stopband is the transition band, which can usually be treated as a "don't care" region in FIR filter designs. The wider the transition band is, the fewer the interpolation filter coefficients that are needed to achieve the same attenuation in the stopband.

Achieving the high stopband attenuations of the interpolation filter is important for performing digital resampling. This is because, after the down-sampling, which is a part of the digital resampling process, the image bands are aliased back to the main signal band as interference. It is necessary to reduce such interference to ensure no or little additional receiver performance loss. The stopband attenuation is highly related to the differences between the passband frequency responses of different subfilters. As will be discussed below, the subfilters of an ideal low-pass filter implemented as an FIR filter bank with the infinite image band attenuations should have the same passband frequency responses after the group delays between them are removed. Moreover, the group delay difference between any two adjacent subfilters should be a constant T_s/L_I.

The frequency bands between the image bands can also be treated as "don't care" regions in the interpolation filter design. Any residual interference and noise in these regions are aliased back into the transition bands around the main signal spectrum and can be rejected by the receiver filter or equalizer. During the filter design, if less effort is spent on the optimization of these transitional and "don't care" regions, it is possible to improve the attenuation of the image bands and to reduce the ripple in the passband for the given number of the filter coefficients. Such improvements, even though they may be small, are still worthwhile to take advantage of when possible.

Properties of the Interpolation FIR Filter
The subfilters of the interpolation filter described above have the following properties.

(1)   Necessary and sufficient conditions of an ideal interpolation filter.

For an interpolation filter to be ideal, the passband frequency response of the ith subfilter, $H_i(f)$, should satisfy

$$H_i(f) = H(f)e^{j2\pi f i T_s/L_I}/L_I, \quad -f_{max} \leq f < f_{max}, \ i = 0, \ 1, \ ..., \ L_I - 1 \quad (7.24)$$

where $H(f)$ is the frequency response of the overall interpolation filter, and f_{max} is the edge frequency of the passband as defined in Section 7.3.1. Namely, the ith subfilter should provide the sampling phase advances of its output by iT_s/L_I relative to the output of the zeroth subfilter. Note that the frequency region considered in (7.24) is only a part of the interpolation filter's frequency response, which is from $-0.5L_I/T$ to $0.5L_I/T$.

It can be shown that (7.24) *is true if and only if the filter's gains in the signal's image bands are equal to zero*, i.e., *the filter provides infinite attenuation in these bands*. The proof of the above conclusion is relatively simple. We leave it as an exercise for the interested reader.

(2)   The mean of the subfilters of the interpolation filter.

The frequency responses of the interpolation filter's subfilters satisfy

$$\sum_{i=0}^{L_I-1} H_i(f)e^{-j2\pi f i T_s/L_I} = H(f), \ -f_{max} \leq f < f_{max} \quad (7.25)$$

where $H_i(f)$ and $H(f)$ are the same as those in (7.24). The relationship given by (7.25) shows that, after the differences between the group delays of $H_i(f)$, $0 < i < L_I-1$, and $H_0(f)$ are removed, the average of the frequency responses of all of the subfilters is equal to the frequency response of the overall interpolation filter in passband. This property can be proved directly from the expression of FIR filter's Fourier transform.

(3)   Relationship between subfilter passband variation and stopband attenuation.

When the stopband attenuations are finite, there will be differences between the passband frequency responses of different subfilters after the differences between their group delays are removed. We can characterize the variations between the subfilters passband responses by the *normalized mean square error between the subfilter passband frequency responses and their mean*, or simply the *MSE of subfilter variation*, defined by

$$\sigma_v^2 = \frac{1}{L_I} \sum_{i=0}^{L_I} \int_{-f_{max}}^{f_{max}} \left| L_I H_i(f) e^{-j\omega i T_s/L_I} - H(f) \right|^2 df \Big/ \int_{-f_{max}}^{f_{max}} |H(f)|^2 df \qquad (7.26)$$

Because the output of a subfilter is a down-sampled version of the output of the interpolation filter, the nonzero stopband frequency responses are aliased back to the passband. Intuitively, the MSE of the subfilter variation should be proportional to the aliased residual energy in the stopbands. Thus, it is inversely proportional to the average stopband attenuation, denoted by $\overline{A}_s = 1/\overline{g_s^2}$, where $\overline{g_s^2}$ is the mean of the squared filter gain in the stopband, and is proportional to the number of the stop bands minus 1, i.e.,

$$\sigma_v^2 \approx C(L_I - 1)/\overline{A}_s = C(L_I - 1)\overline{g_s^2} \qquad (7.27)$$

While it is difficult to prove analytically, (7.27) with $C = 1$ appears to agree well with numerical evaluations. For example, the normalized MSE of the subfilter variation, σ_v^2, is about -40 dB for Example 7.3 in Section 7.4.2 from the numerical evaluation, while the average stop band attenuation is about 45 dB. This heuristic formula could be useful for evaluating the variations between the subfilter frequency responses if needed.

However, while it is true that the variation of the subfilters' passband response has a negative impact on the SNR of the samples at the resampler output, the exact effect depends on many factors. The MSE of the subfilter variations can only be used as a rough estimate of the additional interference introduced into the resampled signal samples during interpolation.

7.4.2   Examples of Interpolation Filter Design

As interpolation filters are commonly FIR low-pass filters, any suitable filter-design algorithm can be used. For FIR filter design, the commonly used approaches are impulse response truncation, windowing-based design methods, and design methods based on optimization procedures [11]. In this section, we consider two popular design algorithms with frequency-domain weighted optimization: the equal-ripple algorithm based on Remez exchange and the least squares algorithms. The popularity of these algorithms is due to their flexibility for optimizing interpolation filter parameters to achieve the best possible receiver performance. Below we provide a brief description of these two algorithms. Examples of the interpolation filter designs that use the LS algorithms will also be provided.

Interpolation Filter Design Based on Equal Ripple Approximation
The FIR filter design with equal-ripple error approximation performs the frequency-domain optimization. Its objective is to design an FIR filter that approximates a given frequency-domain prototype, e.g., the one shown in Figure 7.8, based on the equal-

ripple criterion. An iterative procedure is used to minimize the largest errors in the stopband regions and/or the largest errors (ripples) departing from the prototype in the passband region. In other words, this algorithm performs the min-max optimization.

The most popular tool for equal-ripple FIR filter design is the Parks–McClellan algorithm [16], which is based on the Remez exchange algorithm published by E. Y. Remez in 1934 [17]. One desirable property of this algorithm is that it can apply different weighting factors to different frequency bands. Thus, it is possible to adjust the magnitude of ripples in different passbands and stopbands. This property is especially important to meeting the exact requirements of the interpolation filter for resampling of the received signal. This algorithm was applied to the design of the interpolation filter used in the early applications of digital rate conversion for timing synchronization in echo-cancellation modems as described in [7].

Interpolation Filter Design Based on the Least Squares Criterion
Another popular algorithm for the interpolation filter design is the algorithm based on the optimization according to the LS criterion. Similar to the equal-ripple criterion, the LS FIR filter design algorithm also minimizes the error between the target FIR filter and a frequency-domain prototype. However, instead of minimizing the maxima of the errors, the LS-based algorithm minimizes the total error power. In addition, it also has the flexibility of weighing different frequency band to achieve controlled passband and stopband error characteristics [18]. As a result, the errors in these bands can be adjusted individually by choosing different weighting parameters.

The objective of the weighted LS (WLS) filter design algorithm is to minimize the weighted error power, or L^2 (Euclidean) norm, i.e.,

$$\varepsilon = \int_0^\pi W(\omega)|H(\omega) - A(\omega)|^2 d\omega \tag{7.28}$$

where $H(\omega)$ is the frequency response of the FIR filter, $A(\omega)$ is a prototype function of $H(\omega)$, and $W(\omega)$ is a weighting function. To perform the FIR filter design, $A(\omega)$ is usually a piecewise linear function, which may include "don't care" regions, and $W(\omega)$ is usually constant in each of these regions. Thus, the integral in (7.28) can be decomposed into the sum of the weighted error powers of these regions and the "don't care" regions are excluded. Moreover, $H(\omega)$, the Fourier transform of the FIR filter coefficients h_i's, can be expressed as a function of these coefficients. Thus, ε is minimized by taking the partial derivatives of the sum of the integrals with respect to h_i's, setting them to zero and solving the resulting linear equations. A detailed description of the WLS algorithm and its derivation can be found in [19].

Since receiver performances are determined by the interference power due to the noise and distortions in interpolated samples, the total error power is a more meaningful metric than the maximum error magnitude for such applications. In addition, the LS algorithm computes the optimal FIR coefficients in close form rather than through iterative calculation as in the equal-ripple algorithm. As a result, the LS-based FIR filter design algorithm is popular in interpolation filter designs for timing synchronization. Below, we provide two examples of interpolation FIR filter designs based on the WLS design algorithm.

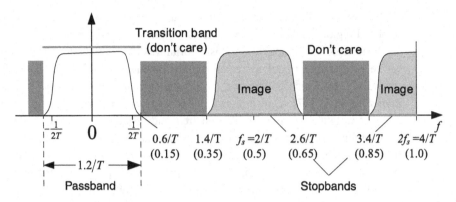

Figure 7.9 Filter design prototype of Example 7.3

Examples of Interpolation Filter Designs and Characteristics

Example 7.3 One scenario, which is often encountered in designing polyphase filter bank for timing synchronization, is that the conversion ratio is close to one. For example, this is the case in echo-cancellation modems, since the local and remote transmitters usually have the same nominal symbol rate $1/T$ and $f_s = 2/T$.

Let us consider a target filter bank, which consists of four subfilters, for performing interpolation. Using the prototype for the interpolation filter design given in Figure 7.8, the Nyquist frequency is equal to $4/T$. If the normalized excess bandwidth is equal to 0.2, the passband in the positive frequency is from zero to $0.5(1 + 0.2)/T = 0.6/T$. The stopbands are on top of the image bands. The first stopband is from $1.4/T$ to $2.6/T$. The second stop band is from $3.4/T$ to $4/T$. Normalized by the Nyquist frequency $4/T$, the width of the passband is equal to 0.3 and centered at zero. In the positive frequency region, the normalized stopbands are from 0.35 to 0.65 and from 0.85 to 1.0. The transition band is from 0.15 to 0.35 and the frequency band between the two image bands, i.e., from 0.65 to 0.85. Both of them are treated as "don't care" regions. This forms the prototype for the interpolation filter design shown in Figure 7.9.

For interpolation of the signal samples, the maximum passband ripple is assumed to be less than 1 dB peak-to-peak (p-p). In addition, the attenuation of the image band signals should be greater than 40 dB. An FIR filter that satisfies these conditions is designed by using the LS FIR filter design routine *firls* [20] available in Matlab and Octave. The frequency response of the filter is shown in Figure 7.10.

The parameters input to the *firls* routine are n, F, A, and W, i.e., the number of coefficients minus 1, the frequency vector, the magnitude vector, and the weighting vector, respectively.

There are a number of methods for estimating the required number of the FIR filter coefficients. Unfortunately, none of these methods is very accurate. An initial guess of the number of the coefficients can be obtained by using the empirical formula:

(a)

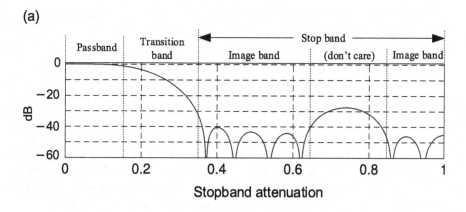

(b)

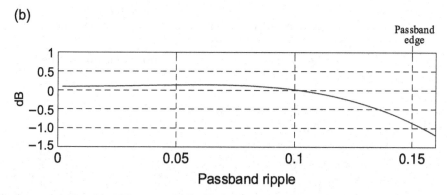

Figure 7.10 Frequency response of interpolation filter in Example 7.3

$$n = 1.5 \times f_s / B_{trans} \qquad (7.29)$$

where B_{trans} is the width of the transition band and f_s is the sampling frequency of the input samples of the filter. The final number can be determined during the filter design by trial and error. In this example, the sampling frequency is equal to $8/T$ and the width of the transition band is equal to $0.8/T$. The calculated number of coefficients is equal to $1.5 \times [8/0.8] = 15$. This number turns out to be a good guess. Given that there are four subfilters, the number of coefficients should be a multiple of 4, i.e., a total of 16 coefficients. Unfortunately, *firls* can only generate an odd number of filter coefficients. We choose $n = 14$, and the actual number of coefficients is equal to $n + 1 = 15$.[3] A zero is added to the end of the generated coefficient vector so that each subfilter has four coefficients.

The frequency vector F contains the pairs of the corner frequencies normalized by the Nyquist frequency $4/T$, of all of the bands, excluding the "don't care" regions. In this example, we have $F = [0, 0.15, 0.35, 0.65, 0.85, 1.0]$.

The gain vector A contains the gains at the corresponding corner frequencies in F. It contains only 0 and 1 for the low-pass filter. Ones are at the edges of the passbands, and

---

[3] When n is not too small, one less coefficient will not significantly affect the performance of FIR filters.

zeros are at the edges of stopbands. In this example, the gain vector used in the filter design is A = [1, 1, 0, 0, 0, 0].

The last parameter is the weighting vector W. To design the interpolation filter using *firls*, different weighting factors can be assigned to different bands. Since each band has only one weighting factor, the length of W is half of the lengths of vectors F and A. The weighting vector that we used in the design is W = [1, 60, 10], which was determined empirically during the design.

Figure 7.10(a) shows the overall frequency response of the filter designed by *firls*, and Figure 7.10(b) shows the details of the passband response. As can be seen from the figures, this design meets the original design goals. From the numerical evaluation, the average normalized variation of subfilter passband response is about -40 dB. Moreover, there are additional degradations when used in conjunction with the linear interpolation discussed above. If such a design cannot meet the receiver performance requirement, the subfilter length and/or the number of subfilters can be increased.

A simple Matlab/Octave m-script file that performs the interpolation filter design is given below:

```
n = 14;
F = [0 0.15 0.35 0.65 0.85 1];
A = [1 1 0 0 0 0];
W = [1 60 10];
b = firls(n, F, A, W);
freqz(b, 1);
```

Example 7.4 We now consider a hypothetical receiver that uses an XO with the natural frequency of 26.00 MHz as the local frequency reference. The transmitter symbol rate is equal to 3.84 MHz with an excess bandwidth of 22 percent. The sampling clock is generated by dividing the XO clock by 4, i.e., with a nominal clock rate of 6.5 Mhz. The interpolation filter is implemented as a filter bank with 8 subfilters. Therefore, 7 zeros are inserted between every two adjacent input samples. As a result, the effective sampling rate after interpolation is equal to $8 \times 6.5 = 52$ MHz. The spectrum of the zero-inserted sequence is similar to that shown in Figure 7.8, except that there are 7 image bands (4 in positive frequency region) instead of 3. As in Example 7.3, only the positive frequency region is considered.

For the given parameters, the passband in the positive frequency region is from zero to $1.22 \times 3.84/2 \approx 2.34$ MHz. The transition band is from 2.34 MHz to $6.5 - 2.34 = 4.16$ MHz, i.e., 1.82 MHz. The stopband starts from 4.16 MHz to 26 MHz, which is the Nyquist frequency. Inside the stopband, there are 4 image bands at $i \times 6.5 \pm 2.34$ MHz, $i = 1, 2, 3$, and from 23.66 to 26 MHz. The energies in the image bands must be rejected. The region between any two adjacent image bands can be treated as "don't care" in the filter design.

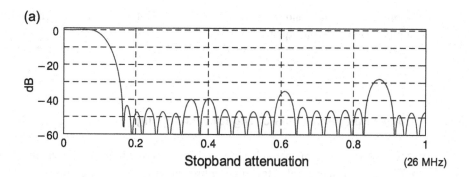

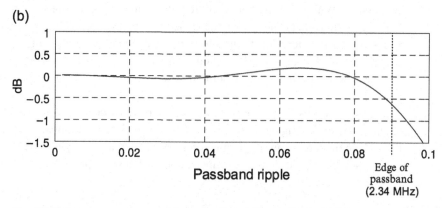

Figure 7.11 Frequency response of the interpolation filter in Example 7.4

To limit the distortion introduced into the received signal samples, the maximum passband ripple is chosen to be less than 1 dB p-p. In addition, the average attenuation of the image band signals is greater than 45 dB. The basic filter design procedure is the same as the one described in Example 7.3. Since the transition band is narrower, to achieve similar inband ripples and stopband attenuations, each subfilter has more coefficients than that in the previous example. Based on the empirical formula given by (7.29), a rough estimate of the number of the filter coefficients is equal to $1.5 \times 52/1.88 = 42$. Because there are 8 subfilters, the coefficient number could be equal to either 40 or 48.

When the same design procedure described in the previous example is used, it is determined empirically that an FIR filter with 47 coefficients, i.e., 6 per subfilter with an added zero, is appropriate. The F vector containing the normalized passband and stopbands frequencies is [0 0.09 0.16 0.34 0.41 0.59 0.66 0.84 0.91 1.0].

When designing the filter, it was found that the attenuations are different in different stopband regions if the same weight is used. To make the attenuations relatively evenly distributed, we split the first stopband into two and assign different weighting factors to the two bands. Specifically, the F and W vectors are

$F = [0\ 0.09\ 0.16\ 0.22\ 0.22\ 0.34\ 0.41\ 0.59\ 0.66\ 0.84\ 0.91\ 1.0]$ and
$W = [1\ 50\ 20\ 5\ 2\ 1]$.

The frequency-domain characteristics of the designed filter are shown in Figure 7.11, which meet the design requirements. If we slightly relax the inband ripple and image band rejections, 39 filter coefficients, i.e., 5 coefficients per subfilter, would be adequate.

7.5   Other Implementations of Digital Resampling

The polyphase filter bank–based method described above provides a flexible, efficient, and analytically tractable approach to performing digital resampling/rate conversion. While this method is preferable in many applications, there are other implementations that can perform the same task and may provide certain advantages under certain practical constraints. In particular, while the computational complexity of the filter bank–based method is low, its control structure may not be straightforward to implement in some cases, e.g. on the pure hardware/ASIC-based platforms.

In this section, we describe two popular alternatives of the polyphase filter bank–based approach for digital resampling. They are particularly suitable for implementations based on pure hardware platforms for generic digital resampling applications. The comparison between different implementations will also be provided.

7.5.1   Farrow's Digital Variable Delay Controller

In [8], a digital variable sampling time-delay controller was proposed by C. W. Farrow for the application to the sampling rate conversion in echo-cancellation modems. It is commonly called *Farrow's delayer*. The central idea is to design an FIR filter with variable coefficients to approximate a delay $(D+\tau)T_s$ in real time, where D is a predefined constant and τ is a variable in the range between 0 and 1. Ideally, the frequency response of such an FIR filter should be equal to $e^{-j\omega(D+\tau)}$.

Each of the filter coefficients of the Farrow's delayer is in the form of a polynomial of τ, such that,

$$h_k(\tau) = \sum_{m=0}^{M-1} c_{k,m}\tau^m, k = 0, 1, ..., K-1 \qquad (7.30)$$

where $h_k(\tau)$ is the kth coefficient of the FIR filter and is a function of τ, and $c_{k,m}$ is a set of precomputed coefficients.

The frequency response of the variable delay filter with the coefficients $h_k(\tau), k = 0, 1, \ldots, K-1$, is given by

$$H(\omega,\tau) = \sum_{k=0}^{K-1} h_k(\tau)e^{-j\omega k} = \sum_{k=0}^{K-1}\sum_{m=0}^{M-1} \tau^m c_{k,m}e^{-j\omega k} \qquad (7.31)$$

which is preferably equal to $e^{-j\omega(D+\tau)}$. The coefficients are determined by minimizing

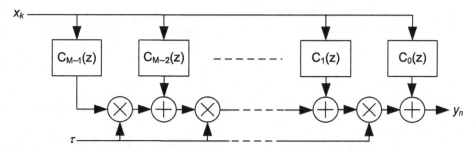

Figure 7.12 Implementation of Farrow delayer

$$\int_0^{2\pi}\int_0^1 \left|H(\omega,\tau) - e^{-j\omega(D+\tau)}\right|^2 d\tau d\omega = \int_0^{2\pi}\int_0^1 \left|\sum_{k=0}^{K-1}\sum_{m=0}^{M-1} \tau^m c_{k,m} e^{-j\omega k} - e^{-j\omega(D+\tau)}\right|^2 d\tau d\omega$$

(7.32)

with respect to $c_{k,m}$ and by letting $D = (K - 1)/2$. A set of such coefficients with $K = 8$ and $M = 4$ were computed and shown in [8] for the targeted applications.

The transfer function of the FIR filter with the coefficients $h_k(\tau)$, $k = 0, 1, \ldots, K-1$, can be expressed as

$$H(z,\tau) = \sum_{k=0}^{K-1} h_k(\tau)z^{-k} = \sum_{k=0}^{K-1}\sum_{m=0}^{M-1} c_{k,m}\tau^m z^{-k} = \sum_{m=0}^{M-1}\tau^m \sum_{k=0}^{K-1} c_{k,m}z^{-k} = \sum_{m=0}^{M-1}\tau^m C_m(z)$$

(7.33)

It leads to a simple hardware architecture for implementation as shown in Figure 7.12.

Compared to the digital resampler implementation based on the filter bank with linear interpolation described in Section 7.3.3.1, Farrow's delayer requires more numerical operations if DSP/software is used for its implementation. At the same time, Farrow's delayer has only one control input. Its design and implementation on pure hardware/ASIC platforms can be quite simple.

It should be noted that the optimization criterion of Farrow's delayer is the minimum variance of the difference between the frequency response of the filter and a constant delay response with the constraint that the coefficients are in polynomial form. No effort is made to achieve the best uniformity of the filter frequency response across the entire delay range. As a result, it is not clear how much the filter's passband characteristics will vary from one delay to another. In contrast, the passband variation of the filter bank–based interpolator can be determined by the stopband attenuations. Thus, for the latter, the variation can be analytically estimated and controlled when the resampler is designed.

7.5.2 Resampling with Arbitrary Delay Using Lagrange Interpolation

Lagrange interpolation is one of the most popular methods to perform interpolation in numerical analysis. By definition, it constructs new data points inside the range of the

Table 7.1 Coefficients of the Lagrange Interpolators of Up to Third Order

	κ_0	κ_1	κ_2	κ_2
$N = 1$	$1-D$	D		
$N = 2$	$(D-1)(D-2)/2$	$-D(D-2)$	$D(D-1)/2$	
$N = 3$	$-(D-1)(D-2)(D-3)/2$	$D(D-2)(D-3)/6$	$-D(D-1)(D-3)/2$	$D(D-1)(D-2)/6$

given data points. Lagrange interpolation can be used to perform digital resampling as shown below.

Digital Resampling Using Lagrange Interpolation

For $N + 1$ given digital samples $x_i = \mathrm{x}(iT_s)$, $i = k$, $k-1$, ..., $k-N$, an Nth order polynomial can be constructed such that [21]

$$f_L(x) = c_{N-1}x^N + c_{N-2}x^{N-1} + \ \cdots\ + c_1 x + c_0, \tag{7.34}$$

which satisfies

$$f_L(iT_s) = x_i, \quad \text{for } i = k, k-1, \ ..., \ k-N, \tag{7.35}$$

Lagrange interpolation generates sample y_n with delay τ relative to kT_s and equal to $f_L(kT_s - \tau)$, which can be expressed as a linear combination of x_i, $i = k$, $k-1, ..., k-N$, such that

$$y_n \cong f_L(kT_s - \tau) = \sum_{j=0}^{N}\ \prod_{l=0,\ l\neq j}^{N} \frac{\tau/T_s - l}{j - l}x_{k-j} \cong \sum_{j=0}^{N}\ \prod_{l=0,\ l\neq j}^{N} \frac{D - l}{j - l}x_{k-j} \cong \sum_{j=0}^{N} \kappa_j(D)x_{k-j} \tag{7.36}$$

where $D = \tau/T_s$ is the delay τ normalized by the sampling interval T_s.

Thus, the generated sample y_n with delay τ is a linear combination of the nearby input samples, and the coefficients $\kappa_j(D)$ are functions of the normalized delay D. The coefficients of the Lagrange interpolators of up to third order are listed in Table 7.1.

To obtain the best resampling results, the generated sample is chosen to be near the center of the segment of the input samples [1]. Moreover, the delay range of the interpolator output only needs to be equal to T_s. Any sample with delay outside this range can be generated by shifting the samples in the input delay line and adding samples at one end or the other of the delay line. Thus, D can be expressed as $\lfloor N/2 \rfloor + d$, where $0 \le d < 1$ for N to be odd or $0.5 \le d < 0.5$ for N to be even, and $\lfloor N/2 \rfloor$ is the integer part of $N/2$. The coefficients of the Lagrange interpolator expressed in terms of the fractional delay d are shown in Table 7.2.

As mentioned in Section 7.3.3, the linear interpolation used for improving sampling phase resolution of the polyphase filter bank–based digital resampler is the same as the first-order Lagrange interpolator by defining $1 - d = \alpha$. For this purpose, linear interpolation should be sufficient. Higher order interpolators can also be used but is likely to be unnecessary.

Table 7.2 Coefficients of Lagrange Interpolators in Terms of Fractional Delay d

	κ_0'	κ_1'	κ_2'	κ_3'
$N = 1$	$1-d$	d		
$N = 2$	$d(d-1)/2$	$-(d+1)(d-1)$	$d(d+1)/2$	
$N = 3$	$-d(d-1)(d-2)/2$	$(d+1)(d-1)(d-2)/6$	$-(d+1)d(d-2)/2$	$-(d+1)d(d-1)/6$

The Optimality of the Lagrange Interpolator for Data Resampling

The use of Lagrange interpolation for digital resampling started as a direct application of the interpolation method widely used in numerical analysis to digital signal processing [1]. From a system and signal processing point of view, it was later shown that the Lagrange interpolator is equivalent to an FIR delayer that has maximally flat response at zero frequency [22].

The concept of the maximal flatness in the FIR frequency response was introduced in [23]. As shown in Section 7.2, the frequency response of an ideal resampler with delay D is $e^{-j\omega D}$, i.e., the magnitude of the spectrum is constant, and the phase is linear with respect to D within the passband. As defined in [23], to achieve the maximal flatness at zero frequency, the frequency response of the digital FIR delayer should satisfy

$$\frac{d^k[H_D(\omega) - e^{-j\omega D}]}{d\omega^k}\bigg|_{\omega=0} = 0, \quad \text{for } k = 0, \ 1, \ ..., \ N-1 \tag{7.37}$$

where N is the number of the FIR coefficients.

It was shown that the coefficients of the FIR filter that satisfy (7.37) have the same form of that given by (7.36) [23]. Therefore, the Lagrange interpolator is an optimal delayer in the sense of achieving the maximal flatness relative to the ideal delayer at zero frequency.

Farrow's Architecture for Lagrange Interpolator Implementation

As proposed in [24], the Lagrange interpolator can be implemented by using the architecture for the implementation of Farrow's delayer shown in [8]. Below we provide an example to illustrate this approach.

Using the coefficients of the Lagrange interpolator with $N = 3$ given in Table 7.2, the interpolator output for a given delay d can be expressed by

$$y_n(d) = -\frac{d(d-1)(d-2)}{6}x_k + \frac{(d+1)(d-1)(d-2)}{2}x_{k-1}$$
$$-\frac{d(d+1)(d-2)}{2}x_{k-2} + \frac{d(d+1)(d-1)}{6}x_{k-3}$$

After simplification and reorganization of the terms, we obtain

$$y_n(d) = d^3\left(-\frac{x_k}{6} + \frac{x_{k-1}}{2} - \frac{x_{k-2}}{2} + \frac{x_{k-3}}{6}\right) + d^2\left(\frac{x_k}{2} - x_{k-1} + \frac{x_{k-2}}{2}\right)$$
$$+d\left(-\frac{x_k}{3} - \frac{x_{k-1}}{2} + x_{k-2} - \frac{x_{k-3}}{6}\right) + x_{k-1} \tag{7.38}$$

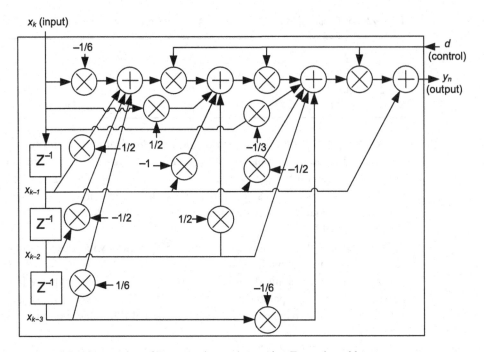

Figure 7.13 Implementation of Lagrange interpolator using Farrow's architecture

The right side of (7.38) can be viewed as the input sample sequence filtered by an FIR filter with transfer function

$$H(z, d) = \sum_{m=0}^{2} d^m C_m(z) \qquad (7.39)$$

where $C_0(z) = z^{-1}$, $C_1(z) = -\frac{1}{3} - \frac{z^{-1}}{2} + z^{-2} - \frac{z^{-3}}{6}$, $C_2(z) = \left(\frac{1}{2} - z^{-1} + \frac{z^{-2}}{2}\right)$, and $C_3(z) = -\frac{1}{6} + \frac{z^{-1}}{2} - \frac{z^{-2}}{2} + \frac{z^{-3}}{6}$.

As (7.39) has the same form as (7.33), the block diagram shown in Figure 7.12 can be directly used for the implementation of the Lagrange interpolator. Figure 7.13 shows the details of such an implementation of the third-order Lagrange interpolator with the formulation of (7.38).

Farrow's forms of the Lagrange interpolators of different orders can be derived similarly. A general formulation of the subfilters in the Lagrange interpolator implementation based on Farrow's architecture was given in [24].

For digital resampling, the third-order Lagrange interpolator is a good choice. As shown in [1, 10], an odd-order Lagrange interpolator, which has an even number of input samples, is more suitable for digital interpolation than an even-order one. Overall, this form of a low-order Lagrange interpolator is quite attractive if the demand for accuracy of interpolation is not very high.

Applications of Lagrange Interpolator to Timing Synchronization
Because the Lagrange interpolators can generate samples with an arbitrary delay, they can be used for the sample timing-phase control in TLLs described in Chapter 6.

Specifically, the desirable timing phase contained in the phase register can be used to generate the input of the delay control variable d. However, since the timing phase nominally increases by T or $T/2$ at the phase controller output from one sample to the next, it is natural to define the increase in the value of the phase register as a time advance rather than delay. If this is the case, this value needs to be translated into the delay d to control the sampling phase of the Lagrange interpolator.

As defined above, the fractional delay d in an odd-order Lagrange interpolator is in the range between 0 and 1. If we denote the value in the timing-phase register of a TLL loop by α_F, which is also in the range between 0 and 1, d can be simply computed as $1 - \alpha_F$. For an even-order Lagrange interpolator, the range of d is between -0.5 and $+0.5$. For the phase value α_F as defined above, d can be computed as $0.5 - \alpha_F$.

Examples of using the Lagrange interpolator in a CDMA DLL will be described in Section 7.7.2.

7.5.3   Comparison of Three Resampler Realizations

The FIR interpolation filter of a polyphase filter bank approximates the ideal sample reconstruction filter, which is optimal according to the Nyquist sampling theorem. For a filter bank that has L_I real subfilters, the frequency response of the FIR filter is periodic with a period of L_I/T_s Hz and symmetric with respect to zero frequency. Therefore, the frequency response of such resampler can be fully specified in the range from zero to $L_I/2T_s$ Hz. This makes the resampler based on an FIR filter bank flexible and analytically tractable. Specifically, it is possible to choose the parameters of the interpolation filter such that the filter can shape the signal frequency-domain characteristics while performing resampling. For example, the interpolator can be designed to serve as a compromise equalizer or to reject the ADC quantization noise if so desired.

This type of digital resampler is especially suitable for the rate conversion of digital communication signals. Unlike digital resampling for other signal processing applications, the passband flatness is less important than the uniformity of the interpolator at different delays. Ideally, if a resampler has infinite attenuation in the stopbands, the passband responses of the subfilters are all identical except for the group delays. Practically, since the attenuations are finite, the passband characteristics change from one subfilter to the next besides the differences in their group delays. Nonetheless, the variance of the passband characteristics between different subfilters is fully controllable by properly designing the FIR interpolation filter to have predefined stopband attenuations.

Both the Farrow's delayer and the Lagrange interpolator are designed to minimize the difference between their frequency responses and that of the ideal delayer with different optimization criteria. Farrow's delayer minimizes this difference under the constraint that the coefficients of the FIR filter are the polynomials of delay τ. By comparison, the Lagrange interpolator is optimal in the sense that the difference between its frequency response and that of the optimal delayer achieves maximal flatness at zero frequency.

However, no optimization efforts were made to shape the passband frequency response and to control the out-of-band responses for either Farrow's delayer or the Lagrange interpolator. For Farrow's delayer, this conclusion was specifically stated in

[8]. The Lagrange interpolator does not have control over its inband and out-of-band characteristics either. Therefore, there is no direct way to control the passband variations at different sampling time delays and no control of the out-of-band characteristics for either of these two resamplers.

When considering the implementation complexity, the polyphase filter bank–based resampler is quite attractive, especially when the application requires high SNR of the interpolated signal samples. For example, the complexity of performing linear interpolation to generate an output sample is only slightly more than that for generating the two required adjacent subfilter output samples. When used in echo-cancellation modems, Farrow's interpolator as proposed in [8] is implemented with four similar subfilters. This is twice as complex as the filter bank–based implementation described in [7].

Low-order Lagrange interpolators, e.g. up to the third order, are particularly attractive for resampling signals with relatively low SNR due to the simplicity of their implementation. For processing high SNR communication signals, the order of the interpolator needs to be increased. If the required order is larger than 6, a Lagrange interpolator is more complex than the implementation of Farrow's delayer and of the filter bank–based resampler.

Note that the complexity of the polyphase filter bank–based resamplers can be evaluated analytically. The complexity does not increase significantly even when the required signal SNR is increased. The relationship between the signal SNR requirement and the complexities of Farrow's delayer and the Lagrange interpolator need to be evaluated numerically or by simulation.

The main advantage of the two interpolators based on Farrow's structure discussed in this section is that they are conceptually simple for hardware/ASIC implementations. Each of them has only one control input, i.e., the required sampling time delay and the entire implementation does not change during operation. These interpolators are particularly suitable for the implementations in ICs designed for generic resampling applications.

The filter bank–based resampler requires one additional control step of selecting and using the two sets of FIR filter coefficients. While this resampler is also implementable using all hardware-based architectures, they are particularly suitable for software-based implementations in DSP/ASIC, especially for high demand and application-specific designs.

All resampler implementations need to perform the control of shifting new input samples into the data delay line at the proper time. There is no significant difference in the implementation complexity among them in this regard.

7.6   All-Digital Timing Control with Digital Resampling

General considerations of timing synchronization implementations and their realization using digital TLLs have been discussed in Section 6.2.4. A high-level block diagram of such a generic implementation was shown in Figure 6.2. The data sampler shown there

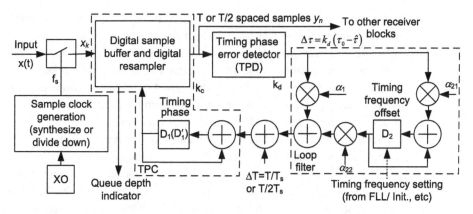

Figure 7.14 A typical all-digital timing control loop employing digital interpolation

performs the timing-phase control. It can be implemented in either analog or digital form. Below we describe an all-digital implementation of TLLs. As an example, the polyphase filter bank–based resampler described above is used as the digital sampler. Other possible implementations will also be discussed.

For applications to timing-phase control in TLLs, the digital resampler generates another sample stream from the sample stream at the local sampling rate. The second sample stream is at a rate that is synchronous to the remote transmitter symbol rate. Therefore, both digital resampling and digital rate conversion essentially mean the same thing, and these two terms can be used interchangeably in this context.

7.6.1 A Typical All-Digital Timing Synchronization Block

The block diagram of a typical all-digital TLL is shown in Figure 7.14. It is similar to the one described in Section 6.2.5 but uses digital resampling for sample timing-phase control. This structure can be used for generating either T or T/2 spaced samples.

The input analog signal $x(t)$ is sampled at a sampling clock with frequency f_s to generate digital samples x_k. The clock is derived from the local XO, through either a sampling-clock frequency synthesizer or a simple frequency divider/multiplier. The sampling frequency f_s can be fixed or adjustable, as long as it satisfies the Nyquist sampling rate of $x(t)$.

The input samples x_k's are stored in a buffer, which may have a buffer queue depth-monitoring function as will be discussed later. The T or $T/2$ spaced signal samples, y_n, generated from the samples x_k's by a digital resampler, are used for timing synchronization and other receiver functions. The rate of the resampler output is synchronous to the remote transmitter symbol rate $1/T$ during normal receiver operations. Below, we will assume that such output samples are at the rate of $1/T$. The operations are similar if the output sampling rate is equal to $2/T$.

A timing-phase error detector (TPD) compares the timing phase of the resampler output to the desired timing phase using one of the techniques described in Chapter 6. The detected timing-phase errors are processed by a loop filter. In the figure, we assume that a second-order loop filter is used. Specifically, the timing-phase errors are scaled and accumulated in the register D_2, which contains an estimate of the timing frequency offset of y_n. During initialization, the register is initialized using the frequency-offset estimate from the initialization and/or carrier synchronization blocks.

The register D_1 inside the TPC of the second-order loop contains the estimate of the desired timing phase of y_n. The output of the loop filter is the sum of the contents of D_2 times constant α_{22} and the TPD output times constant α_1.

The output of the loop filter is accumulated in the register D_1 every time a resampler output is generated, i.e., every T. A constant timing-phase increment $\Delta T = f_s/(1/T) \cong T/T_s$ is also added to D_1 together with the loop filter output to achieve the nominal timing-phase advance per output sample. Note that the TLL updating interval may not be the same as the resampler's output sample interval T.

The register D_1 contains the required timing phase of y_n normalized by T_s, the input sample interval. The contents of D_1 usually represent an unsigned fixed-point fractional number, whose range $[0, 1)$ corresponds to the time interval $[0, T_s)$. The number in D_1 has one or more integer bits, which are stored in register D_1', to handle the integer part of ΔT and the overflow or underflow bit. After accumulation, the value of D_1 can be larger than one due to the integer portion of ΔT and the overflow/underflow bit generated during accumulation. If an overflow or underflow occurs, the fractional number in D_1 wraps around, and the overflow bit is added to, or the underflow bit is subtracted from, the integer part in D_1'. The value of D_1' after the accumulation determines the number of new samples that should be shifted into the input delay line. D_1' is reset to zero before the next accumulation occurs. The contents of D_1 feed the digital resampler to generate the next sample as will be described in Section 7.6.2.

The design, implementation and parameter selection of the all-digital TLL are essentially the same as that of the all-digital PLL described in Section 4.5.2.

After the training stage, the TLL converges and the average sample timing-phase error reduces to zero. Thus, the receiver timing synchronization is achieved.

7.6.2   Implementation of the Digital Resampler for the TPC in TLL

In this section, we consider the implementation of the TLL's TPC using the digital resamplers described in Sections 7.3 and 7.5. To start, we consider the digital resampler implementation based on the polyphase FIR filter bank with linear interpolation.

7.6.2.1   Polyphase FIR Filter Bank–Based Digital Resampler

The block diagram of such a TLL is shown in Figure 7.15. To facilitate the implementation, the number of subfilters of the filter bank is usually chosen to be a number of the power of 2, denoted by 2^J. To perform data resampling, the sampler determines which two of the subfilters should convolve with the input data sample vector. Below we describe the resampling processing when generating data sample y_{n+1} after sample y_n is generated. The operations involved are essentially the same as those described in Section 7.3.3.1.

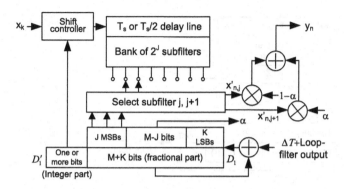

Figure 7.15 TLL phase controller using polyphase filter bank resampler

The register D_1, which contains the timing phase of the output sample, is $M+K$ bits long. The contents of D_1 represent an unsigned fix-point fractional number in the range of $[0, 1)$. To generate sample y_{n+1}, the loop filter output, and the nominal timing increment ΔT are added to the contents of D_1 at time nT. The integer part of the accumulated value, which is equal to the integer part of ΔT modified by the overflow/underflow bit during the addition, is stored in the register D_1', which contains the integer bits of the accumulator. The remainder is stored back into D_1. Denoting the contents of D_1' by m_{n+1}, m_{n+1} new samples are shifted into the input sample delay line. The first element of the delay line becomes $x_{k_{n+1}}$, where $k_{n+1} = k_n + m_{n+1}$. The register D_1' is then reset to zero.

The J MSBs of D_1 determine the subfilter indices i_{n+1} and $i_{n+1}+1$ to generate the outputs for interpolation. The index i_{n+1} is equal to the value of the J MSBs of D_1 as an integer. The $(i_{n+1})^{\text{th}}$ and $(i_{n+1}+1)^{\text{th}}$ subfilters generate $x'_{k_{n+1},i_{n+1}}$ and $x'_{k_{n+1},i_{n+1}+1}$ to be linearly interpolated. As described in Section 7.3.3.1, an exception is made if $i_{n+1} = 2^J-1$. In this case, we have $i_{n+1} + 1 = 0$ and a new sample is shifted into the input delay line to generate $x'_{k_{n+1}+1,0}$ for performing interpolation.

The interpolation coefficient α is equal to the remaining $M-J$ bits as a fractional number with a value between 0 and $1-2^{-(M-J)}$. The value of $M-J$ is determined by the required precision of α.

Moreover, D_1 contains K additional LSBs, which are not involved in the interpolation. These LSBs are needed if the loop filter output is less than 2^{-M} at a low-frequency offset and with a small α_{22} as shown in Figure 7.14 and/or the precision of ΔT is higher than the M bit-long D_1. The purpose and requirement of these additional LSBs have been discussed in Section 4.5.2.

The above procedure is repeated to generate more output samples to be used by the phase error detector and other receiver functions.

7.6.2.2 Farrow's Delayer or Lagrange Interpolator as TPC

When Farrow's delayer or the Lagrange interpolator described in Section 7.5 is used as the digital resampler to serve as the TPC, the TLL phase controller operations are basically the same as those described earlier. Such resamplers are most efficiently

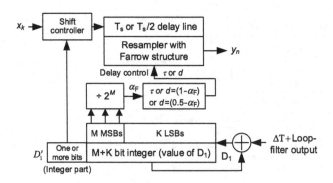

Figure 7.16 TLL phase controller using resamplers based on Farrow's architecture

implemented using the Farrow's architecture and are referred to as *resamplers based on Farrow's architecture.*

When implementing such digital resamplers, D_1 contains $M + K$ bits. The M MSBs represent an unsigned fractional number α_F in the range of $[0, 1)$. They are used by the resampler to control the timing phases of the output samples. The K LSBs are used for phase error accumulation generated by the loop filter as discussed previously.

The fractional number α_F represents the normalized timing phase of the sample to be generated by the resampler as defined in the previous subsection. For the implementation described, any increment in the value of α_F from one sample to the next represents the advance of the timing phase. Since the input to the resamplers based on Farrow's architecture should be the delay of the sampling time, α_F must be converted to the normalized delay τ or d as discussed in Section 7.5. Specifically, τ or d is equal to $1 - \alpha_F$, if Farrow's delayer or an odd-order Lagrange interpolator is used as the resampler, or $d = 0.5 - \alpha_F$, if an even-order Lagrange interpolator is used. Alternatively, it is also possible to modify the part of TLL implementation related to D_1 such that the output of D_1 represents the delay that can be sent to the resampler without translation. A block diagram of using resamplers based on Farrow's architecture is shown in Figure 7.16.

7.6.3 Additional Considerations of Practical Digital TLL Implementations

In this section, we examine a few practical aspects of the implementation of TLLs employing digital resamplers.

Bit-Width Requirements

The general considerations regarding PLL implementation discussed in Section 4.5.2 are applicable to the implementation of an all-digital TLL. Below, we will discuss only a few of the specifics related to timing synchronization.

In the TLL shown in Figure 7.14, the difference between the nominal sampling clock frequency and $1/T$ is compensated by the nominal phase advance ΔT. The frequency offset contained in register D_2 is only used to compensate for the residual sampling frequency offset due to the inaccurate timing clock reference. The required word length of D_2 is determined by the ratio of the XO's maximum relative frequency error, which is usually much smaller than one, to the required timing frequency accuracy. At the same time, the value ΔT is equal to the nominal sampling frequency f_s times T. Since $f_s T$ is on the order of one, the required word length of ΔT is significantly longer than the word length of D_2.

For example, assume that the accuracy of the XO is 100 ppm, the required accuracy of the timing-clock frequency $1/T$ is 0.1 ppm, and the nominal rate conversion ratio is equal to 1.5. It is sufficient if D_2 is 10 bits long. However, the word length of ΔT should be 24 bits long since $2^{-23} > 10^{-7} > 2^{-24}$. In order for the smallest frequency offset to be accumulated without the loss of precision, the word length of D_1 should also be equal to 24 bits.

Synchronization between the Digital Resampler and Real-World Time

When a TLL completely operates in the digital domain, the receiver does not have the concept of the real-world (analog) time. In many communication systems, especially in wireless communication systems, this is not an issue. In such systems, periodic synchronization signals, such as burst or frame synchronization signals, are usually sent by the transmitter to the receiver at a prespecified time. A receiver can use these signals as time marks to synchronize the operation of the resampler to the real-world time.

However, if the data are transmitted continuously without such synchronization signals, e.g., in some wireline communication systems, it is necessary to monitor the availability of the input samples. When the input samples are all converted to output samples, the operation of the resampler should stop until sufficient input samples become available.

Such a monitoring and control function can be achieved by monitoring the available samples in the input delay line. If the delay line is treated as a queue, the monitoring is performed by a queue depth monitor. The resampler operates as fast as it can to generate the output samples and stores them in an output buffer. If there are not enough samples in the input buffer, the resampling operation stops. The operation resumes once there are sufficient samples in the input buffer again.

7.7 Examples of Digital Rate Conversion for Timing Control in Communication Systems

The all-digital TLLs and their implementation considerations described in Section 7.6 are applicable to the timing synchronization block in any digital communication receiver. In this section, we provide three examples.

7.7.1 Application to Timing Control in Echo/Self-Interference Cancellation Modems

Echo-cancellation modems with all-digital timing control appeared on the market in the mid-1980s. They were the earliest commercial modem products employing digital resampling technology. The theory and realizations of the digital rate conversion for timing synchronization in echo-cancellation modems can be found in a number of references including [6, 8, 25]. The digital rate-conversion technology is especially attractive for timing synchronization in echo-cancellation modems because it substantially reduces the number of related hardware components and, thus, is a cost-effective solution.

Digital resampler-based timing synchronization had become the de facto implementation in many generations of wireline full-duplex communication devices and systems employing echo cancellation including ISDN, xDSL, and gigabit Ethernet. In this section, we first briefly describe the operations and implementation of the echo-cancellation modem. The advantages of using digital resampling techniques for performing timing synchronization over previous hardware-based methods will be discussed. Finally, the design and implementation will be presented. These principles and implementation details are applicable to the implementation of wireless communication systems based on the *in-band full-duplex* (IBFD) technology [26, 27, 28]. Since IBFD implementations have attracted considerable attention in the digital communication technical communities recently, what is presented in this section has both historical and current importance.

7.7.1.1 Basics of Full-Duplex Communications with Echo/Self-Interference Cancellation

In a conventional wireline or wireless system, the transmitters and receivers are operated in different frequency bands or at different time slots to avoid the interference to the received signal from the transmitter. In other words, such systems consist of two TDM or FDM half-duplex communication links. Only half of the resources are used for signal transmission in each direction at the same time. To fully use the available resources, in a full-duplex communication system based on echo/self-interference cancellation, the transmission and reception use the same frequency band and at the same time. As a result, such systems have the potential to double the spectrum efficiency compared to systems that are not using this technology.

However, the received signal is usually quite weak due to the channel propagation loss. When both of the transmitted and received signals coexist in the same frequency band at the same time, part of the transmitted signal would present in the received signal and constitute strong interference. The interference is generated either by the transmitted signal leaking through direct paths from the transmitter port to the receiver port or by reflections due to discontinuities in the propagation media. In general, these interferences in the received signals can be modeled as the transmitted signal passing through a linear channel. We will refer to the part of the transmitted signal showing in the received signal as *echoes*.

To eliminate such interference, the receiver emulates an accurate replica of the echoes. The constructed replica is subtracted from the received signal. The operations that generate the echo replica and subtract it from the received signal are performed by the *echo canceller* in the receiver. A high-level block diagram of a receiver with such an echo canceller is shown in Figure 7.17.

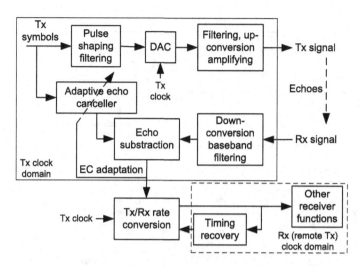

Figure 7.17 High-level block diagram of an echo/self-interference cancellation modem

The replica of the echoes is generated by an adaptive echo canceller, which convolves the local transmitted data symbols with the estimated CIR of the echo channel. The output of the subtractor is used to adaptively update the echo canceller's coefficients. All of these functions operate synchronously to the local transmitter symbol rate. However, the received signal after the echo removal is processed by receiver functions. It must be sampled at the receiver clock rate, which is synchronous to the remote transmitter symbol rate. Thus, it is necessary to convert the received signal samples from the local transmitter clock rate to the remote transmitter clock rate, if these two clocks are asynchronous.

Above, we provided a brief overview of an echo/self-interference cancellation receiver. More details of the principles, operations, and implementations of echo-cancellation receivers can be found in [29, 30, 25]. Below, we will focus on the sample rate conversion methods from the local transmitter clock rate to the remote transmitter clock rate.

7.7.1.2 Analog and Digital Sampling Rate Conversion in Echo-Cancellation Modems

In the early days of the echo-cancellation modem development, such rate conversions were performed by analog means. As shown in Figure 7.18(a), the received analog signal is converted to digital samples by an ADC operated at a rate that is an integer multiple of the local transmitter symbol rate. The echo synthesizer generates the samples of the echo replica, which are at the same rate as the ADC samples.

After the synthesized echo is subtracted out, the received signal samples are almost echo-free and contain only the remotely transmitted signal and additive noise. However, their sampling rate is still synchronous to the local transmitter clock. In order to be used by other receiver functions, it is necessary to convert the rate of the samples to be synchronous to the remote transmitter symbol rate.

To perform analog rate conversion, the echo-canceled received signal samples are converted to the analog domain by a digital-to-analog converter. After sample-and-hold

(a)

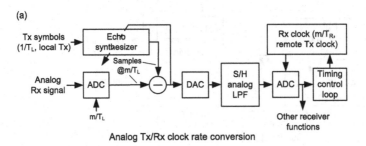

Analog Tx/Rx clock rate conversion

(b)

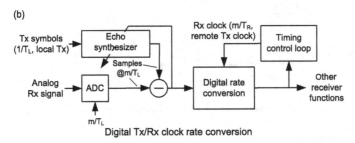

Digital Tx/Rx clock rate conversion

Figure 7.18 Analog and digital rate conversions of echo-cancellation modems

operations, the analog signal generated is low-pass filtered and then resampled by an ADC. The sampling rate is controlled by a TLL and is synchronous to the remote transmitter symbol rate.

It can be seen from Figure 7.18(a) that two ADCs and one DAC are needed in the implementation. Including another DAC in the transmitter, the number of these converters is doubled compared to a modem without echo cancellation. Thus, in the early 1980s, there was a great incentive to employ the digital resampling technology to reduce the number of ADCs and DACs to the level of the modems with no echo cancellation. Moreover, even though the DSPs were still in the infant stage at that time, their processing powers were able to handle this task if the resampler is designed and implemented properly. These are the reasons why digital resampling/rate conversion was considered and implemented in the echo-cancellation modems as its early applications to commercial modem products.

The implementation of the transmitter/receiver sample rate conversion using digital resampling in echo-cancellation modems is illustrated in Figure 7.18(b). The echo-cancellation process is the same as that shown in Figure 7.18(a). The difference between the implementation shown in Figure 7.18(b) and that in Figure 7.18(a) is that the echo-canceled samples, which are at the local transmitter sample rate, are converted to become synchronous to the remote transmitter symbol rate by digital resampling instead of by analog means.

The principles of the digital rate conversion in echo-cancellation modems are the same as those shown in Section 7.6. Specifically, if the remote and local transmitters have the same nominal symbol rate, Example 7.3 in Section 7.4.2 is appropriate for the implementation.

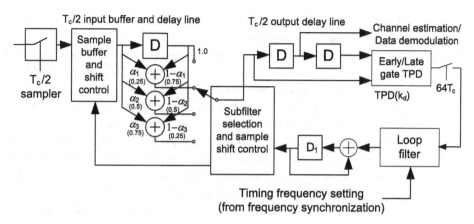

Figure 7.19 A cdma2000 DLL with digital interpolation

7.7.2 Implementation of Timing Control with Digital Resampling in DS-CDMA Receivers

In Section 6.5.4, we described an example of delay locked loops (DLLs) in a cdma2000 RAKE receiver. In the design shown there, it was assumed that the input data signal is sampled at the rate of $8/T_c$, where T_c is the chip interval. When the DLL makes a delay adjustment, it selects the sample closest to the desired timing phase. Thus, the resolution of the adjustment is equal to $T_c/8$. The error caused by the low timing resolution would not cause noticeable performance degradation for the DS-CDMA receivers operating in low SNR environments such as voice communications. Because signal samples used by the DLL are at the rate of $2/T_c$, the rate of the input samples is preferably also equal to $2/T_c$, rather than $8/T_c$, to reduce the cost and complexity of implementation. Below we consider two implementations of the DLL timing-phase control using linear and third-order Lagrange interpolations with the input signal sampled at $2/T_c$.

7.7.2.1 Implementation Using Linear Interpolation

Digital resampling by linear interpolation, i.e., first-order Lagrange interpolation, discussed in Section 7.5.2, can be directly used to generate samples with a $T_c/8$ resolution from $T_c/2$ samples. In low SNR environments, the sample distortion requirement for resampling is not critical. Therefore, the simple linear interpolation of two adjacent $T_c/2$ samples would be acceptable. The RAKE finger shown in Figure 6.15 is modified to incorporate linear interpolation of the input samples with the $T_c/2$ spacing. The block diagram of such a RAKE finger implementation is shown in Figure 7.19.

Linear Interpolation Implemented Using a Polyphase Filter Bank
To generate samples with a $T_c/8$ resolution from $T_c/2$ samples by linear interpolation, the resampler uses an interpolation filter equivalent to an FIR filter bank with four subfilters, each of which has two coefficients, i.e., (0.0, 1.0), (0.25, 0.75), (0.5.0.5), (0.75.0.25). Given that the value of any of these coefficients is a negative power of 2 or

1 minus such a value, the hardware implementation involves only barrel shifters and subtractors, but no multipliers.

The digital resampler–based DLL operates as follows. As in the all-digital timing control loop discussed in Section 7.4.2, the register D_1 contains an M bit fixed-point number, which represents an unsigned fractional number. One out of the four subfilters is selected based on the value of D_1. The two MSBs of D_1 interpreted as an integer determine the index of the subfilter selected. The remaining bits of D_1 are required for timing error accumulation. Two samples in the buffered input delay line convolve with the selected subfilter coefficients to generate an output sample, which is stored in another $T_c/2$ delay line. The samples in the second delay line are used by the TPD, channel estimation, and data demodulation.

Normally, a $T_c/2$ sample is shifted into the input delay line once every output sample is generated. However, if D_1 experiences an overflow after accumulating the loop filter output, the contents of D_1 wrap around and the subfilter index represented by its two MSBs changes from 3 to 0. When this occurs, an additional $T_c/2$ sample is shifted into the input delay line. Conversely, if D_1 experiences an underflow, the index represented by its two MSBs changes from 0 to 3 and no sample is shifted into the delay line. In both cases, the underflow or overflow bit is discarded.

Compensation of Unequal Subfilter Gains

One issue that engineers encountered in practical CDMA receiver implementation using the simple linear interpolator described above was that the gains of different subfilters were not equal. Because many functions of DS-CDMA receivers depend on the estimated average power of the samples, the subfilter gains should be compensated so they are equal to each other to achieve acceptable receiver performance.

The received DS-CDMA signals with low SNR can be modeled as a zero-mean random processes. The power spectral density of the cdma2000 signal at the receiver filter output is approximately in a raised cosine form with 12.5 percent excess bandwidth. As a result, the autocorrelation function of the signal is very close to the sinc function. When the signal is sampled at $2/T_c$, the cross-correlation between the adjacent samples is $\rho \approx \mathrm{sinc}(0.5) = 0.637$. The output linear gains of the interpolation filter coefficients $(\alpha, 1-\alpha)$ can be shown to be

$$g_{\rho,\alpha} = \sqrt{\rho + \left[\alpha^2 + (1-\alpha)^2\right](1-\rho)} \tag{7.40}$$

Relative to the gain of a subfilter filter with a single tap of 1, if $\alpha = 0.5$, we have $g_{\rho,\alpha=0.5} = 0.905$ and if $\alpha = 0.25$ or 0.75, we have $g_{\rho,\alpha=0.25/0.75} = 0.929$. In order to make the output powers of the subfilters equal to each other, the outputs of the subfilters with $\alpha = 0.5$ and $\alpha = 0.25/0.75$ are multiplied by 1.105 and 1.076, respectively.

Analysis of Signal-to-Distortion Ratio

The cdma2000 DLL utilizing linear interpolation described earlier has been used in practical receiver implementations as an alternative to the DLL implementation shown in Section 6.5.4. Using the analytical result given by (7.22), we can estimate the worst-case signal-to-distortion ratio of such an implementation. Since the excess bandwidth of a cdma2000 signal is equal to 0.125 and the sampling rate is equal to $2/T_c$, we have $\omega_x = \pi \times 1.125/2 = 0.5625\pi$. The value of L_I is one as there are no subfilters. By substituting these values into (7.22), we have

$$\gamma_{sd,1} \geq \frac{320 L_I^4}{\omega_x^4} = \frac{320}{(0.5625\pi)^4} \approx 32 \sim 15dB \tag{7.41}$$

The limit given by (7.41) is independent of the number of the sampling phases. If the quantization resolution is limited to $T_s/8$, the signal to distortion ratio will be slightly lower.

For the receiver that uses the $T_s/8$ samples for timing adjustment as described in Section 6.5.4, we have

$$\gamma_{sd,0} \geq 12 L_I^2/\omega_x^2 = 12 \times 4^2/(0.5625\pi)^2 = 64.48 \sim 18.1\,dB \tag{7.42}$$

Hence, the signal-to-distortion ratio of using linear interpolation of $T_s/2$ samples is about 3 dB lower than directly employing $T_s/8$ sampling. Such a ratio should be adequate for CDMA voice services but may not be sufficient for data services.

Implementations Using Farrow's Structure
The linear interpolator described above can also be implemented using the Farrow structure as shown in Section 7.5.2. Namely, the interpolated sample can be computed as

$$y_n = x_k - (1 - \alpha)(x_k - x_{k-1}) = x_{k-1} + \alpha(x_k - x_{k-1}) \tag{7.43}$$

where α takes a value from the set of {0, 0.25, 0.5 and 0.75}.

However, the gain correction factor given by (7.40) would still be necessary.

7.7.2.2   Implementation Using the Third-Order Lagrange Interpolator

To improve the signal-to-distortion ratio at the output of the interpolator, instead of linear interpolation, the second-order or third-order Lagrange interpolator described in Section 7.5.2 can be employed. Below, we consider an implementation using the third-order interpolator similar to the one shown in Figure 7.16. The structure of the DLL is the same as that shown in Figure 7.19 with the exception that the Lagrange interpolator shown in Figure 7.13 is used as its TPC. Specifically, the M MSBs, e.g., $M = 5$, of the register D_1, treated as a fractional number in the range of $[0, 1-2^{-M}]$, represents the sampling phase α_F. The input to the Lagrange interpolator, i.e., the sampling delay d, is equal to $1 - \alpha_F$. The remaining LSBs of D_1 are for the accumulation of the timing-phase errors generated from the DLL loop filter.

Normally, one new sample is shifted into the input sample delay line of the interpolator per output sample. If an overflow occurs during the DLL update, the value of α_F changes from close to one to close to zero, and an additional new sample should shift into the input delay line. Conversely, if an underflow occurs, the value of α_F changes from close to zero to close to one. In this case, no new input sample will be shifted into the delay line. These operations are the same as the DLL implementations described earlier and in Section 6.5.4.

7.7.3   Digital Rate Conversion for OFDM Timing Synchronization in a Multimode Receiver

It has been shown in Section 6.6.2 that OFDM receivers are insensitive to a timing frequency offset. Even if a sampling frequency offset is as large as 5 to 10 ppm, an LTE receiver performance with the largest DFT size, i.e., 2048, still may not be noticeably affected. Nonetheless, digital resampling/rate conversion is still valuable to the implementations of OFDM receiver timing synchronization when flexibility and higher sampling frequency accuracy are required. Below we discuss a few possible applications of digital rate conversion for OFDM receivers.

The digital samples input to the DFT for performing the OFDM demodulation should be at the OFDM channel sampling rate f_s. In many cases, however, it may not be convenient and/or desirable to directly generate such digital samples by operating ADCs at the nominal rate of f_s. For example, to generate the digital samples at the sampling rate f_s from an XO with a nominal frequency of f_{XO}, a clock rate converter will be needed. Traditionally, the sampling clock is generated by a frequency synthesizer as described in Section 4.6.2. Alternatively, the OFDM samples at the desired sampling rate can also be generated by digital rate conversion. The latter approach provides flexibility in modern multimode communication systems. It is particularly advantageous when full or partial software-defined receiver architectures are employed.

The implementation of the receivers in multimode devices, e.g., devices that support several different technologies such as WCDMA, cdma2000, and LTE, is another example of the possible applications of digital rate conversion. In such devices, it is desirable to use a single sampling clock with a frequency higher than, or equal to, the highest required sampling rate of the received signal among all of the systems. Rate converters based on digital resampling can be used to generate samples at the desired sampling rate for each system. In addition to its flexibility, a digital rate converter can also reject the ADC quantization noise during the process when necessary and appropriate.

An additional advantage of this approach is that the planning of combatting RF interference can be simplified. Using a different sampling frequency for each system would result in different tone interference called *RF spurs*. To simplify spur planning and control, digital samples with the same initial sampling rate can be generated for received signals from different systems. Digital sampling rate conversion is preferred in generating the desired sampling rates from the samples with the initial common rate because no spurs at new frequencies will appear after the digital rate conversion.

In such a multisystem environment, digital resampling is used to generate the OFDM samples at the desired sampling rate. Its application to the generation of the LTE samples at the proper sampling rate is straightforward. Below, we consider a hypothetical case where a single ADC sampling clock is used in an LTE/WCDMA receiver.

LTE/WCDMA Receiver with a Common ADC Sampling Clock

In many GSM/WCDMA receivers, XOs with a nominal frequency of 26 MHz are often used. In an LTE/WCDMA dual-mode receiver, it is possible to sample the input signal at the nominal frequency of such an XO. When operating in the WCDMA mode, digital down-rate conversion can be used to bring down the sampling rate by a factor of 4, i.e., to 6.5 MHz. One popular form of such a down-sampling rate converter is the two-stage half-band FIR filter [2, 3] due to its simplicity and efficiency. From the samples at 6.5 mega-samples/second (MS/s), the samples required by the WCDMA receiver can be generated by the digital resampling techniques described above. For example, the rate conversion can be done by an FIR filter bank–based resampler that uses the interpolation filter shown in Example 7.4 in Section 7.4.2. Moreover, the sample rate conversion can be combined with the digital TLL as shown in Section 7.6.

For a receiver that is capable of receiving LTE signals with 20 MHz bandwidth, the data samples input to the DFT of the receiver is usually at the sampling rate of $f_s = 30.72$ MHz when the DFT size is equal to 2048.[4] Since the data subcarriers occupy 18 MHz of the spectrum, a sampling rate higher than 20 MHz would satisfy the Nyquist sampling rate requirement for such LTE signals. Therefore, the digital samples at 26 Ms/s can be directly converted to samples at 30.72 Ms/s as required by the DFT operations. The design and parameters of such an implementation are outlined below.

We choose the number of subfilters to be equal to 8 and use the linear interpolator described in Section 7.3.3.1 to generate the DFT input samples. The data signal bandwidth normalized by the sampling frequency is equal to $18/26 \approx 0.769$. Based on the analysis given in Section 7.3.3.2, it can be shown that the average signal-to-distortion ratio achievable by using linear interpolation should be higher than 46 dB.

With eight subfilters, seven zeros are inserted between every two adjacent input samples to increase the sampling frequency to 208 MHz. The Nyquist frequency of the zero-inserted samples is equal to 104 MHz. The design procedure described in Section 7.4.2, Example 7.4, allows an interpolation filter with 47 or 48 coefficients, i.e., six coefficients per subfilter, to be used. The ripple in the frequency band that contains the active subcarriers should be less than 1 dB peak to peak, and the average attenuation of the image bands can be greater than 45 dB. Distortion caused by linear interpolation is about 46 dB as computed according to (7.22). Given these two distortion levels and seven stopbands, the overall SNR would be greater than 37 dB at the rate converter output. With a noise margin of 6 dB, the worst-case SNR degradation is equal to 1 dB. Thus, such a rate converter can be used for a targeted LTE receiver when a 31 dB received signal SNR is acceptable.

The rate conversion implementation from the nominal XO frequency to the DFT sample rate is similar to the timing locked loop shown in Figure 7.14 and Figure 7.15. For a fixed conversion ratio, the normalized timing-phase increment

---

[4] As the number of active subcarriers is equal to 1200 in the 20 MHz LTE signal, it is possible to use 23.4 MHz for the sample rate if the DFT size is reduced to 1536.

per output sample is a constant equal to $\Delta T = 1/r$. If necessary, the frequency offset due to the inaccuracy of the XO can be corrected by the rate conversion based on the estimated carrier frequency offset obtained from the carrier synchronization block as discussed below.

XO Frequency Error Compensation

Even though the OFDM receiver performance is insensitive to the sampling frequency offset as shown in Section 6.6.2, it may be desirable to reduce the offset when possible. This is especially important for OFDM systems with large FFT sizes.

In the OFDM receiver that uses an XO as the frequency reference, an estimate of the frequency offset caused by the XO's inaccuracy can be provided by the carrier synchronization block. The remote transmitter's carrier and the timing clock frequencies are usually locked, or very close to each other. As a result, the estimated local XO frequency offset can be used for correcting the timing frequency offset.

Let us first consider the ideal case where the XO's native frequency is exactly equal to its nominal frequency $\tilde{f}_{XO}$. The sampling frequency $\tilde{f}_{s,XO}$ is synchronous to the remote transmitter's channel sample frequency f_s with a frequency conversion ratio $r = f_s / \tilde{f}_{s,XO}$. The timing-phase increment per output sample is equal to $T_{s,T} \cong 1/f_s = \tilde{T}_{s,XO} (\tilde{f}_{s,XO}/f_s) = \tilde{T}_{s,XO}/r$, where $\tilde{T}_{s,XO} = 1/\tilde{f}_{s,XO}$. Practically, however, if the XO's actual frequency $f_{s,XO}$ has an error relative to the nominal value $\tilde{f}_{s,XO}$ given by

$$\delta f_{s,XO} = 1 - f_{s,XO}/\tilde{f}_{s,XO} \tag{7.44}$$

we have $\tilde{T}_{s,XO} = T_{s,XO}(1 - \delta f_{s,XO})$. The timing-phase increment per output sample normalized by the input sample interval $T_{s,XO}$ is equal to

$$\Delta T = T_s / T_{s,XO} = (1 - \delta f_{s,XO})/r \tag{7.45}$$

Assuming that the receiver has achieved carrier synchronization when generating the digital samples, the carrier synchronization block generates an estimate of the XO's relative frequency error $\hat{\delta f}_{XO}$. The required timing-phase increment normalized by $T_{s,XO}$ can be computed from (7.45) by using $\hat{\delta f}_{XO}$ in place of $\delta f_{s,XO}$. After the correction, the timing frequency error would be greatly reduced. If necessary, a timing-frequency error detector can be implemented for the compensation of the residual timing frequency error.

A block diagram of the sampling rate converter with the XO frequency error compensation is shown in Figure 7.20. Its operation should be clear from the figure based on the previous discussion. The requirement for the word length of D_1 is determined by the frequency-offset accuracy required by the OFDM receiver and could be less demanding than that required by the single-carrier systems.

The example described above is only one way to apply digital rate conversion for the OFDM receiver sample generation. There are many alternatives such as the FIR or IIR structures as described in the literature, including [3, 10, 31]. Modem designers can select the algorithms and architectures proper for the implementations under their practical constraints.

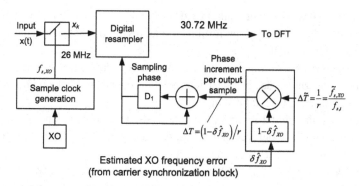

Figure 7.20 Sampling rate conversion with compensation

7.8 Summary and Remarks

In order to perform timing synchronization, it is necessary to adjust the sample timing phase so that the generated samples are synchronous to the remote transmitter symbol clock and at the optimal timing phase. In the early days of digital communications, timing-phase adjustments were accomplished by adjusting the sampling time of the ADC sampling clock in the analog domain. Since the mid-1980s, timing synchronization gradually migrated to digital implementation by using the digital resampling technology. In recent years, digital resampling has become the de facto method for implementing timing-phase adjustment in most digital communication receivers. This chapter covered the theoretical fundamentals and practical aspects of digital resampling technology in the context of timing synchronization.

This chapter first presented the Nyquist sampling theorem and analog signal reconstruction from sample sequences to establish the theoretical foundation of digital resampling. Digital resampling using the interpolation filter was then introduced as a direct application of the analog signal reconstruction. The most popular form of its implementation, the polyphase FIR filter bank, was presented, and various theoretical and practical aspects were discussed in detail. Two other popular forms of resampler implementations, Farrow's delayer and the Lagrange interpolator, were also presented. Their characteristics and details of their implementations were discussed and compared with the FIR filter bank based resampler.

The basics of all-digital timing synchronization using digital resamplers and the practical aspects of their implementations were also considered. Finally, we presented three hypothetical design examples of timing synchronization implementations for practical communications systems. Timing synchronization with the digital resampler in echo/self-interference cancellation, DS-CDMA, and dual-mode LTE/WCDMA receivers were described.

As a summary of this chapter, a few noteworthy remarks are provided below.

- The theoretical foundation of digital resampling is the Nyquist sampling theorem.
- A sample sequence satisfying the Nyquist sampling rate can be used to reconstruct the original analog signal using the interpolation filter with no loss of information.
- Digital resampling is a special case of signal reconstruction with the interpolation filter.

- The polyphase FIR filter bank provides a flexible and efficient implementation of the interpolation filter for performing digital resampling/rate conversion. It is also analytically tractable.
- An arbitrary resampling timing phase and/or conversion rate can be achieved by performing zero-order or linear interpolation of the subfilter outputs of the FIR filter bank–based digital resampler.
- Other popular digital resampling methods used for receiver timing synchronization are Farrow's delayer and the Lagrange interpolator. Both of them can be implemented using Fallow's architecture.
- The ideal digital resampler should have a flat frequency response with a constant group delay in its passband band that contains the signal of interest.
- For the digital resampler used for rate conversion of received signal samples in digital communications, it is more important to have the same frequency response at different delays than to have ideal flat frequency response inside the passband.
- The frequency responses of the passband and transition bands of the resamplers based on FIR filter banks are fully controllable.
- The variations of the passband frequency responses at different delays can be controlled by properly designing the stopband attenuations.
- A polyphase FIR filter bank–based digital resampler can be specifically designed for attaining required signal characteristics at its output to achieve the desired receiver performance.
- Such digital resamplers are especially suitable for digital receivers that are fully or partially based on the software-defined architectures. They can also be implemented using hardware/ASIC-based architectures.
- Both Farrow's delayer and the Lagrange interpolator try to achieve a flat frequency response of the passband with different optimization criteria. These methods have no direct control over the transition band characteristics and the minimizing variations of the frequency responses at different delays.
- Farrow's delayer and the Lagrange interpolator are particularly suitable for general IC hardware designs for generic digital interpolation applications.
- All three types of digital resamplers can be used for the implementation of digital timing synchronization in digital receivers as was shown in the given examples.
- In multisystem environments, receivers can employ digital resampling to generate the required sampling rates for different systems from an initial common sampling rate to simplify RF interference planning and spur control.

References

[1] R. W. Schafer and L. Rabiner, "A Digital Signal Processing Approach to Interpolation," *Proceedings of the IEEE*, vol. 61, no. 6, pp. 692–702, 1973.

[2] M. G. Bellanger, G. Bonnerot, and M. Coudreuse, "Digital Filtering by Polyphase Network: Application to Sample Rate Alteration and Filter Banks," *IEEE Transactions on Acoustics, Speech, and Signal Processing*, vol. ASSP-24, no. 2, pp. 109–14, 1976.

[3] R. E. Crochiere and L. R. Rabiner, "Interpolation and Decimation of Digital Signals – A Tutorial Review," *Proceedings of IEEE*, vol. 69, no. 3, pp. 300–31, 1981.

[4] M. G. Ballanger, J. L. Daguet, and G. P. Lepagnol, "Interpolation, Extrapolation, and Reduction of Computation Speed in Digital Filters," *IEEE Transactions on Acoustics, Speech, and Signal Processing*, vol. ASSP-22, no. 4, pp. 231–5, 1974.

[5] R. Crochiere and L. Rabiner, *Multirate Digital Signal Processing*, Englewood Cliffs, NJ: Prentice Hall, 1983.

[6] F. Ling, "Digital Rate Conversion with a Non-Rational Ratio for High-Speed Echo-Cancellation Modem," in IEEE International Conference on Acoustics, Speech, and Signal Processing, 1993. ICASSP-93, Minneapolis, MN, USA, 1993.

[7] Dept. of R&AD, Codex Corp, "Old Codex Memos on Digital Timing Synchronization, 1985," August 20, 2015. [Online]. Available: www.slideshare.net/FuyunLing/old-codex-memo-on-digital-timing-synchronization-1985. [Accessed 23 August 2015].

[8] C. W. Farrow, "A Continuously Variable Digital Delay Element," in IEEE International Symposium on Circuits and Systems, 1988, Espoo, Finland, 1988.

[9] F. M. Gardner, "Interpolation in Digital Modems I. Fundamentals," *IEEE Transactions on Communications*, vol. 41, no. 3, pp. 501–7, 1993.

[10] L. Erup, F. M. Gardner, and R. A. Harris, "Interpolation in Digital Modems. II. Implementation and Performance," *IEEE Transactions on Communications*, vol. 41, no. 6, pp. 998–1008, 1993.

[11] A. V. Oppenheim and R. W. Schafer, *Digital Signal Processing*, Englewood Cliffs, NJ: Prentice Hall, 1975.

[12] H. Nyquist, "Certain Topics in Telegraph Transmission Theory," *AIEE Transactions*, vol. 47, pp. 617–44, 1928.

[13] Wikipedia, "Interpolation," [Online]. Available: https://en.wikipedia.org/wiki/Interpolation. [Accessed 27 September 2015].

[14] J. G. Proakis, C. M. Rader, F. Ling, and C. L. Nikias, *Advanced Digital Signal Processing*, New York: Macmillan, 1992.

[15] 3GPP TS 25.104, "Third Generation Partnership Project; Technical Specification Group Radio Access Network; Base Station (BS) Radio Transmission and Reception (FDD)," 3GPP, Valbonne, France, 2009.

[16] T. W. Parks and J. H. McClellan, "A Program for the Design of Linear Phase Finite Impulse Response Filters," *IEEE Transactions on Audio and Electroacoustics*, vols. AU-20, no. 3, pp. 195–9, 1972.

[17] E. Y. Remez, "General Computational Methods of Tchebycheff Approximation," *Atomic Energy Commission Translation*, no. 4491, pp. 1–85, 1957.

[18] A. Tarczynski, G. D. Cain, E. Hermanowicz, and M. Rojewski, "WLS Design of Variable Frequency Response FIR Filters," in Proceedings of the 1997 IEEE International Symposium on Circuits and Systems, Hong Kong, 1997.

[19] I. Selesnick, "Linear-Phase FIR Filter Design by Least Squares," 9 August 2005. [Online]. Available: http://eeweb.poly.edu/iselesni/EL713/firls/firls.pdf. [Accessed 23 February 2016].

[20] MathWorks, "firls – Least Square Linear-Phase FIR Filter Design," [Online]. Available: www.mathworks.com/help/signal/ref/firls.html. [Accessed 24 August 2015].

[21] T. I. Laakso, V. Valimaki, M. Karjalaimen, and U. K. Laine, "Splitting the Unit Delay: Tools for Fractional Delay Filter Design," *IEEE Signal Processing Magazine*, vol. 13, no. 1, pp. 30–60, 1996.

[22] O. Herrmann, "On the Approximation Problem in Nonrecursive Digital Filter Design," *IEEE Transactions on Circuit Theory*, vol. 18, no. 3, pp. 411–3, 1971.

[23] E. Hermanowicz, "Explicit Formulas for Weighting Coefficients of Maximally Flat Tunable FIR Delayers," *IEEE Electronics Letters*, vol. 28, no. 20, pp. 1936–7, 1992.

[24] V. Valimaki, "A New Filter Implementation Strategy for Lagrange Interpolation," in Proceedings IEEE International Symposium Circuits and Systems (ICAS-95), Seattle, Washington, USA, 1995.

[25] F. Ling, "Echo Cancellation," in *Adaptive System Identification and Signal*, London: Prentice Hall International (UK), 1993, pp. 407–65.

[26] A. Sahai, G. Patel, and A. Sabharwal, "Pushing the Limits of Full-Duplex: Design and Real-Time Implementation," in Rice University Technical Report TREE1104, June 2011. [Online]. Available: http://arxiv.org/abs/1107.0607 in. [Accessed January 2014].

[27] Y. Choi and H. Shirani-Mehr, "Simultaneous Transmission and Reception: Algorithm, Design and System Level Performance," *IEEE Transactions on Wireless Communications*, vol. 12, no. 12, pp. 5992–6010, 2013.

[28] F. Ling, "Achievable Performance and Limiting Factors of Echo Cancellation in Wireless Communications," in IEEE Information Theory and Applications Workshop (ITA), 2014, San Diego, CA USA, 2014.

[29] M. M. Sondhi and A. J. Presti, "Application of Automatic, Transversal Filters to the Problem of Echo Suppression," *Bell Systems Technical Journal*, vol. 45, no. 10, pp. 1847–50, 1966.

[30] V. G. Koll and S. B. Weinstein, "Simultaneous Two-Way Data Transmission over a Two-Wire Circuit," *IEEE Transactions on Communication*, vol. 21, no. 2, pp. 143–7, 1973.

[31] P. P. Vaidyanathan, *Multirate System and Filter Banks*, Englewood Cliffs, NJ: Prentice Hall, 1992.

Index

Printed in the United States
by Baker & Taylor Publisher Services